Piling Engineering

SECOND EDITION

W.G.K. FLEMING, BSc, PhD, MICE
Chief Engineer
Cementation Piling and Foundations Ltd
Rickmansworth

A.J. WELTMAN, MSc, DIC, MIMM, FGS
Consulting Geotechnical Engineer
Stroud

M.F. RANDOLPH, MA, PhD
Professor of Civil Engineering
University of Western Australia

W.K. ELSON, Dip. CE, PhD, MICE
Senior Lecturer in Civil Engineering
Brighton Polytechnic

Blackie A & P
Glasgow and London

Halsted Press, an Imprint of
John Wiley & Sons, Inc.
New York and Toronto

Published by
Blackie Academic & Professional, an imprint of Chapman & Hall,
Wester Cleddens Road, Bishopbriggs, Glasgow G64 2NZ, UK

Chapman & Hall, 2-6 Boundary Row, London SE1 8HN, UK

Blackie Academic & Professional, Wester Cleddens Road, Bishopbriggs, Glasgow G64 2NZ, UK

Chapman & Hall GmbH, Pappelallee 3, 69469 Weinheim, Germany

Chapman & Hall Japan, ITP-Japan, Kyowa Building, 3F, 2-2-1 Hirakawacho, Chiyoda-ku, Tokyo 102, Japan

DA Book (Aust.) Pty Ltd, 648 Whitehorse Road, Mitcham 3132, Victoria, Australia

Chapman & Hall India, R. Seshadri, 32 Second Main Road, CIT East, Madras 600 035, India

Copublished in the United States and Canada by HALSTEAD PRESS, an imprint of John Wiley & Sons, Inc., New York and Toronto. Orders from the USA only should be sent to John Wiley & Sons, 1 Wiley Drive, Somerset, New Jersey 08873.

© 1992 Blackie & Son Ltd

First published 1985, this edition 1992
Paperback edition 1994, reprinted 1994

Typeset by Thomson Press (India) Limited, New Dehli
Printed in St Edmundsbury Press, Bury St Edmunds, Suffolk

British Library Catolguing in Publication Data
Piling Engineering.-2nd ed.
    I. Fleming, W.G.K.   II. Weltman, A.J.
    III. Randolph, M.F.
    642.1
    ISBN 0-7514-0194-3

Library of Congress Cataloging-in-Publication Data available
Piling Engineering / [edited by] W.G.K. Fleming . . [et al.] 2nd ed.
    p.   cm.
    Includes bibliographical references and index.
    ISBN 0-470-21825-8
    1. Piling (Civil Engineering)   I. Fleming, W.G.K.
TA780.P494   1992
624.1'54-dc20   91-31618
                                            CIP

Preface

When the first edition of this book was published in 1985, its aim was to provide an up-to-date text for engineers who were faced with the several problems of carrying out site investigations where piling was an envisaged foundation solution, discovering what piles were available on the market, designing a system which would be adequate for their requirements, appreciating the problems which could be encountered in the construction phase, testing the product for soundness and behavioural characteristics, and dealing with the difficult matter of specifying a solution which was not only adequate, but also economic.

Piling has always been a subject which has fascinated engineers and called forth the inventive spirit of those involved in its many aspects. It is not surprising therefore, that after some six years, a variety of improvements can be made to the text to include some of the latest changes in technique and advances in technology.

The structure of the first edition has been retained, but each chapter has been up-dated as necessary. The chapters dealing with the design of single piles and of pile groups have been extensively re-written and improved, and the chapter dealing with load testing has been extended to deal more fully with the theory of dynamic testing, which has become more popular in recent years. Elsewhere, the text incorporates new piling techniques and recognizes, for example, the increasing use of continuous-flight augered piling which is now widely monitored by computers on board the piling machines. Retaining wall design has also advanced by the introduction of new computer programs, and a variety of Standards relating to the whole field of piling have been revised.

It is hoped that the reader will find the book a useful reference, giving insight into the current technology of the subject. My thanks are due to my co-authors for contributing their time, effort and experience to the work, and we would each wish to express our thanks to the colleagues with whom we work for the experience which they readily share and to our respective families who have tolerated the expenditure of time to complete the work.

W.G.K.F.

Contents

4   Design of single piles   92

5   Design of pile groups   179

List of symbols

A	Area of pile (base or shaft)
B	Breadth of pile group
b	Breadth of H pile
c	Recoverable pile movement (during driving)
c_u	Undrained shear strength
d	Pile diameter
E	Young's modulus
E_p	Young's modulus of an equivalent solid pile of the same radius
$(EA)_p$	Cross-sectional rigidity of the pile
$(EI)_p$	Bending rigidity of the pile
e	Exponent in calculation of group (stiffness) efficiency
F	Coefficient in flexibility matrix
f	Unit friction
G	Shear modulus of the soil
G_l	Shear modulus of the soil at depth l
G_0	Shear modulus at ground surface
H	Lateral load
I	Influence factor
K	Earth pressure coefficient
k	Efficiency of pile driving hammer; stiffness of pile, cap or foundation; coefficient of subgrade reaction
l	Embedded length of the pile
m	Rate of increase of soil shear modulus with depth
M	Applied moment
n	Number of piles; rate of increase with depth of coefficient of subgrade reaction
P	Axial force, usually at working load
p_u	Ultimate lateral force per unit length of pile
Q	Axial force, usually ultimate capacity
q	Pressure (e.g. end-bearing) or resistance (e.g. cone)
R	Resistance (pile driving); correction factor, settlement or deflection ratio
r	Radius, radial co-ordinate
r_m	Maximum radius of influence of pile under axial loading
s	Centre to centre pile spacing; set of pile during driving

T	Transformation matrix
u	Lateral displacement
w	Axial displacement
z	Depth, co-ordinate perpendicular to ground surface
α	Ratio of skin friction to undrained shear strength; interaction factor
β	Ratio of skin friction to effective overburden pressure; departure angle between direction of loading and line joining pile centres; angle of retained soil surface to horizontal
δ	Interface angle of friction.
ζ	Parameter in solution for axial pile response
η	Ratio of radius of pile base to that of pile shaft; efficiency of pile groups
θ	Rotation of pile head under lateral loading
λ	Parameter in axial capacity calculation; pile-soil stiffness ratio, E_p/G_l
μ	Parameter in solution for axial pile response
ν	Poisson's ratio for the soil
ξ	Ratio of shear modulus at depth l, G_l, and that below the pile base, G_b
π	Mathematical constant
ρ	Parameter giving relative homogeneity of the soil
σ	Normal stress
τ	Shear stress
ϕ'	Angle of friction for soil
ψ	Angle of rake of pile

Principal subscripts

b	denoting base of pile
c	characteristic or critical (see solution for lateral pile response); also cone or cap
f	foundation
h	horizontal
o	denoting value at the pile shaft
p	pile
r	radial
s	denoting shaft of pile
t	denoting top of pile
v	vertical
N	Standard Penetration test result (blows/300 mm)
c'	Effective cohesion
c_w	Wall adhesion
α	Inclination of wall to horizontal
β	Angle of retained soil surface to horizontal
H	Height
F	Wall rotational stability factor
R	Resultant force

1 Introduction

Since man first sought to establish secure dwellings and to cross streams and rivers where fluctuating water levels gave an element of uncertainty to his travel arrangements, the driving of robust stakes or piles in the ground has provided a means whereby the hazards of living could be reduced.

Various peoples in different parts of the world found it convenient to dwell by lake shores where food, water and easy transport were readily to hand and where water levels remained reasonably constant. Evidence of piled settlements has been found on the borders of lakes in, for example, Switzerland, Italy, Scotland and Ireland.

It is believed that some of these settlements were in use about 4000 years ago and they were sometimes of considerable size, as on the shores of Lake Geneva opposite Morges. In another settlement at Robenhausen, it has been estimated that over 100 000 piles had been used. In this case the piles were beneath a covering of peat moss which it is reckoned would have taken at least 2000 years to form. Recent archaeological excavations elsewhere in Switzerland and in France have confirmed that piles were used widely for housing along marshy lake shores.

At Lough Drumkeery in Co. Cavan, Ireland, about 30 000 ancient piles including primitive sheet piles were found in 1863. These consisted of birch and oak, and the oak piles had been carefully pointed and driven to a depth of about 3 metres*.

The first historical reference to piling appears to be by Herodotus, the Greek writer and traveller who is sometimes referred to as the 'father of history' and who lived in the fourth century B.C. He records how an African tribe, the Peonions, lived in dwellings erected on lofty piles driven into a lake bed. The piles were driven under some kind of communal arrangement but after a time a law had been made that when a man wished to marry, he had first to drive three piles. Since the tribe was polygamous, the number of piles installed was considerable. This system provided a unique and practical method of founding an expanding settlement.

*The Lake Dwellings of Ireland, W.G. Wood-Martin, Longman Green & Co., London, 1886.

One of the earliest uses of piling appears to have been by the Phoenicians who used sheet piles for dock and shore constructions in connection with their expanding sea trade. In effect the sheet piles appear to have been derived from the skill of boat builders in the planking of ships.

The cedars of Lebanon were exported to the Egyptians who were great sailors and builders but were without suitable timber—soon cedar wood sheet piles were used to enable them to sink wells. Indeed the sustained demand for cedar was such that in time the great cedar forests of the area were reduced to only small remnants.

Greek and Roman engineers used piles for shore works at many places along the Mediterranean coast, but in this situation they encountered difficulties with the destructive 'teredo navalis' so that few of their structures have survived in this environment to give archaeological evidence. In Britain, a Roman bridge spanned the Tyne at Corbridge, about 20 miles west of Newcastle on Tyne, using piles to support the construction. These were discovered when an old bridge was taken down in 1771, and it was found that they were of black oak and about 3 metres in length. Much evidence exists elsewhere, testifying to the skill of the Romans in solving difficult foundation problems. The Emperor Trajan built a bridge over the Danube, the foundations of which were exposed during the eighteenth century. When examined it was said that the surface of the timber was petrified to a depth of about 20 mm, and that the timber beneath this surface layer was completely sound. About A.D. 0 Vitruvius, a Roman architect, wrote a treatise *De Architectura*, and described in it a method of sheet piling for use in creating dams and other water-retaining structures, again testifying to the skills of his generation.

The city of Venice came into being, it is believed, as refugees began to seek security on the islands of the Lagoon of Venice from the barbarian invaders who pillaged the remnants of the Roman empire. The position of the city, in an easily defended location, was determined in the early 8th century and piles were driven for the support of all the buildings which were gradually constructed as the city became more and more prosperous.

Amsterdam, founded about 1000 years ago, was built almost entirely on piled foundations of 15 to 20 metres length; the piles in this case were sawn off level and capped with thick planks. The Romans by contrast often capped their piles with a mixture of stone rubble and concrete.

It is of interest to note that Creasy (*An Encyclopedia of Civil Engineering*, 1861) says that in Holland piling and capping by planking was still in use, with rough stones rammed between the planks, but that timber and stone used together did not produce 'solidity and duration'. He comments that pile driving has for its object the consolidation of a soil which is not sufficiently compact and he recommends the test loading of piles, with load being kept in place for long enough to remove any doubts about the security of the piles.

Little progress in the art of driving piles seems to have been made between

Roman times and the beginning of the 19th century. The 'Pile Engine' of the early 1800s can have varied little from the Roman type. For a small engine a timber frame was used to suspend the 'ram' which was actuated by a single rope, divided at its end into as many small ropes or 'tails' as required. At each 'tail' a man worked and the method was known as the 'whip ram' method. Small rams would have a mass of about 120 kg and often consisted of oak strongly hooped by iron. Coulomb stated that for this kind of work a man's effort was normally the equivalent of raising a 75 kg weight through a height of 1000 m in one day. For man operated piling frames in about A.D. 1700 it was said that 20 men could operate a 360 kg ram, the procedure being to raise the ram 25 or 30 times in succession over a period of about 1 minute. There was then an equal rest period, and the 'volley' was repeated, but it was said (somewhat sadly) that a further minute was often wasted. Larger engines raised their heavier cast-iron rams, which could weigh up to $1\frac{1}{2}$ tonnes, by a winch, which might be powered by men or horses. When the ram reached a prearranged height a scissor catch at the ram head compressed and disengaged it, so that it slid down its guide to produce the necessary impact.

At this time, as no doubt for hundreds of years previously, it was fully appreciated that timber piles subject to wetting and drying cycles were much less durable than piles which were permanently submerged. Where permanent submergence was not possible, the piles above water were often coated with tar. Iron shoes were also fitted to piles where driving involved penetration into stony layers.

It was in the nineteenth century that, as in many other engineering fields, big changes began to take place both in terms of materials and motive power. Metal piles, mostly in the form of cast-iron pipes, became available in the mid-1830s and these were used for important structures on account of their durability. Screw piles were first used for the foundations of the Maplin Sands Lighthouse in the Thames Estuary in 1838, the inventor being Alexander Mitchell. In 1824 Joseph Aspdin patented his hydraulic cement, later known as Portland cement on the grounds that its appearance bore a resemblance to Portland stone. While in his initial patent he apparently withheld some manufacturing details, by the middle of the century hydraulic cement had become widely available and before the end of the century Coignet and Hennebique, French competitors, had introduced the successful reinforced concrete which was later to be brought to Britain by L.G. Mouchel.

Steam power had first been applied to the driving of piles in Britain by John Rennie in 1801 or 1802 at the Bell Dock at the entrance to the London Docks. The ram was hoisted by an 8-hp engine, constructed by Boulton and Watt, and this engine, or one similar to it, was used again at Hull docks in 1804. In 1843, Nasmyth produced a radical departure from this type of machine, introducing a revolutionary type of hammer. In this new hammer the ram was attached to the lower end of the piston rod and the whole weight of the steam cylinder acted on the pile. It achieved striking rates of 80 blows per minute and was used for example at the site of the Newcastle High-Level Bridge to drive 70-ton piles, the initial supervision of the works being carried out by Robert Stephenson. In 1846 a compressed-air hammer was proposed by Clarke, Freeman and Varley, and in 1849 Clarke and Motley produced a vacuum hammer. This was indeed an era of invention and engineering excitement.

In about 1870, a pile driver powered by gunpowder was launched upon the world by a Mr Shaw of Philadelphia. The machine was claimed to have several advantages over any other type available at the time, not least being that it required only a man and a boy to work it. A cast steel cap or 'mortar' rested on the pile head, in which was formed a recess sufficient to contain a few ounces of gunpowder. Mounted on a frame above this anvil or mortar was a steel drop-weight with a projecting plunger, designed on falling to strike the gunpowder in the recess and cause an explosion. This action then threw the dropweight back up into its retainers ready for the next strike. The function of the man was to pull a lever which released the dropweight, while the boy hung on a ladder adjacent to the pile head and threw in the charges of gunpowder as required. The operating rate achieved was 15 to 20 blows per minute, but if it became any faster, the heat of the mortar became sufficient to cause premature explosions, which apparently were not good for the morale of the boy.

A machine of this type was brought to England to the Severn Tunnel works, but it proved unsuccessful in the sandy conditions which had previously baffled other piling machines. It was then brought to London to St. Katherine's Dock to work for the East London Railway but here its use was again short-lived. After only three charges of gunpowder had exploded, it caused such a state of alarm among the dock authorities that they ordered its removal forthwith. The same, or a similar machine, was then used to drive piles for the third bridge over the River Elbe at Dresden and was highly praised for efficiency, economy and speed. Piles of 1.8 m to 2.5 m length were driven at a rate of up to 30 per day using 60 gunpowder cartridges per pile. The cost per pile for driving was reported as '8 shillings and 6 pence'.

In 1897, A.A. Raymond patented the Raymond pile system and was, it appears, the first person to develop a practical economical cast-in-place concrete pile. In 1903 R.J. Beale also developed a method of driving a steel pipe, plugged at its lower end, and of subsequently filling it with concrete and then withdrawing the tube, while in 1908 E. Frankignoul, a Belgian, invented an early version of the Franki driven-tube pile which was later developed to provide expanded bases and which became widely known and successful throughout the world. At about the same time the first precast concrete driven piles appeared and although the originator is not known, the earliest drawings of precast concrete piles in the UK appear to have come from the Hennebique Company.

The use of steel I-beam piles originated in the United States before 1900, when fabricated sections were used for highway bridges in Nebraska, but after

1908 the Bethlehem Steel Co. produced rolled H-sections which quickly captured the market for this type of pile.

The use of steam-operated hammers continued throughout the first half of this century for driving piles of all types but declined in the post-war period after 1946, and in-place diesel-operated hammers became popular. Drop hammers, operated from diesel winches and crane drums, have nevertheless remained to the present time because of the basic simplicity of the method. Now diesel hammers have lost some of their former popularity for environmental reasons, and it would appear that hydraulically-powered hammers are finding increased usage because of relative efficiency and easy variable control.

Bored piles, formed by percussion boring tools, have certainly been known and used in the United Kingdom since the early 1930s, but it would be wrong to think of bored piling as a development of the 20th century. Long before the term 'bored pile' was coined, and before hydraulic cements came into common use, the 'well foundation' was used in many countries for the support of major structures. Such foundations have been dug in India for example for hundreds of years, with the stone foundation being carried up from the base of each excavation or boring. The Taj Mahal, which was built in the period 1632 to 1650, made extensive use of this type of foundation, and throughout the Mogul period from 1526 this type of support was used for many bridges across deep scourable river beds.

The same technique proved particularly effective for railway bridges when the railways were first introduced into India and was equally successful in England where three railway bridges across the Thames in London at Charing Cross, Cannon Street and Victoria Station were founded on well-type foundations in the mid-nineteenth century, casting what were in effect large-diameter piles within hand-dug caissons.

With the advent of hydraulic cements, however, it was no longer necessary to restrict the dimensions of well foundations to those which would allow workers to descend and construct a pier from the bottom of the excavation upwards, since concrete could be mixed at the surface and poured in. This meant that for relatively light loads small-diameter piles could be constructed by percussion boring equipment as used for well sinking, and so the art of bored piling developed.

Bored piling offered certain advantages over driven types for at least part of the market, since a small tripod machine could be used to suspend the boring tool and could at the same time provide enough effort to enable relatively short lengths of steel tube, which were necessary to retain soft and unstable upper ground strata, to be removed after concrete placing was complete. When compared with the relatively large machines of the same period needed to insert driven piles, the equipment was small and could readily operate on sites to which the larger machines could not gain easy access.

In its early days, the method was probably a good deal less reliable than the methods used to install driven piles, but as companies built up experience, reliability improved. Nevertheless it was still difficult to cope with water inflows at depth, and in order to overcome this problem the 'pressure pile' was invented. In this system an airlock was screwed on to the top of the steel lining tube which was used to stabilize the hole. This was done at the end of the excavation stage, and using appropriate air pressures within the tube, water was expelled from the boring. Concrete was then placed in batches through the airlock and thus the segregation which might have occurred had concrete been poured directly into water, or by means of an unreliable bottom-opening skip, was avoided. The method proved rather cumbersome and has now fallen into disuse. Instead concrete is now poured into water-filled boreholes using tremie pipes.

Today, although small tripod or similar percussion boring equipment remains, the majority of bored piles are formed using rotary augering machines. This results from rapid development of this type of plant since about 1950. Machines are now available to bore a wide range of pile diameters from about 150 mm to over 2 metres, while yet other equipment is designed to enable concrete or grout to be injected to form a pile through the hollow stem of a continuously flighted boring auger as the laden auger is withdrawn from the soil.

At the same time driven piles have developed considerably and two main forms exist for land-based work, namely the precast jointed pile and the cast-in-place pile. The former type of pile replaces to a large extent the long precast piles which were common in the early part of this century and requires much smaller and lighter handling and driving plant, while the latter offers a cheap and reliable pile where the pile penetration required is not more than about 18 metres.

Steel H piles and tubular steel piles are used for limited applications. The H section pile causes minimal soil displacement and can stand up to fairly heavy driving conditions while the tubular pile finds application in many shoreline works where both direct load and bending capacity are often needed. Very large and long tubular piles are used extensively in major offshore works such as the founding of oil platforms at sea. The cost of steel is however such that these types of pile find only occasional application for inland jobs.

The design of piles and pile groups has advanced steadily in recent times with much of this work being carried out by engineers who specialize particularly in foundation engineering. Effective stress methods are being developed for individual pile designs, while computer-based techniques for assessing the settlement behaviour of large and complex pile groups are finding increasing application.

The construction of piles today is also carried out for the most part by specialist contractors who exercise considerable skill in adapting methods of working to enable sound piles to be formed even in the most forbidding ground conditions, although it should still be borne in mind that every technique used can encounter difficulties and unforeseen problems at times. The skill of a particular contractor often depends on his particular experience and this is only gained over many years, perhaps using a range of methods.

In modern piling practice site investigation has become an essential precursor to making sensible decisions both regarding pile design and the choice of appropriate construction methods, and is all the more important because of the wide range of methods and equipment now available. Financial savings made at the investigation stage are often followed by the unwelcome discovery of adverse conditions which can eventually impose an extra burden of cost far exceeding the earlier savings which may have been made.

Clearly it is essential that, with ever-increasing pile stresses which are consequent upon the strong incentives which exist to use materials economically, piles must be made both sound and durable. The integrity of different types of pile in a wide range of ground conditions has been investigated and the CIRIA reports PG2[1] and PG8[2] give some insight into the potential problems which can arise in certain circumstances. At the same time the development of integrity testing methods now provides a means for detection of major pile construction problems, and this in turn implies that pile construction can reach higher standards than ever before[3].

Good specification of piling works is also important, and with reference to several of the common types of piling, information on this is provided in the Institution of Civil Engineers publication *Specification for Piling*. This should at least be a standard reference work for those preparing such documents for any land-based contract. So much in piling depends on an exercise of judgment by an experienced engineer that those who undertake this type of work must remain both well informed and close to the realities of dealing with one of nature's most variable materials, the ground. Performance in the field equally requires alert supervision and sound understanding of the processes.

While the evaluation of methods and the advancement of engineering knowledge have changed the piling industry in important ways throughout its history and while the pace of change has been particularly rapid in recent years, it should still be realized that whatever the equipment or methods employed, a great deal of reliance must be placed on individuals whose skill, experience and honest endeavour eventually turn engineering concepts into dependable reality. It has ever been so.

[1]Thorburn, S. and Thorburn, J.Q. *Review of problems associated with the construction of cast in place piles.* CIRIA Report PG2, 1977.
[2]Healy, P.R. and Weltman, A.J. *Survey of problems associated with the installation of displacement piles.* CIRIA Report PG8, 1980).
[3]Weltman, A.J. *Integrity testing of piles: a review.* CIRIA report PG4, 1977.
Sliwinski, Z.J. and Fleming, W.G.K. The integrity and performance of bored piles. *Advances in Piling and Ground Treatment for Foundations,* I.C.E., London, 1984.

2   Site investigation for piling

2.1 Planning

The planning of a site investigation for piling works is essentially similar to that for shallow foundations, apart from the obvious requirement for greater depth of exploration. The main additional factors which should be considered are the practical aspects of pile installation.

The first stage in planning a site investigation comprises a desk study of the available information. This subject is comprehensively discussed by Dumbleton and West (1971), Weltman and Head (1983) and BS 5930 (1981). It should also be noted that the Eurocode 7 drafts are in preparation which relate to site investigation. In remote and overseas locations little published information may be available and in this instance, more emphasis must be placed on the exploratory work. The second stage should include a site inspection in order to confirm, as far as is possible, the data collected during the desk study, and to make as many additional preliminary observations as possible.

The investigation should take into account details of the foundation design if known, including the tolerance of the structure to settlement, since this can dictate the extent of the investigation. Where these details are unknown it is important that the exploratory work provides sufficient information to permit decisions concerning the type of foundation, as well as the actual foundation design, to be made. The major requirement of the investigation in terms of the design is to provide comprehensive information over the full depth of the proposed foundations and well below any possible pile toe level. This is necessary to permit flexibility in the pile design and to be certain that the anticipated strata thicknesses exist. Of equal importance is a thorough appraisal of the groundwater conditions, including its occurrence and response during boring operations.

For the case of straightforward pile design in well-documented homogeneous deposits, it is generally possible to make firm proposals for the number and depth of boreholes at the planning stage, although some aspects of the programme of exploratory work, including the field and laboratory testing, can usually only be outlined. As the complexity of the proposed piling works increases, however, the more flexible the approach to the investigation has to be to allow adequate feedback and analysis of information as it is obtained. This is particularly true where there are large variations in design loads for heavily loaded piles or for piles into rock-head which varies in level or which has a weathered profile. For complex projects, a two-stage approach may be the most cost effective. The second stage is then based on the preliminary findings of the first. Good communications must exist between the site investigation contractor and client in order to tailor the exploration to the requirements of the foundation design and the practical aspects of pile installation vis-à-vis the prevailing soil conditions. Where a piled foundation is proposed for an extensive project, piling trials including pile testing can be cost effective and should be included as part of the investigation. Reference may be made to the ICE publication *Specification for Ground Investigation with Bills of Quantities* (1989) for additional guidance. Detailed information of the ground conditions reduces the possibility of encountering unexpected problems due to adverse soil or groundwater conditions. In this respect, it is essential that all the available factual information is passed on to the contractor at the tender stage since the offer to carry out the work will be made on the basis of the information provided.

Other specific aspects of the investigation may include matters such as the seismic risk of the area, aggressive soil conditions (especially in fills) and the possibility of aquifer pollution if piles are to penetrate contaminated ground. Consideration should also be given to acceptable levels of noise for the area, the extent of existing foundations adjoining the site and the sensitivity of neighbouring structures and buried services to vibration and soil displacement. For piled foundations the investigation should include comprehensive details of trafficability (which can vary with seasonal changes in the weather) and access (including headroom) for heavy plant. Water supplies, the facilities for waste disposal and site drainage would also be relevant to the investigation, since the installation of deep foundations can produce large quantities of spoil and slurries.

In the following sections, only those aspects of site investigation which relate to piling works are discussed.

2.2 Depth of exploration

The extent of the exploration is not limited solely by the zone of influence caused by the loaded piles, but is dependent on other factors such as considerations of general stability, the necessity to understand the overall groundwater regime, highly heterogeneous soil conditions or the effects of piling on adjacent structures. The zone of influence is a good starting point, since it quantifies to some degree the lateral and vertical extent of the stressed zone. In the case of a piled raft, significant stresses are transferred to a considerable depth (see Figure 2.1), and it is recommended that the depth of

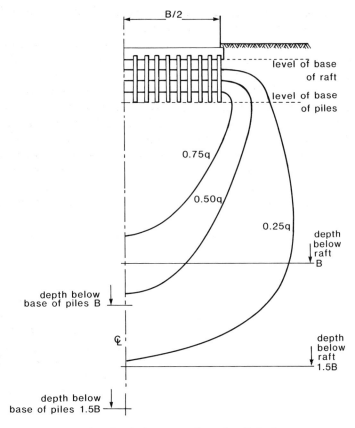

Figure 2.1　Vertical stress envelopes for piled raft

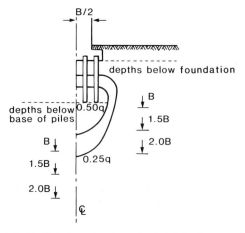

Figure 2.2　Vertical stress envelopes for isolated pile group

exploration should be at least one and a half times the width of the loaded area. In the special circumstance where the geological conditions or rockhead level are well known, such depths may be reduced. As a general guide the distance is measured below the base of the pile group, and the loaded area is calculated allowing a spread of one in the horizontal direction to four in the vertical direction over the estimated length of the pile. For isolated pile groups (under a single column for instance), the stressed zone based on the spread of load concept is relatively much shallower (see Figure 2.2), but a single proving borehole should be taken to a depth in excess of $2B$ in order to ensure that 'punching' through a relatively strong stratum into a weaker stratum cannot occur. In addition, the intensity of loading at the tip of end-bearing piles may be high, and a rock-bearing stratum should be adequately investigated particularly where heavy column loads are to be transmitted to single or small groups of highly loaded piles.

Where piles are to be founded in a bearing stratum of limited thickness (such as a gravel layer overlying clay) it is essential that a large number of bores or probings are employed—possibly spaced on a close grid, to determine whether the bearing stratum thickness and level is maintained over the proposed area of loading. In these cases, detailed exploration may be restricted to a limited number of locations, permitting the calibration of a lower cost penetration method which may be used to provide the necessary coverage. It is important in this situation that the competence of the underlying stratum to sustain the transmitted stresses is established. Where piles are to be founded in a rock-bearing stratum, the likely variability of rock-head level should be assessed from a geological point of view and the investigation should be sufficient to establish rock-head contours in adequate detail and to obtain profiles of strength and weathering.

It should be noted that the strict geological definition of bedrock (and therefore rock-head level) may not concur with the views of a piling specialist, and this point should be made clear in the dissemination of information. In ground conditions which are known to be of a heterogeneous nature, a qualified geologist or engineering geologist should be involved with the work from an early stage. Even if piles are to be taken through heterogeneous soils to found in underlying bedrock, variations in groundwater conditions, or the occurrence of lenses of silt, sand, large boulders or open work gravel may jeopardize the successful installation of driven or bored cast-in-place piles and these conditions should be fully investigated.

2.3 Groundwater conditions

The investigation of groundwater conditions is particularly pertinent to piling works. The findings may influence the choice of pile type and it is essential that the piling contractor has the data necessary to assess the rate of water inflow that may occur during the installation of conventional bored piles.

Observations of the ingress of water including the levels at which water seeps and collects should be made and noted on the site investigation logs in such a way that they may be related to the casing and borehole depths. These water level observations will be unlikely to reflect the true hydrostatic conditions, but the information is of use to piling contractors tendering for bored or drilled piles.

Some fine soils of low permeability permit fairly rapid boring and excavation without seepage of water for the period of time that the bore is open, although the groundwater level may be near the ground surface. This is typical of silty clays, such as London Clay. However, the possibility of the occurrence of silt or fine sand lenses within these soils should be recognized and they should be detected if present, since such layers may result in rapid accumulation of water in the bore. Where these conditions are anticipated, appropriate measures can be taken, but the contractor will require the information at the tender stage.

In many situations pile bores in homogeneous clay may be similar in length to the clay layer thickness, and with natural variation in bed levels, there can be a risk of a pile boring extending into underlying strata. The underlying beds may be permeable, and will probably be under a considerable head of water. The site investigation must delineate the interface between the clay and underlying beds with some accuracy, to minimize the risk of penetrating such aquifers during pile construction.

The heads of water recorded by observation in boreholes, or measurement of levels in piezometers may be subject to seasonal variation, and ideally measurements should be collected over a period of months where the levels are critical. The period should of course include the 'wetter' parts of the year. This is not always possible, but the topography and geology of the surrounding area should enable an assessment of any potential problem to be made. A special case concerns tidal variation. Accurate estimates of maximum groundwater levels may be essential if, for instance, piles are to be constructed by the bentonite method, when a positive head of slurry has to be maintained during construction. Tidal variation in groundwater (which can occur in permeable soils) at the proposed location of the piles has therefore to be monitored during spring tides in order to obtain a realistic level.

Whilst measures can be taken to combat erosion of the concrete in bored piles by flowing groundwater immediately after casting, the possibility of this occurrence must be known beforehand, and openwork, highly permeable gravels under a seepage head should be detected. Their presence may be suspected in glaciated valleys for instance, where coarse granular material may be found in outwash fans. The situation could also arise in alluvial gravels if there is a marked absence of finer material. Simple permeability tests can be useful in the assessment of this problem.

Chemical attack of steel or concrete piles may occur due to dissolved salts in acidic groundwater. Groundwater samples should be taken from boreholes in such a way that they are uncontaminated or diluted by water added during boring or drilling. Special samplers may be required to obtain water samples at given levels. In made ground or backfill such as quarry or mine waste, extremely corrosive conditions may be present and this situation should be adequately investigated. Industrial waste may also give rise to corrosive ground conditions and in built-up areas, the previous use of the ground should be established. Chemical tests on soil samples may be carried out to give additional information.

2.4 Sampling, in-situ testing and laboratory testing

In certain respects, decisions regarding the optimum programme of sampling and testing may only be made after some knowledge of the ground conditions

Table 2.1 Typical in-situ tests and their application to pile design

In-situ test	Application to pile design
Vane test	For measurement of in-situ undrained strength in soft to firm clays. Results may be applied to the estimation of downdrag on a pile shaft. The remoulded strength of sensitive clays may be relevant to overall stability problems associated with pile driving in soft clays.
Standard penetration test (SPT)	Investigation of thickness of bearing strata. Direct application of N value in empirical formulae for pile load capacity. Estimation of the angle of friction ϕ', in granular deposits, for use in pile design. Crude estimates of cohesions in stiff clays or matrix dominant fills. A tentative relationship between N value and modulus for use in pile/soil systems has been proposed.
Static cone test	Direct application of sleeve friction and point resistance to the design of driven piles. Estimation of the shear strength of clays and production of detailed soil profiles.
Pressuremeter	Estimates of modulus of soil for possible application in pile design, but results may not be strictly appropriate. Estimates of shear strength in weak rock.
Stressprobe pressuremeter	Gives a measure of the undrained shear strength of stiff clays and especially developed for over-water use.
Camkometer	Determination of shear strength parameters in clays, silts and sands. Shear modulus (with reservations regarding its applicability as for Pressuremeter), lateral stress and pore-water pressures can be measured.
Plate bearing test (over a range of depths)	Gives shear strength and modulus in all soil types. The shear strength parameters are highly relevant to pile design, as similar volumes of soil are stressed. The modulus values may not be quite so relevant as the pile installation can markedly influence this parameter. Important that this test is done carefully.
Simple permeability tests	For estimation of flow in permeable gravels or fissured rock. This may be a factor relevant to the selection of pile type. The tests can also be used in weak rock, to indicate highly fissured or fractured zones.

at the site has been gained, and some conclusions regarding the possible piling alternatives have been reached. This situation emphasizes the need for a flexible approach and the value of preliminary studies to facilitate the initial assessment. In many cases, the selection of a suitable pile type is controlled by external factors such as cost, vibration or noise level, and there is no simple way to relate a pile type to the prevailing soil and groundwater conditions.

Guidelines are given in BS5930, Section 4.3, for sample spacing, the size of bulk samples, types of sampling equipment and the intervals between in-situ tests, and these are applicable to the recommendations given in to the following notes.

Some situations may arise where special tests may be required or the intensity of sampling or testing may have to be increased to provide sufficient data on particular aspects. An example would be the estimation of rock modulus in a weak fissured rock, where, for instance, plate bearing tests combined with intensive large-diameter rotary diamond coring and possibly other in-situ tests may be necessary. Such an approach would be very much dependent on the costs involved in the additional exploration compared with the costs of deeper, or larger (possibly over-conservative) pile foundations.

The usual forms of cable tool boring equipment and rotary drilling equipment are appropriate, but in view of the additional depth of boring likely to be necessary, larger than usual starting diameters for boreholes and drillholes may be prudent. This permits 'stepping down' at an intermediate depth in order to maintain a reasonable rate of progress. Normal sampling and testing methods suffice, but there is a bias towards in-situ static and dynamic penetration tests, especially the SPT and static cone test, since pile design by empirical methods can be based directly on the results. Table 2.1 summarizes typical in-situ tests which may be carried out, and their application to pile design.

2.5 Methods of exploration for various soil types

In the following sections, specific soil types are described with appropriate methods of exploration. Whilst the ground conditions may well closely approximate those described, intermediate or mixed conditions are quite likely, and a combination of methods would then be employed. Pile design is related to the soil conditions, and if established methods of design are to be employed, the investigation should provide appropriate parameters.

2.5.1 Stiff fissured clays

In these soils, bored cast-in-place piles or low-displacement driven piles are usually employed. At the present time pile design is generally based on values of undrained strength obtained from triaxial compression tests on 'undisturbed' open-drive samples, using the 'α' adhesion factor method. Scattered shear strength values are usual as a result of the fissures and of sample disturbance, and a sufficient number of tests should be carried out to permit selection of a mean or upper and lower bound strength profiles without too much guesswork.

In fissured clays, sample size and orientation influence the results of strength tests. If the 'α' method is employed, using α values recommended by Skempton (1959), undrained triaxial tests should be carried out on 38mm diameter specimens cut from open-drive U100 samples. Due care should be exercised to relate the strength obtained from undrained tests to the mass behaviour of the soil below a pile base where the operational strength will generally be lower than that obtained from laboratory tests. Estimates of modulus of deformation from laboratory tests are very approximate, and even the results obtained from plate bearing tests may not be wholly representative of the installed pile.

In very stiff fissured clays with a high silt content, open-drive sampling becomes impracticable and in-situ dynamic or static penetration tests such as the SPT or the static cone penetrometer may be employed. Pile design from the results of these tests is highly empirical although it is generally adequate for a preliminary test pile. A number of correlations have been proposed between the SPT 'N' value and the cohesion of silt and hard clays. These are given below.

$$c_u = 20N \text{ kN/m}^2 \text{ (Meyerhof)}$$
$$c_u = 13N \text{ kN/m}^2 \text{ (Terzaghi and Peck)}$$
$$c_u = 7N \text{ kN/m}^2 \text{ (Reese, Touma and O'Neill)}$$
$$c_u = 4 \text{ to } 6N \text{ kN/m}^2 \text{ (Stroud and Butler)}$$

Clearly the range of values is large, and the correlations should not be employed in isolation, but checked against any available field and laboratory test results for the particular clay being investigated. Corroboration is particularly necessary for the higher c_u derivations. It should be noted that UK practice tends towards the use of lower ratios compared with typical US practice. This could have come about as a result of a number of factors, such as typical SPT field methods and may not relate to differences in the soil. The relationship between 'N' and c_u proposed by Stroud and Butler (1975) is frequently adopted for use in UK non-sensitive clays where other measurements are not available. It is discussed further in section 2.5.2 in connection with glacial tills. Where the static cone penetrometer is used to estimate soil strength, the cohesion, c_u, may be predicted from the relationship

$$q_c = Nk\ c_u + p_0$$

where q_c denotes the loading on unit area of the cone (kN/m^2)

Nk denotes the cone factor

c_u denotes the undrained shear strength (kN/m^2)

and p_0 denotes the total overburden pressure (kN/m^2).

Estimation of shear strength by this method has been used extensively in

North Sea glacial clays, where electrical cones are used (see Figure 2.3). Cone factors, Nk, of 25 are recommended by Marsland (1976), who also noted that the ultimate bearing capacities measured by the cone were greater by factors of between 1.2 and 3.0 than those obtained from plate bearing tests. Further information on cone testing may be found in Meigh (1987).

Of the latest methods available for the determination of deformation and strength parameters from in-situ measurements, there is the self-boring

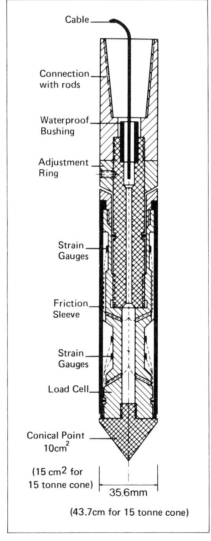

Cable

Connection with rods

Waterproof Bushing

Adjustment Ring

Strain Gauges

Friction Sleeve

Strain Gauges

Load Cell

Conical Point 10cm^2

(15 cm^2 for 15 tonne cone)

35.6mm

(43.7cm for 15 tonne cone)

Figure 2.3 Cross-section of electric static cone. Illustration courtesy of Fugro Ltd., Hemel Hempstead pressuremeter such as the 'Camkometer' or the Marchetti spade dilatometer. Further information on pressuremeter testing may be found in Mair and Woods (1987).

Where pile design is to be based on effective stress methods, the sampling programme should provide material for consolidated undrained tests, with measurement of pore-water pressure in order to estimate c' and ϕ'.

2.5.2 *Cohesive glacial till*

Tills are generally overconsolidated, and in some cases fissured. The notes under section 2.2 are therefore relevant to these materials. However, there should be increased awareness of the probability of soil variations which could be of considerable significance in pile selection and installation. Detailed logging of boreholes and samples is therefore necessary. Whilst progress in boring through glacial soils may be slow, it is imperative that these deposits are investigated in depth. If boulders are encountered, they should be proved for at least 3 m, and preferably deeper. Bedrock should not be assumed unless there is some corroborative evidence. When it is not possible to penetrate boulders by cable tool methods, rotary diamond coring can be used, although this then precludes further soft ground sampling and some types of in-situ test.

Open-drive sampling and in-situ testing become more difficult as the proportion of coarse clastic material in a till increases. Similarly, problems may arise in the pile installation. In these situations, it is important to establish a grading for the till, and particle size distribution tests on suitably sized bulk samples should be considered. It is difficult to secure samples which are free from sampling bias in a glacial soil of this type, unless the deposits are at shallow depth in which case an excavated trial pit may be considered. The depth of piled foundations usually precludes this approach and the advice of geologists or engineers familiar with the local ground conditions should be sought. The site investigation borehole logs should record chiselling time and the rate of borehole and casing advance, since this may give valuable information regarding the difficulties of installing small-diameter temporary casings for bored cast-in-situ piles. Resort may be made to rotary diamond coring using air flush, which is successful in some situations, although this method may be extremely costly and core recovery may be poor. The static cone penetrometer has proved a useful method of predicting shear strength in glacial tills in the special environment of the North Sea. As discussed in section 2.5.1, a cone factor, Nk, of 25 has been suggested by Marsland (1976). Other authors recommend factors of between 15 and 30 (Toolan and Fox, 1977) and 15 and 20 (Kjekstad *et al.*, 1978). An empirical relationship between Nk and the cone resistance intercept on a q_u-depth profile is proposed by Thorburn *et al.* (1981).

Less satisfactory, but more economical, is the SPT. Stroud and Butler (1975) recommend correlation coefficients between the N value and the

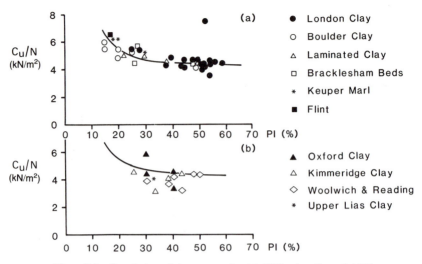

Figure 2.4   Correlation of shear strength with SPT value (Stroud, 1974)

undrained shear strength, c_u, of between 4 and 6 and relate the $N–c_u$ factor to the plasticity index of the clay (PI), as shown in Figure 2.4.

Whilst in-situ tests overcome the difficulty of sampling in cohesive tills containing granular material, the presence of the clastic components interferes with some of the tests, and every effort should be made to secure undisturbed samples for laboratory tests. If the recovered material is too mechanically disturbed for triaxial testing, moisture content, PL and LL determinations should be made. Note too that there is a minimum shear strength for remoulded clays based on the Skempton-Northey plot (Skempton and Northey, 1952) of liquidity index against undrained shear strength (see Figure 2.5), which gives a lower bound to the strength of the material. Points representing test results which fall below this line are probably from excessively disturbed samples, or from material which has dried out, the plot being for saturated clays. Envelopes for the undrained cohesion of matrix dominant tills for the 'CL' and 'CH' Casagrande sub-group classification according to moisture content are presented in Figure 2.6. Estimates of the shear strength from various sources can be combined to give upper and lower bounds, and factors of safety selected accordingly.

The notes made in connection with the observation and recording of groundwater conditions are especially pertinent in glacial soils, and great care should be taken to relate the observed conditions to the stratigraphy. To this end, a number of suitably disposed piezometers should be installed to record the hydrostatic head in selected strata.

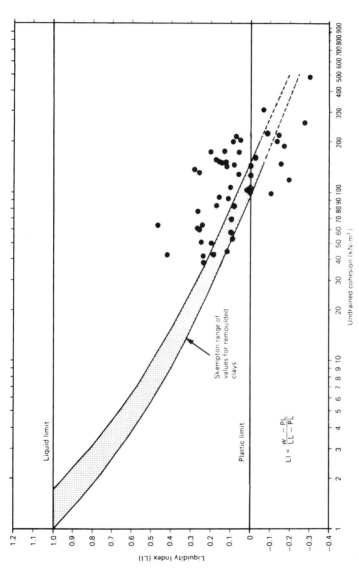

Figure 2.5 Relationship between Liquidity Index and undrained cohesion showing typical range of values in glacial till. CIRIA Survey (1977). Illustration courtesy of CIRIA, London

2.5.3 *Fine-grained granular deposits*

For the purpose of establishing parameters suitable for pile design, deposits of silts and sands are generally investigated by means of in-situ tests. Dynamic tests such as the SPT are sometimes used and the resulting N values employed in estimating pile capacity. However the use of the static cone penetrometer is generally advantageous in thick deposits of silts or fine sands to give a more reliable ground profile. The method also overcomes difficulties which are commonly met in these soils where water pressures cause piping into the base of open boreholes giving rise to low N values, although the SPT drive can be extended in such soils and the penetration recorded for each 75 mm. Values of N derived in this way are not however strictly comparable to those obtained in the normal way. For a given relative density, the N value is dependent on the effective overburden pressure, the overconsolidation ratio of the deposit, and, for derived parameters such as angles of internal resistance, the grading and shape of the particles. Correction to the N value

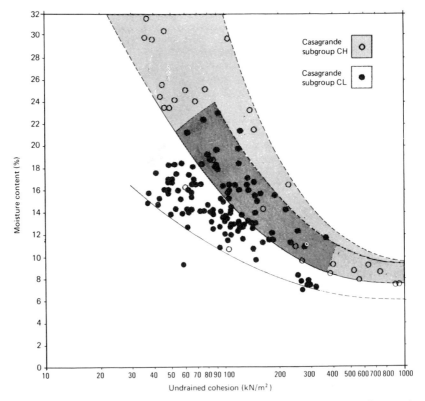

Figure 2.6 Relationship between undrained cohesion and moisture content for matrix-dominant till. Illustration courtesy of CIRIA, London

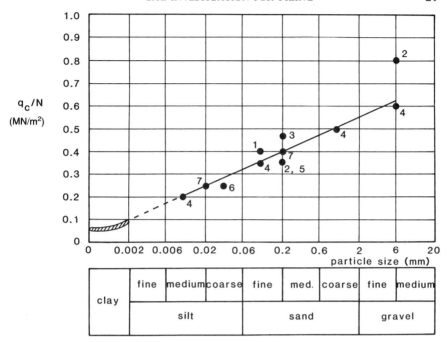

Figure 2.7 Correlation of cone resistance with SPT value. 1, Meigh and Nixon; 2, Meyerhof; 3, Rodin; 4, Schmertman; 5, Shulze and Knausenberger; 6, Sutherland; 7, Thorburn and McVicar

for the effective overburden pressure has been given most attention by a number of researchers. This is discussed further in section 2.5.4 in connection with medium- and coarse-grained granular deposits.

The modern electric cone penetrometer has overcome the problem of rod friction which was present with the mechanical type, and no correction is required for cumulative rod weight. The cone provides a measure of end resistance (q_c) and sleeve friction (f_s), which can be used for the design of driven piles. Where the static cone is employed, and a correlation is required with SPT N values (or vice versa), approximate conversion between the two parameters is possible using the relationship between cone resistance (q_c) and N for soil according to particle size as proposed by a number of authors. These are summarised in Figure 2.7.

It should be recognized that dense silts may result in high penetration resistance to driven piles without the realisation of correspondingly high bearing capacities and their presence should be adequately established. Accurate gradings of such deposits are not generally required for assessment of pile performance, and representative samples are in any event difficult to secure. Water pressures should be known in such strata especially where bored piles may be employed, as severe loosening of these materials can occur during boring where there are seepage pressures. Ameliorative measures to control loosening during continuous-flight auger boring are currently under development. Stand-pipes should be installed, or a piezometer used to monitor the water pressures, in specific strata.

2.5.4 Medium- and coarse-grained granular deposits

Exploration for pile foundations in this type of material is often carried out using the SPT, but the heavier capacity static cone tests are equally suitable. In-situ permeability tests, and bulk sampling for grading analyses, provide useful and sometimes essential additional information. It is general practice to employ empirical methods of pile design based on the results of penetration tests. Where the size range is large, and includes a finer fraction (as may be the case in glacially derived materials) the results of penetration tests may be misleading, and care should be exercised in the selection of representative values. In such soils, groundwater seepage pressures may cause loosening in the base of boreholes, and the penetration values may be affected if suitable precautions are not taken. These would include topping up boreholes with water, the use of small-diameter shells to clean out the base of the borehole, and reducing the rate of retrieval of the shell from the base of the borehole. Notwithstanding these devices, some reduction in N value may still occur. Whilst low values may lead to over-conservative design, it is also possible that unexpectedly heavy driving may be experienced with displacement piles and full penetration may not be possible.

Correction to the N value for effective vertical stress at the depth of the test may be carried out. This should not be applied to correct for poor technique—where tests have been carried out in loosened ground for instance, and it is recommended that the correction is only applied where consistent values have been obtained over a series of tests. A comprehensive review of the SPT by Skempton (1986) gives corrections which may be applied to N for the effective vertical overburden pressure, and also for the test as currently carried out in the UK, compared with the test when the relationship between N and relative density, D_r, was originally proposed by Terzaghi and Peck (1948). The overburden pressure relates the N value of the test to a reference value at a stress level of 100 kN per square metre (C_{N1}), as shown in Figure 2.8. A revision to the relationship between the N value and relative density, using a standardized N_{60} value as given by the modern test with an automatic trip hammer is provided in Figure 2.9, for $(N_1)_{60}$ values, these being N_{60} values corrected for overburden pressure.

In terms of the parameters which may be derived from the N value, Stroud (1989) presents charts relating the angle of internal friction with $(N_1)_{60}$. Figure 2.10 shows a relationship between the effective angle of friction (ϕ') less the constant volume angle of friction (ϕ'_{cv}) for a range of relative density and Figure 2.11 gives a relationship between the constant volume angle of friction

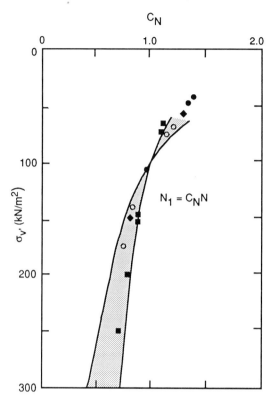

Figure 2.8 Correction for overburden pressure: (○) fine sands fill; (●) fine sands NC; (◆) fine sands OC; (■) coarse sands NC. Shaded region indicates laboratory tests (Marcuson and Bieganousky, 1977; Gibbs and Holtz, 1957); unshaded region indicates field test data. After Skempton (1986)

and particle shape. Stroud additionally considers the effects of overconsolidation ratio on the effective angle of friction.

These relationships enable a value of ϕ' to be derived from 'N' via the relative density, D_r.

Further causes of inaccurate N values are tests which are carried out in material below the base of a borehole which is mechanically disturbed to an excessive depth by overshelling, or testing partially or totally within the borehole casing. Double range tests, i.e. driving the tool an additional 300 mm, may reveal this occurrence, exhibited either as a marked increase in resistance over the final range, or a sudden drop respectively. In dense granular soils, it is often impracticable to use the SPT shoe, as it is too easily damaged. In these particular soils, the SPT 60° cone can be substituted without significantly influencing the results.

Figure 2.9   Effect of relative density, based on field data. After Skempton (1986)

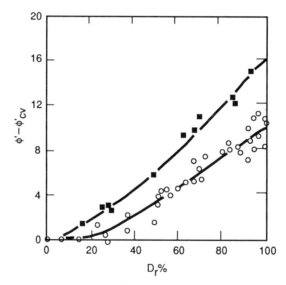

Figure 2.10   Variation of $\phi' - \phi'_{cv}$ with relative density: ($\bigcirc$) triaxial; ($\blacksquare$) plane strain; mean effective stress (p') = 150 to 600 kN/m²; ϕ' measured as secant value. After Bolton (1986)

   Chiselling depths and times should be recorded on the borehole logs (with the weight of the chisel) since this may be the only evidence available to provide an indication of the maximum boulder size present. If rotary diamond coring is used to penetrate a boulder, there may be some difficulty drilling

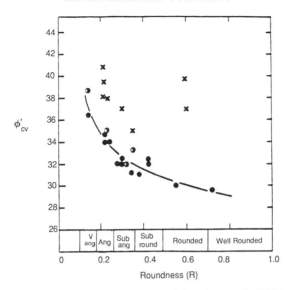

Figure 2.11   Relationship between particle shape and ϕ'_{cv} (after Youd, 1972) based on triaxial tests: (●) uniformly-graded quartz; (◑) uniformly-graded feldspar or feldspar quartz mixtures; (×) well-graded

below the boulder in purely granular material. A rock roller 'tricone' bit may permit progress, but at the expense of anything other than minimal geotechnical data. Every effort must be made to penetrate dense layers and to carry the investigation to the full depth to ensure that weaker or less dense layers at depth are fully explored. In order to install bored piles through, or into these strata, the groundwater conditions should be thoroughly investigated as discussed in section 2.3. The permeability of the deposits should be established by field permeability tests and the values should be related to the particle size distribution as seepage conditions are particularly relevant if non-displacement piles are selected for the foundation.

2.5.5 Normally consolidated clays

These clays are not employed as bearing strata but influence pile design as a result of 'downdrag' or by the effects they may have on the construction of bored cast-in-situ piles. The magnitude of the downdrag may be estimated from the undrained cohesion of the clay using the results of direct penetration vane tests, self-boring pressuremeters, or from laboratory triaxial compression tests on carefully taken undisturbed samples. It is generally necessary to employ thin-walled piston samples for this purpose. Such samples may also be used for laboratory determination of the shear strength of the clay in terms of effective stress to permit the calculation of downdrag by effective stress methods.

Where the undrained strength of these clays is low, there is a possibility of producing ground failure and deformed pile shafts under the head of concrete employed in the construction of bored cast-in-situ piles. A sufficient number of in-situ tests or laboratory tests should be carried out to establish a strength profile so that this possibility may be examined. Clays of a 'very soft' consistency are potentially troublesome especially in association with highly organic and peat layers or at a clay/gravel interface.

2.5.6 Weak rocks

Some information relating to the strength/density of weak or weathered rock may be obtained from static or dynamic penetration tests. The profiles may show broad changes in rock strength and may be useful in estimating the penetration of driven piles. Pile design from the results of dynamic penetration tests is generally not straightforward, especially in chalk and weak or weathered Triassic strata. In chalk, guidance on the use of the SPT is given by Hobbs and Healy (1979). It is now generally recognized that the use of the 60° cone, rather than the shoe, can significantly affect the N values obtained, and the shoe should be used for correlation with other work and to provide N values for use in empirical formulae. The sample obtained from the SPT tool is not representative.

In many cases penetration results are insufficient on their own, and it will be necessary to carry out rotary diamond coring in a diameter large enough to ensure as close to 100% core recovery as possible, even in weak friable rocks. Open-drive samples (U100), even if possible to obtain, do not yield representative material. Weathered sedimentary rocks frequently exhibit erratic strength profiles and weak zones below less weathered beds must be detected if they are present.

Percussive drilling combined with penetration resistance measurement may give some guide to the driving resistance of driven piles, but significant zones may be missed and there is generally little substitute for adequately detailed logs prepared from drill cores. Laboratory tests on small lengths of core are unlikely to yield information directly applicable to the pile design, but may be useful as an aid to the rock description and to establish a rock grade.

The results of field tests on rock cores by means of the point load test may be used in a similar way. Additional information on the in-situ rock mass may be obtained from permeability tests (packer tests for instance) which reflect the fracturing, fissuring and jointing of the rock structure. Where detailed information of in-situ rock condition or rock grades is required, especially in formations such as chalk or weak Triassic strata, a purpose-excavated shaft is recommended. Additional information can be obtained by carrying out plate bearing tests at selected horizons as the shaft is excavated.

Settlement of piled foundations on rock may be estimated from in-situ modulus measurements using the pressuremeter or by plate bearing tests carried out over a range of depths, though with adequate geological information, settlement may be shown to be too small to justify any extensive investigation. Pressuremeter testing is comparatively cheap, but suffers the disadvantages of possible disturbance of the rock when drilling the pocket for the probe, and 'sample' size and orientation effects. The plate bearing test is costly, but can give superior information.

Groundwater information is particularly important for bored or drilled piles, and adequate notes should be made of water encountered during drilling operations. Some qualitative information may be obtained from the quantity of water/mud returns if a fluid flushing medium is being used, and water levels in drilled holes should be recorded if the borehole is left standing for any length of time. In continuous aquifers, simple standpipes may be adequate to establish the hydrostatic head, but in more complex ground, piezometers will be required to isolate water pressures associated with a particular structure.

2.6 Backfilling site investigation boreholes

On sites likely to be piled, the backfilling of site investigation boreholes should be carried out with special care. This is particularly important in the case of bored piles where the loss of a pile shaft due to accidental connection to water-bearing strata via a badly backfilled borehole may occur. This may be accompanied by loss of ground in granular deposits. In an underreamed pile base, the likelihood of striking a site investigation borehole is increased and the consequences of a delayed collapse could be serious in a large-diameter pile where men may be working in the base. It is recommended that a weak grouted cement backfill is employed or, if suitable material is available, heavily compacted clay plugs may be formed at frequent intervals in the bore. For future reference, an accurate survey of the boreholes should be carried out on completion of the work.

3 Basic piling methods

In this chapter, the main pile types and methods of installing them are described, with details of currently available plant, pile types and proprietary methods.

The two basic methods of installing piles are well known, namely driving them into the ground, or excavation of the ground, usually by boring, and filling the void with concrete. Other methods have evolved to cope with certain ground or groundwater conditions, or to be more economic over particular (usually restricted) load capacities. Typical of this type of development are methods which involve a combination of driven and bored techniques. To install these piles, a casing is driven into the ground followed by in-situ concreting. Other more specialized techniques exist, such as the injection of grout into the ground. Within each category, variations exist associated with the particular proprietary method in use.

3.1 Pile types

In order to categorize the various types of pile and their method of installation, a simple division into 'driven' or 'bored' piles is often employed. This is adequate in many situations, but does not satisfactorily cope with the many 'hybrid' types of pile in use. A more rigorous division into 'displacement' or 'non-displacement' piles overcomes this difficulty. In the displacement (generally driven) pile, soil is moved radially as the pile shaft enters the ground. There may also be a component of movement of the soil in the vertical direction. Granular soils tend to become compacted by the displacement process, and clay soils may heave, there being little immediate volume change as the clay is displaced.

Piles of relatively small cross-sectional area, such as 'H' section piles, are termed 'low displacement piles', and the effects of compaction or soil heave are reduced. This can be advantageous if long lengths of pile are to be driven through granular deposits, or if the piles are at close centres, or if clay heave is a problem.

In the non-displacement (generally bored) pile, lateral stresses in the ground are reduced during excavation and only partly reinstated by concreting.

Problems resulting from soil displacement are therefore eliminated, but the benefit of compaction in granular soils is lost.

The displacement of the soil by a pile during installation is therefore a fundamental property, and its recognition in any classification of pile type is clearly advantageous. Little-used types such as screw piles can also be covered by the (low) displacement classification, whereas they could not be correctly termed 'driven piles'.

The two main categories of pile types may be classified further according to whether preformed units are used, and whether the preformed unit is used as temporary support for the ground and withdrawn during concreting. For non-displacement piles, factors such as pile diameter and underreaming are introduced to the classification, as they have a bearing on the method of installation, and particularly in the type of plant employed.

A scheme of classification of bearing pile types is presented in Figure 3.1. A further distinction is added in this scheme which relates to casting of concrete against the soil, which can occur with both displacement or non-displacement piles. This factor may be significant when ground or groundwater conditions are aggressive to concrete, or where water flows through highly permeable ground and there is a possibility of cement and fines removal from fluid concrete.

3.2 Displacement piles

Figure 3.1 shows the main types of displacement piles to comprise either totally preformed piles or piles in which a displacement method is used to form a void which is then filled with concrete.

3.2.1 *Totally preformed displacement piles*

These piles are either tubular or solid sections which are driven into the ground. The hollow tubular types may be formed from steel or concrete, and the solid sections from steel, timber or concrete; the latter are precast and may also be prestressed. Ease of transportation, handling on site and length adjustment has led to popularity of the precast jointed type. Some micro piles are also included in this category. A summary of the main types of preformed displacement pile is given below.

Totally preformed displacement piles

(a)	Precast concrete	: full length reinforced (prestressed)
		: jointed (reinforced)
		: hollow (tubular) section
(b)	steel	: 'H' section
		: tubular section
		: other

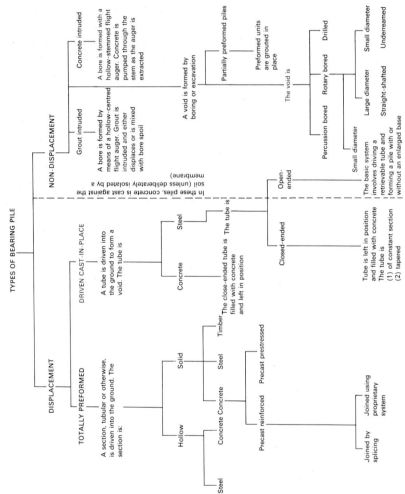

Figure 3.1 Classification of bearing pile types. After Weltman and Little (1977)

3.2.2 *Precast reinforced concrete piles*

Precast concrete piles are now usually of the jointed type, unless a large contract with a more or less constant depth of piling makes it economical to precast the piles on site, thus overcoming the transportation problem. Typically, precast non-jointed piles are of square section and of generally larger dimensions than jointed types, i.e. 400 to 600 mm square to work at 150 to 200 tonnes respectively. Extending precast piles which do not have joints is a lengthy process, involving breaking down the projecting pile head to provide a suitable lap for the steel and casting concrete to form the joint, butting piles together in a steel sleeve with an epoxy cement, or inserting dowel bars into drilled holes and using an epoxy cement to form the joint. Good alignment of the pile sections is required to prevent excessive bending stresses developing on subsequent re-driving.

There are some benefits from prestressing concrete piles. Tensile stresses which can be set up in a pile during driving are better resisted, and the pile is less likely to be damaged during handling in the casting yard and when being pitched. Bending stresses which can occur during driving are also less likely to produce cracking. However, the ultimate strength in axial compression is decreased as the level of prestress is increased, and the prestressed piles are therefore more vulnerable to damage from striking obstructions during driving. They are also difficult to shorten and special techniques have to be employed. As a result they are most suitable for a constant-length application.

Load capacities for prestressed concrete piles are given in Table 3.1 (after Weltman and Little, 1977).

3.2.3 *Jointed piles*

Jointed piles are not prestressed, but as each unit is a shorter length, generally 1 to 13 m, although even shorter sections are available from some manufac-

Table 3.1 Load capacities for prestressed concrete piles for normal driving conditions. After Weltman and Little (1977).

For square section of side—mm	Pile capacity (kN)	
	Ratio wt. hammer/pile	
	0.9	0.6
250	590	530
300	850	760
350	1160	1040
400	1500	1350
450	1900	1700

Note: Grade 40 (BS 8110) concrete for pre-tensioned tendons has been assumed (i.e. characteristic strength of 40 N/mm^2 at 28 days) turers, handling stresses are much lower, and there is no danger of overstressing from this cause. The piles are joined by various means —bayonet fittings or wedges being typical. The joint should be as strong as the pile, and preferably of the same moment of resistance. The fit of the joints should be good, as energy is lost through them, and increased hammering is then required to drive the pile. Angularity produced by the joints can lead to high bending stresses, and any joints which produce this effect should be rejected, with a misalignment of 1 to 300 considered a permitted maximum. Two methods of joining the pile segments are given in Figure 3.2.

The piles are usually of square section, but other forms, such as hexagonal sections are also produced. Typical capacities range from 700 to 2500 kN with pile cross-sections usually in the range 250×250 to 400×400 mm. Sizes above and below this range are however produced. In suitable ground conditions, piles have been driven to depths of up to 100 m, although the vast majority of typical applications do not exceed 30 m or so.

The units are cast from a high-cement-content mix with a water/cement ratio usually of around 0.4. The minimum compressive strength of cubes is generally around 60 N/mm² at 28 days. Pile units with square cross-sections will usually have a main reinforcement bar at each corner; a 12 to 20 mm dia. deformed bar with a yield stress of 600 N/mm² being typical. Hexagonal section piles characterized by the 'Herkules' type have six main tensile reinforcement bars. An advantage of a hexagonal pile cross-section is a slightly better distribution of stress over the pile section during driving, and possibly less susceptibility to damage from torsional stresses. Transverse reinforcement in the form of 5 mm dia. wire 'spiral' controls longitudinal cracking, sometimes with additional bars near the pile ends. The joints are normally formed from a steel plate with integral starter bars projecting into the pile. A fixture in the casting formwork sets the plate square to the pile, but strict quality control of the squareness of the plate and concrete strength are required to produce successful jointed piles. Concrete cover should be the minimum to provide protection from corrosion, as large thicknesses of cover can lead to spalling. Accurate positioning of an adequate number of spacers is therefore an important aspect of casting the pile units. Hard steel points can be fixed at the toe of jointed piles for limited penetration into weak rock. The piles are generally driven with a drop hammer of 3 to 4 tonnes weight, although hydraulic hammers, vibrators or diesel hammers are sometimes used.

3.2.4 Shoes for concrete piles

Steel or cast-iron shoes with pointed or flat ends are generally fitted to concrete piles for driving through coarse granular soils or (weak) rock. In uniform clays or sands, they are not usually necessary. The pile point may be flat or tapered.

(a)

(b)

(c)

Figure 3.2. Two methods of joining pile segments. (a) Detail of the 'Herkules' pile joint. Photograph courtesy of A. Johnson Construction Company Ltd., Glasgow. ((b) and (c)) Locating and driving home the locking pin in the corner slots of the Westpile's Hardrive pile endplates. Photographs courtesy of Westpile Ltd., Uxbridge

3.2.5 *Hollow tubular-section concrete piles* (*cylinder piles*)

Hollow concrete tube piles have been produced in diameters to sustain loads in excess of 2 MN, typically 900 to 1400 mm in diameter. The concrete tubular sections would be post-tensioned and make efficient use of reinforced concrete, having twice the moment of resistance of a solid concrete pile of equivalent weight. The large-diameter tubular concrete pile has been used in marine applications to take large combined axial and bending loads. During driving, a point may be used to close the end of the pile for penetrating hard strata, but the pile may also be driven open-ended. In either condition, large soil displacements occur, and the pile is not suitable for high-density application unless in a loose soil. In marine applications the corrosion resistance of the high-quality concrete from which the piles are constructed is generally adequate, but in freeze-thaw conditions, the long-term resistance may suffer. The piles have been driven to depths of 60 m, but because of difficulties in cutting off any excess lengths of pile, or the costs of extending them, they are best suited for driving to a predetermined level. Tubular concrete piles are not in common use at the present time.

3.2.6 *Preformed steel piles*

Totally preformed steel piles comprise 'H' and 'I' sections, and also some tubular sections which are not filled with concrete after driving. Because of fears of rapid corrosion, there has been a reluctance by some engineers to employ steel piles, but in natural, undisturbed ground, free from contaminants, the corrosion rates of steel are low. The advantages of steel piles are (a) ease of handling, (b) little likelihood of over-stressing during lifting or pitching, and (c) ease of extension or reduction in length to cope exactly with variations in driven length. They are more suitable than concrete piles for driving through thin layers of harder strata, or natural obstructions, but against this, slender steel piles such as 'H' sections tend to wander off line and this is exacerbated if the piles are joined by welding. The tendency for slender piles of this type to wander off line during driving has been investigated by Hanna (1968). The uneven pressure distribution across the toe of a pile as the soil is displaced from the pile base is implicated as the cause. Other reasons, such as a form of dynamic instability, have also been cited. Length for length, steel piles tend to be more costly than concrete piles, but good load-carrying capacity for a given weight of pile can reduce driving costs. Slender 'H' section piles have a low volume displacement and this may be a useful property for high-density piling in soils which tend to heave or become over-compacted. In some ground conditions, such as chalk or weak siltstones or sandstones, there may be a problem in achieving a suitable resistance with 'H' section piles. In chalk, this is believed to be the result of the shaft whipping as it is driven (Hobbs and Healy, 1979). In the other rocks, George *et al.* (1976) recommend adding wings to 'H' piles to increase their surface area.

Figure 3.3 Typical cross-section of an 'H' section pile. Courtesy of British Steel

Typical cross-sections and dimensions of an 'H' section Universal Bearing pile are given in Figure 3.3 and Table 3.2.

3.2.6.1 *Hollow steel piles.* Hollow steel piles are of tubular or box type sections. They overcome some of the problems encountered with the flexibility of slender 'H' sections, but are large displacement piles as the open end tends to plug. Heave of the ground and the possibility of over-compacting granular deposits so that piling becomes progressively more difficult may preclude their use in high-density applications. In loose granular deposits, the latter effect may be beneficial. Ground heave can be reduced by grabbing out soil which enters the pile base during driving, and possibly preboring to some depth. As hollow piles are driven through cohesive deposits, the internal friction of the plug of soil entering the pile base will exceed the bearing capacity of the soil at the pile tip, and displacement will then occur unless the plug of soil is grabbed out. Soil displacement at depth may not lead to heave at the surface, and whether or not the plug is grabbed out depends on the tendency of the soil to heave, and the total depth of driving required, as removing the plug facilitates pile penetration. The depths at which these various effects occur are however difficult to judge and control.

Hollow piles may be driven closed-ended through cohesive soil containing cobbles and small boulders, but heave is then more likely to occur.

This type of steel pile performs well in resisting impact and bending loads, and the availability of large-diameter sections enables considerable loads to be carried. This combination of properties has led to their extensive use in marine structures, where long freestanding lengths of pile are common in, for example, jetties, piers and mooring dolphins.

Box section piles are typically formed by welding two standard sheet pile sections, edge-to-edge, to form a box. The properties of piles formed in this way, using Larssen sheet piles are given in Table 3.3, with typical dimensions identified in Figure 3.4. Permitted maximum pile loads will be dependent on the grade of steel employed, provided suitable ground reaction is present. Greater load bearing capacity may be obtained by using Frodingham double box piles. In a practical application, with extended lengths of pile above a

Table 3.2 Dimensions and properties of BSC 'H' section piles

Serial Size	Mass	Depth of Section D	Width of Section B	Thickness Web and Flange T	Root Radius r	Depth between fillets d	Ratio $\frac{D}{T}$	Area of Section	Moment Inertia Axis X-X	Moment Inertia Axis Y-Y	Radius Gyration Axis X-X	Radius Gyration Axis Y-Y	Elastic Modulus Axis X-X	Elastic Modulus Axis Y-Y
mm	kg/m	mm	mm	mm	mm	mm		cm²	cm⁴	cm⁴	cm	cm	cm³	cm³
356 × 368	174	361.5	378.1	20.4	15.2	290	17.7	222.2	51134	18444	15.2	9.11	2829	976
	152	356.4	375.5	17.9	15.2	290	19.9	193.6	43916	15799	15.1	9.03	2464	841
	133	351.9	373.3	15.6	15.2	290	22.5	169.0	37840	13576	15.0	8.96	2150	727
	109	346.4	370.5	12.9	15.2	290	26.9	138.4	30515	10901	14.8	8.87	1762	588
305 × 305	223	338.0	325.4	30.5	15.2	247	11.1	285.0	52817	17570	13.6	7.80	3126	1080
	186	328.4	320.5	25.7	15.2	247	12.8	237.0	42625	14108	13.4	7.70	2597	881
	149	318.2	315.5	20.6	15.2	247	15.4	190.0	33040	10869	13.2	7.60	2075	689
	126	312.4	312.5	17.8	15.2	247	17.6	161.3	27526	9013	13.1	7.50	1760	576
	110	307.9	310.3	15.4	15.2	247	20.0	140.4	23580	7689	13.0	7.40	1532	496
	95	303.8	308.3	13.4	15.2	247	22.7	121.0	20111	6529	12.9	7.30	1324	424
	88	301.7	307.2	12.3	15.2	247	24.5	112.0	18402	5959	12.8	7.30	1220	388
	79	299.2	306.0	11.1	15.2	247	27.0	100.4	16400	5292	12.8	7.26	1096	346
254 × 254	85	254.3	259.7	14.3	12.7	200	17.8	108.1	12264	4188	10.7	6.22	965	323
	71	249.9	257.5	12.1	12.7	200	20.6	91.0	10153	3451	10.6	6.10	813	268
	63	246.9	256.0	10.6	12.7	200	23.3	79.7	8775	2971	10.5	6.11	711	232
203 × 203	54	203.9	207.2	11.3	10.2	161	18.0	68.4	4987	1683	8.54	4.95	489	162
	45	200.2	205.4	9.5	10.2	161	21.1	57.0	4079	1539	8.46	4.90	408	133

Note: Current sizes and properties may require revision to suit EC steel grades.

Table 3.3 Larssen box piles: dimensions and properties

Section	B mm nom.	H mm nom.	d mm	Mass per metre kg	Section Area cm²		Least Rad.of Gyr cm	Approx Perim. mm	Moment of Inertia cm⁴		Section Modulus cm³	
					Steel only	Whole Pile			About X-X	About Y-Y	About X-X	About Y-Y
6 W	525	252	7.8	89.4	113.8	1122	9.33	1652	10028	39280	796	1400
9 W	525	300	8.9	102.0	129.9	1344	11.43	1748	16976	43812	1132	1562
12 W	525	349	9.0	120.8	153.9	1589	13.00	1872	25992	55442	1490	1970
16 W	525	391	10.5	136.6	174.0	1776	14.80	1952	38126	60920	1950	2164
20 W	525	447	11.3	154.6	196.9	2005	16.63	2084	54474	70120	2438	2482
25 W	525	502	12.1	175.8	223.9	2221	18.33	2164	75240	80082	2998	2830
32 W	525	502	17.0	207.3	264.1	2224	18.02	2168	96584	85806	3848	3032
3	400	299	14.1	124.4	158.6	1020	11.22	1596	19954	30559	1348	1399
4A	400	428	15.7	148.0	188.7	1413	13.82	1772	47055	36039	2199	1649
6(122 kg)	420	502	22.0	244.0	310.6	1794	14.77	2080	106780	67756	4237	2920
6(131 kg)	420	502	25.4	262.0	333.6	1794	14.47	2080	118588	69859	4725	3011
6(138.7 kg)	420	502	28.6	277.4	353.2	1794	14.27	2080	128448	71392	5117	3077

Figure 3.4 Typical cross-section of a Larssen double box pile. Courtesy of British Steel

sea bed for example, the loads may be limited by the slenderness ratio of the effective length of pile above the supporting soil.

Spiral welded tubular steel piles, formerly produced by BSP, are available from Tex Steel Tubes Ltd. in an extended range of sizes, from 250 mm diameter to 2.1 m diameter. The wall thicknesses vary from 3 mm in the smaller diameters to 19 mm in the larger sizes, but a range of thicknesses is available in each tube diameter.

3.2.7 Timber piles

For modest loads and piling depths of about 12 m, timber piles are quite suitable, provided certain precautions are taken. They are infrequently used in the UK, although they are still in common use in Scandinavian countries. In the UK the 3M Civil Engineering Company offers timber piles of up to 400 × 400 mm cross-section.

Working loads are unlikely to exceed 500 kN, partly because of difficulties in driving piles to give sets appropriate to greater working loads and partly

Table 3.4 Lengths of timber piles

Timber	Square section side (mm)		Max. length approx. (m)
Douglas fir	(sw)	400	15
Pitch pine	(sw)	500	15
Greenheart	(hw)	475 +	18 +
Jarrah	(hw)	250	9
Okan	(hw)	250	9
Opepe	(hw)	250	9

sw   softwood
hw   hardwood

because of the generally smaller cross-sections of timber piles and lower compressive strengths. Stresses should be checked according to CP112 for compressive loads parallel to the grain. Pile cross-sections are usually square although circular sections can be used. Section sizes depend on the type of timber; BS 8004 quotes the cross-sections and lengths given in Table 3.4. The lengths of timber piles quoted in Table 3.4 are somewhat greater than the typical pile length quoted at the start of this section, to allow sufficient spare length for trimming after driving.

3.2.7.1 *Driving timber piles.* Drop hammers are usual for driving timber piles. Damage through overdriving manifests itself in 'brooming' of the pile head, cracking of the shaft and unseen damage to the pile tip. The driving process crushes and separates the fibres at the head of the pile—hence the 'broom' effect. A steel band around the pile head may restrict this damage to manageable proportions, but timber piles are not particularly suitable for driving through dense strata, or strata with obstructions. A steel or cast-iron shoe is generally attached to the tip of the pile. For heavy driving, the weight of the piling hammer should be equal to the weight of the pile. When the driving is easy, it should be not less than half the weight of the pile. Particular care should be taken to make sure that the pile heads are square, that the hammer blows are axial to the pile, that the pile is properly set up in the leaders, and not strained to bring it vertical after a bad start to the drive. If damage occurs, and the hammer weight is suitable for the driving conditions, it may be necessary to de-rate the pile (i.e. reduce the design load of the pile), so that smaller drops and fewer blows can be used for its installation.

3.2.7.2 *Preservative treatment.* Both softwoods and hardwoods should be treated. Although some hardwoods such as greenheart have slight inherent resistance to decay, it is not usual to rely on this for long-term protection. The procedures set out in British Standards 913 or 4072 should be followed. Note that if cracking of the pile shaft occurs, or there is damage at the head that extends below the cut-off level, the preservative treatment will not be effective and the trimmed end of the pile should be treated. Where softwoods are used as bearing piles, it is normal to trim these below the groundwater level (for the driest season) and extend the shaft as necessary with a concrete or possibly suitably corrosion-protected steel section.

3.2.7.3 *Quality and tolerances.* Timber for piles is specially selected for straightness of grain and should also be free from defects such as shakes which would weaken it and render it liable to fail during driving, or impair preservative treatment. Piles machined to a section should not be out of straightness by more than 25 mm over their length. Round piles may be out of straightness by up to 25 m over a 6 m chord.

3.2.7.4 *Marine conditions.* Timber piles have long been favoured for construction of small jetties on account of their energy-absorbing qualities. They are, however, open to attack above the mud line from a variety of marine organisms, and the marine borers *Limnoria* and *Teredo* are usually mentioned in this context, for which arsenates and creosotes are recommended preservatives respectively. Greenheart has some natural resistance to attack. In shallow, warm waters, pile attack is more rapid than in cold, rough water conditions.

3.2.8 *Driven cast-in-place displacement piles*

Driven cast-in-place displacement piles are suitable for a wide range of loads. The pile may be of two forms, one employing a driven temporary tube to form the void which is then filled with concrete as the tube is withdrawn, and the other employing a driven tube to form a permanent casing. The former variation is most commonly used in the lower load range, whilst the latter can be used to support considerable loads in suitable ground conditions. As the load-supporting medium is generally the concrete infill, methods of driving have been evolved which do not over-stress the non-load-bearing forming tube. Thus this type of pile is often bottom driven using an internal hammer to reduce driving stresses on the tube, rather than being top driven. A steel plate closes the end of the pile, and the hammer works on a plug of 'earth' dry concrete in the base of the pile tube. Cased piles, formerly manufactured by BSP, but now made by Tex Steel Tubes Ltd., provide a typical example, for which working loads, hammer sizes and pile sizes are given in Table 3.5. Reinforcement for these piles is limited to starter bars at the top of the pile, unless large column stresses are to be resisted. Other forms employ a mandrel which distributes the driving stresses to the casing over a large area, and provides some internal support to the tube. An example of this system is the Westpile's 'Shell' pile (see Figure 3.5). Pile sizes and design loads for the Westpile's 'Shell' pile are given in Table 3.6.

In those piles in which the tube is retrieved, it is necessary either to bottom drive the tube with a plug that may be driven out, or to top drive the tube and close off the bottom end of the pile with an expendable blanking plate or 'shoe'. When the tube is withdrawn, the shoe remains at the base of the pile. In the former system, a plug of dry-mix concrete or gravel is compacted in the base of the tube using an internal drop hammer. The hammer may then be used to drive the tube into the ground via the compacted plug. At the required depth the tube is restrained at ground surface, and the plug driven out to form an enlarged base.

This type of 'driven cast-in-place' pile is typified by the 'Franki' pile. Concrete infill may be a zero slump mix, deposited in the tube from a skip and compacted by the internal drop hammer as the temporary tube is withdrawn. Alternatively a higher workability mix can be used, and compacted by an

Table 3.5 Sizes, drop hammer weights and normal design loads for cased piles driven with internal drop hammer, (From BSP data; now Tex Steel Tubes Ltd.)

Internal dia. of casing (mm)	Weight of internal drop hammer (tonnes)	Design load (tonnes)		Casing thickness (mm)	Welded plate shoe thickness (mm)
		Ordinary soil*	Rock etc.†		
254	0.75	15	20	3.3	9.5
305	1.25	30	35	3.3	9.5
	1.75	(40)§	45		
356	2.0	40	50	3.3	12.5
	2.75	(50)§	65		
406	2.5	50	60	3.7	12.5
	3.5	(60)§	85		
457	3.0	65	80	4.0	16
	4.0	(75)§	100		
508	4.0	80	100	4.4	16
	5.5	(90)§	130		
559	5.0	100	125	4.8	19
610	6.0	120	150	5.3	19

*Ordinary soil; sand, gravel or very stiff clay.
†Rock etc.; rock, very dense sand or gravel, very hard marl or hard shale.
§Subject to individual consideration.
The above table is normally applicable for pile lengths up to about 15/18 m. If the piles are driven in two lengths, the upper length may be of lesser thickness. For longer piles an increased casing thickness will be required for the lower portion of the piles.

Figure 3.5 Westpile's Shell piling system. Photograph courtesy of Westpile Ltd., Uxbridge

Table 3.6 Leading dimensions and
nominal load capacities of Westpile's
Shell Pile

O.D. Shell (mm)	Nominal working load (kN)
340	450
400	650
440	800
530	1200

Note: All shells are of 914 mm length.

electric vibrator clamped to the top of the tube. In both cases, reinforcement is installed in the pile shaft before concreting, widely spaced bars being necessary to allow the low workability mix to penetrate to the exterior of the pile. If the pile is to resist compressive forces only, the reinforcement may be restricted to the upper section.

Alternative methods of forming the shaft are to employ a precast member, or a permanent casing, seated in the enlarged plug. These methods can be helpful when downdrag forces or aggressive ground conditions are present. The plug would then consist of an inert aggregate, and the shaft suitably coated or wrapped to resist downdrag forces or the adverse ground chemistry as appropriate.

Typical Franki pile dimensions range from 275 mm to 650 mm diameter, with design loads from 35 to 200 tonnes. Maximum lengths are of the order of 30 m, although the length usually used is less than 20 m.

3.2.8.1 *Concrete mixes for driven cast-in-place piles.* The mixes used for driven cast-in-place piles vary with the type of compaction applied, as discussed in the previous section. The mixes should not however contain less than 300 kg/m^3 of cement. It is suggested in BS 8004 that the mean compressive stress over the total cross-section of the pile under working load should not exceed 25% of the specified 28-day works cube strength. Some increase in working compressive stresses is possible with permanent steel casing.

3.2.8.2 *Damage to adjacent piles during driving.* Driven cast-in-place piles can be damaged by driving adjacent piles too close or before the concrete has reached a suitable strength. The piles may be damaged by lateral forces or by tensile forces, as the ground heaves. This type of movement can separate shell piles. When it is suspected that pile damage of this type has occurred it may be decided to carry out a pile load test or integrity test as a check. On a pile which is cracked by this type of damage, a load test may yield an apparently satisfactory result; the long-term performance of the pile may however be impaired if the steel reinforcement is exposed in the cracks. To lessen the risk of cracking caused by soil movements, a minimum spacing of $5D$, centre to centre, is often employed when driving adjacent piles when the concrete is less than 7 days old. Integrity tests can however provide sufficient information to modify this rule.

During installation of cast-in-place piles with relatively thin bottom-driven permanent steel casing, collapse of the tube can occur from lateral soil displacement, if the piles are driven at centres that are too close. This has sometimes resulted in the loss of the hammer at the base of the pile, when the collapse occurs above the hammer as the pile is driven. The occurrence is more likely, however, when driving piles inside a coffer-dam. Where this problem is encountered, and there is no way to reduce the piling density, preboring may be considered as a method of reducing the effect over the upper part of the pile. At the design stage, if high-density piling is unavoidable in soils prone to heave such as stiff clays, a low displacement 'H' section pile may be selected as more suitable. For driven cast-in-place piles of the 'Franki' type, the multi-tube technique described by Cole (1972) can be employed. All piles within 12 diameters of each other are considered to form a part of a group, and are driven (and if necessary, re-driven) to final level before basing out and concreting. The same principle may be employed for shell type piles, but it is most important that the shells do not separate during the operation. The amount of uplift of the shells may be checked against the driving mandrel if this is lowered to the bottom of the pile.

3.2.8.3 *Driven expanded piles.* A new form of low displacement driven pile, the Burland 'wedge pile', has recently been patented. Whilst the principle of the wedge pile is not restricted to any particular pile cross-section, in its current form, it comprises a cruciform cross-section (see Figure 3.6a). This is fabricated from four steel angle sections, tack welded back-to-back at their edges, and forms a closed sleeve. The sleeve is fitted with a shoe and driven (or jacked) into the ground to the required depth. A tapered cruciform expander mandrel is then forced down the entire length of the sleeve, splitting the tack welds. An expansion of about 10% of the width of the pile occurs, which increases the shaft friction to a significant degree. The mandrel is withdrawn, leaving a set of spacers behind, which maintain the pile in the expanded condition in order that it may be grouted for permanence (Figure 3.6b).

Pile sizes and load capacities may be varied over an extremely wide range, from small cross-sections of the order of 100 mm to very heavy piles in excess of 1 m across. Piling depths vary according to the section and ground conditions, but there are no special limitations.

3.2.9 *Helical (screw) cast-in-place displacement piles*

Whilst the entry of this type of pile into the ground is effected by means of a type of auger, the process involves compaction rather than removal of the

Figure 3.6 Details of Burland 'wedge' pile system: (a) expander mandrel passing into sleeve; (b) expander mandrel showing recesses for cylindrical spacers. Courtesy of British Steel

soil and, in this respect, the piles are of a displacement type. In forming the pile, a heavy-duty single-start auger head with a short flight is screwed into the ground to the required depth. The auger head is carried on a hollow stem which transmits the considerable torque and compressive forces required, and through which the reinforcement cage is inserted after completion of the insertion process. The end of the hollow stem is sealed with a disposable tip. Following placement of the reinforcement, concrete is poured through this tube, from a hopper at its head. As concrete filling takes place, the auger is unscrewed and removed, leaving behind a screw-threaded cast-in-place pile. By virtue of the combined rotation and controlled lifting rate applied at the extraction stage the 'threads' are of robust dimensions. The sequence of pile construction is shown in Figure 3.7.

This method of forming a pile is known as the Atlas Piling System, and is marketed by Cementation Piling and Foundations in the UK, in association with N.V. Franki S.A of Belgium. A purpose-designed, track-mounted rig provides hydraulic power for auger rotation and the application of downward force, and is fitted with a crane boom for handling reinforcement and concrete skips. For a given pile size and volume of concrete, pile capacities are greater than for traditionally constructed bored piles, although the restricted diameter of the reinforcement cage may be a disadvantage if the pile is required to resist high bending stresses. The system does however combine many of the advantages of a displacement pile with the low noise and vibration characteristics of a bored pile. It will operate in most cohesive and granular

Figure 3.7 Atlas piling system: process

strata to a maximum depth of 22 m, providing piles ranging in diameter from 360 mm to 560 mm.

3.3 Methods of pile driving

The methods of driving piles fall into the following categories:

  (i) dropping weight
 (ii) explosion
(iii) vibration
 (iv) jacking against a reaction.

3.3.1 Drop hammers

The dropping weight, or drop hammer, is the traditional method of pile driving and is still employed. A weight approximately equal to that of the pile is raised a suitable distance in a guide and released to strike the pile head. For single tubular piles a free-standing timber trestle support is sometimes employed to support the pile above ground, provided that an internal drop hammer can be used. Alternatively, tubular guides known as 'chimney pots' can be used. Normally, purpose-made rigs will hoist the pile into position,

Leader mast head

Rope head wheels

Pile rope

Hammer cage

Drop hammer

Hammer hydraulic ram
Pile helmet

Flexible hydraulic pipe

Inner leader mast

Two stage leader Hydraulic ram

Pile

Outer leader

Rear braces

Upper raking hydraulic ram

Leader foot boom

Two hydraulic winches

Leader foot boom hydraulic ram

Lower raking hydraulic ram

Leader mast foot

Akerman H12 base machine
mounted on 750 mm crawlers

Figure 3.8 Highly developed 'Banut 400' piling rig, crawler mounted, hydraulically operated drop hammer machine, with forward, backward and side-to-side raking facilities. Overall height of rig 18.75 m max., 11.53 m min. Max. pile length, 15.0 m. (Hercules Piling Ltd)

support it during driving and incorporate a guide for the drop hammer (Figure 3.8). The rig may be winched on skids or rollers, or is on tracks (Figure 3.9). A crane chassis may form an alternative base machine, and pile guide frame, or 'hanging leaders', supported from the crane jib and strutted from the base of

Figure 3.9   Roller-mounted pile driving rig. Photograph courtesy of CIRIA, London

the chassis (Figure 3.10). These rigs permit raking of the pile with relative ease, and in suitable ground rapid progress can be made.

In guiding the pile, a balance has to be struck between providing suitable directional control, and allowing some freedom for slight pile movement within the guides, particularly at the base of the frame. Twist or slight lateral shifts often occur as a pile is driven. If this is prevented by a tight-fitting gate at the base of the pile frame, or worse, by inclining the guides after driving has started, the pile may be overstressed in torsion or bending. The resultant pile damage may not be apparent at the surface. Slight out-of-verticality rarely affects the pile performance, and deviation from a specified pile location can often be tolerated. A positional tolerance of 75 mm and a maximum of 1 in 75 on axial line for vertical piles are typically specified figures. A deviation from line of 1 in 25 is generally allowed for raking piles. These tolerances accord with those given in the ICE *Specification for Piling* (1987). A check calculation should be made, or proof supplied by the piling contractor, that at the maximum tolerable inclination and eccentricity, the pile is not overstressed in bending when under load.

Drop hammer weights generally vary from 0.5 to 2 times the pile weight, with drops usually in the range of 200 mm to 2 m. Since the peak stresses at the pile head can be greatly increased if the hammer strikes the pile eccentrically, a

Figure 3.10   Pile driving rig with hanging leaders. Photograph courtesy of RDL Munro Piling

long narrow hammer is preferable, as there is more chance of the blow being axial and it has better impact characteristics. Some increase in hammer drop is usually required for raking piles, the driving energy being reduced by approximately 10% for a 1 in 3 rake. Distress at the pile head is more likely to be the result of using a hammer that is too light and hence needing an excessive drop, than using a hammer that is too heavy. The effect of increasing the ratio of hammer weight to the pile weight on pile driving efficiency is shown in Figure 3.11, for a 15.2 m long, 355 mm square-section concrete pile. In this example the pile weight is 4.7 tonne. The energy available from the hammer varies from a few percent at a hammer/pile weight ratio of 0.32 to almost 50% at a ratio of 1.06. Inertia losses and the temporary compression losses are

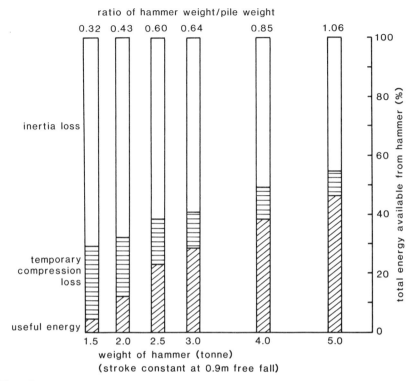

Figure 3.11 Energy distribution during driving of a precast reinforced concrete pile for various hammer weights. After Packshaw (1951)

reduced as the ratio increases to provide the increase in available energy. The maximum stresses at the top of a pile may be calculated from the relationships given by Broms (1973), as follows.

For a precast concrete pile $\sigma_h = 3\sqrt{h_e}$ MPa
For a steel pile (with wooden packing) $\sigma_h = 12\sqrt{h_e}$ MPa
For a steel pile (without wooden packing) $\sigma_h = 18\sqrt{h_e}$ MPa
For a timber pile $\sigma_h = 1.2\sqrt{h_e}$ MPa

where σ_h denotes the maximum stress, and h_e (cm) denotes the equivalent height of free fall of the hammer. Note that the velocity of striking the pile head, v_0, is given by

$$v_0 = \sqrt{2h_e g}.$$

The effect of the mass of the hammer is to alter the length of the stress wave produced in the pile. The length L_w of the stress wave is given approximately by the relationship

$$L_w = \frac{3W_h}{W_p/L}$$

where W_h denotes the total weight of the hammer, and W_p/L denotes the weight of the pile per unit length.

During driving, the ground conditions will dictate the way in which the compressive stresses produced at the pile head by the hammer blow are reflected from the pile tip. Under easy driving conditions, the compressive stress is reflected from the toe as a tensile wave. When the pile length is somewhat greater than half the length of the stress wave, there is a likelihood of tensile stresses appearing towards the top of a pile. These can be damaging to precast concrete piles, and when driving such piles in soft ground the hammer drop should be reduced, especially if using a light hammer to drive a long pile.

When heavy driving conditions are encountered, the initial compressive wave will be reflected from the pile toe as a compressive wave, but depending on the dynamics of pile/hammer contact, this compression wave can be further reflected as a tension wave at the pile head if the hammer is not in contact with the pile head at the time the returning wave arrives. Tensile forces can therefore occur under heavy driving conditions, which can be particularly damaging to concrete piles. Large compressive stresses occur at the pile toe and head when obstructions or dense strata are encountered, which may be sufficient to cause steel piles to deform or, if driving is protracted, fatigue can occur with similar results.

Variants of the simple winch-raised drop hammer include rams raised by steam or compressed air or hydraulically, free falling from the top of the stroke (single-acting hammers), and hammers having pressure applied on the

Table 3.7 Approximate minimum hammer sizes for driving bearing piles based on design load of pile for drop hammers and single-acting hammers. (BSP data.)

	Approximate minimum hammer mass	
Design load of pile (kN)	Steel piles (tonne)	Concrete piles (tonne)
400	—	2
600	2	3
800[1] to 900[2]	3	4
1500	5	—
2250	8	—
3000	10	—

[1] Concrete pile
[2] Steel pile

Figure 3.12   A Junttan hydraulic hammer driving precast segmental piles. Photograph courtesy of Cementation Piling and Foundations Ltd., Rickmansworth

downstroke (double-acting hammers). Double-acting air hammers are not generally suitable for driving concrete piles. However, large single-acting hammers have been developed with maximum net hammer weights approaching 300 tonnes, producing rated energies of up to 2.5 MNm using steam pressure (Vulcan 6300). By using a fluid cushion, the impact of the hammer is controlled in the larger pile hammers of the Hydroblok type, and a sustained push rather than a sharp peak of stress is transmitted to the pile at each blow. This type of hammer blow is better for driving through soils where the tip resistance of the pile is low as in clays. Hydraulic hammers tend to be more efficient and quieter in operation than other types, and are tending to be used in place of simple drop hammers and diesel hammers in many applications. They are versatile and may be used on a wide variety of bearing piles, including raking piles, and usually with only slight modification, may be used under water. They are also suitable for pile extraction. A typical hydraulic hammer is shown in Figure 3.12.

Approximate drop hammer and single-acting hammer sizes to drive various sizes of piles based on the design load of the pile are listed in Table 3.7.

Hydraulic hammers of the BSP type are typical of hammer sizes normally

Table 3.8 Typical characteristics of selected BSP hydraulic hammers

	HH3	HH5	HH7	HH11	HH16	HA30
Ram weight (tonne)	3	5	7	11	16	30
Blow rate at max. stroke (bpm)	46	40	38	38	30	28
Maximum potential energy (Kgm)	3600	6000	8400	13 200	19 200	36 000
Engine power (BHP)	150	150	150	292	292	750

Notes: 1) Engine power is typical figure (1 BHP = 0.75 kW)
2) Ram stroke is variable on all hammers between 0.2 and 1.2 m

employed. A summary of the leading characteristics of these hammers is given in Table 3.8.

3.3.2 *Pile driving by diesel hammer*

Early experiments in driving by the explosive energy of gunpowder are related in Chapter 1. The modern equivalent is the use of the diesel hammer to produce rapid, controlled explosions. In operation, the diesel hammer employs a ram which is raised by explosion at the base of a closed cylinder. Alternatively, in the case of the BSP double-acting diesel hammer, a vacuum is created in a separate annular chamber as the ram moves upward, and assists in the return of the ram, almost doubling the output of the hammer over the single-acting type. The ram weight is less than that of a drop hammer, but the more rapid action can make up for this loss in efficiency. The action of a diesel hammer of the Delmag type is best suited to driving piles which derive the greater part of their resistance at the pile tip, such as occurs when driving through granular deposits. The hammer is more versatile than a drop hammer, as it can be set on top of a pile without guide rails (leaders), attached to a crane or piling rig. For raking piles, however, hanging leaders or a pile frame are required.

A characteristic of diesel hammers is the way in which the energy output increases with driving resistance. In a very soft soil, there may be insufficient resistance to pile movement to cause satisfactory compression in the ram cylinder to produce an explosion, and a smaller hammer may be necessary in the early stages of driving for this reason.

Detailed hammer specifications should be obtained from manufacturer's data, but summary specifications are given in Table 3.9(*a*) and (*b*). The figures

Table 3.9(a) Summary specification for BSP double-acting diesel hammers

Hammer size	B15	B25	B35	B45
Ram wt. (kg)	1500	2500	3500	4500
Energy per blow (kg.m.) at 80 blows/min	3790	7320	8050	11400
Maximum rake	1 in 1	1 in 1.25	1 in 1.5	1 in 2
Suitable for driving steel bearing piles of design load* (kN)	600 to 900	1500	2250	3000
Suitable for driving concrete bearing piles of design load* (kN)	400 to 600	800	—	—

*Allowing F of S of 2.0, and for vertical piles. Hammer sizes must be increased for raking piles.

Table 3.9(b) Summary specification for Delmag diesel hammers

Hammer size	D8–22	D12	D15	D22–23	D30–23	D36–23	D46–23	D62–22	D80–12
Ram wt. (kg)	800	1250	1500	2200	3000	3600	4600	6200	8000
Energy per blow (m.k.g.)	2400 to 1300	3125	4125	6700 to 3350	9100 to 4550	11 500 to 5750	14 600 to 7300	22 320 to 11 160	27 200 to 17 440
Maximum rake*	1 in 2	1 in 1	1 in 3	1 in 3 to 1 in 1	1 in 3 to 1 in 1	1 in 3 to 1 in 1	1 in 3 to 1 in 1	1 in 2 to 1 in 1	1 in 2
Suitable for concrete bearing piles of wt. (kg) (depending on ground conditions)	2500	4000	4500	6000	8000	10 000	15 000	25 000	30 000

*For rakes of 1 in 1, the hammers are uprated by fitting an upper cylinder extension.

relate to plant in first-class condition, and for this reason, may not be appropriate in some cases to the actual conditions obtaining. Using pile driving analysers, it has been found that the energy delivered by a diesel hammer in good condition rapidly diminishes in use, unless carefully maintained in first-class condition. Site conditions are not always conducive to such levels of maintenance.

3.3.3 Vibratory methods of pile driving

Vibratory hammers are usually electrically powered (but may be hydraulically powered), and consist of contra-rotating eccentric masses within a housing attached to the pile head. Power requirements are up to 250 kVA, usually supplied by a mobile generator. The majority of pile vibrators run at low frequencies, typically 20 to 40 Hz. At these frequencies, neither the exposed length of pile nor the soil will be in resonance. Sound propagation is low, and in cohesionless soils good rates of progress can be realized. During the driving progress, the granular soil immediately adjacent to the pile is effectively fluidized, and friction on the shaft is considerably reduced.

In cohesive soils, fluidization will not occur, and vibratory pile driving methods are not generally as effective.

3.3.3.1 *Resonance pile driving.*

If the frequency of vibration is increased up to perhaps 100 Hz, the pile will resonate longitudinally, and penetration rates can approach 20 metres per minute in loose to moderately dense granular soils, falling to 5 m per minute or less in dense granular soils. At these frequencies non-cohesive soils are fluidized to the point where the frictional resistance on the pile shaft is reduced to close to zero, and the pile driving energy is delivered to the pile toe.

This method of pile installation lacks thorough investigation. It is potentially very effective, but there is a tendency for the driving equipment to suffer from the damaging effect of the severe vibration associated with the large energy inputs required to drive bearing piles. Noise and vibration propagation can be high, leading to settlement in nearby structures.

3.3.4 *Jacking methods of pile installation*

By jacking (usually hydraulically) against a reaction, lengths of pile, either in short units or in continous lengths, may be forced into the ground. The method tends to be restricted to micropiling (see section 3.6), as the reaction loads are then provided by the structure being underpinned. Other, portable reaction systems may be used to jack against in situations other than underpinning. A special application is the 'Taywood Pilemaster' system, in which sheet piles can be forced into the ground against the reaction provided by adjacent piles. Jacking methods are exceptionally quiet and vibration-free in use and by monitoring the pressure in the hydraulic system an indication of the pile failure loads can be obtained during installation.

3.4 Non-displacement piles

By the excavation of a borehole in the ground, a pile can be produced by casting concrete in the void. Some soils, such as stiff clays, are particularly amenable to the formation of piles in this way, since the borehole walls do not require support, except close to the ground surface. In unstable ground, such as gravels, the ground requires temporary support, from casing or a bentonite slurry. Alternatively, the casing may be permanent, but driven into a hole which is bored as the casing is advanced. A different technique, which is still essentially non-displacement, is to intrude a grout or a concrete from an auger which is rotated into a granular soil, and hence produce a grouted column of soil. The three non-displacement methods are therefore

(i) bored cast-in-place piles
(ii) partially preformed piles
(iii) grout or concrete intruded piles.

Particularly in pile types (i) and (ii), little or even negative soil displacement occurs during boring, and indeed, granular deposits may be loosened by the boring action. This is in contrast to a displacement pile, which produces (sometimes beneficial) ground compaction in granular deposits. On the other hand, the rough and irregular interface between the pile and the soil which is formed during the boring action tends to improve load transfer, and the difference in skin friction between the two types is less than it might be. In type (iii) the auger is usually made to run full throughout the process, and relaxation of the walls of the bore may be inhibited.

3.4.1 *Bored cast-in-place piles*

The range of pile sizes and hence design loads is considerable for this form of pile. The excavation process generally (but not exclusively) produces a borehole of circular cross-section, and piles are referred to as small-diameter when less than 600 mm, and large-diameter when greater than this nominal size. The smaller diameter piles tend to be bored by tripod rigs employing percussion methods, and the larger size by rotary or percussive grabbing methods. Nominal diameters (mm) for bored piles are as follows:

(i) 300, 350, 400, 450, 500, 550, 600 (small-diameter)
(ii) 750, 900, 1050, 1200, 1350, 1500, 1800, 2100 (large-diameter).

3.4.2 *Small-diameter percussion bored cast-in-place piles*

In this type of pile, the boreholes are formed by rigs similar to those employed for site investigation, using equivalent techniques. The tripods and equipment are heavier, however, and minimum pile diameters start at around 300 mm, which is about the maximum for a site investigation borehole. The percussion tools and screwed casing are very similar. In cohesive soils, a 'clay cutter' is used, often weighted to improve penetration. The winch for raising the cutter and spoil to the surface may be driven by diesel or compressed air. The hole is advanced by repeatedly dropping the clay cutter in a cohesive soil, and winching the cutter to the surface with its burden of spoil. The cutters may consist of an open cylinder, with a hardened cutting edge or sometimes a cruciform section. Clay is extruded past the cutting ring, and adheres to the cruciform blades and may be pared away at the surface, or may pass into the body of the cylindrical type of cutter, and be shaken out at the surface. To assist the cutter in stiff clays, a little water may be added to the borehole. It is unavoidable that some of this water will be drawn into the pore space in the clays as they swell on stress relief from the boring action, but the reduction in soil strength does not appear to be significant from this cause, and it is more important to bore and concrete the shaft rapidly.

Tripod rigs are light and easily transported and are therefore very suitable

Figure 3.13 Typical cutting tool used in construction of small-diameter bored cast-in-place piles. After West (1972)

Figure 3.14 Shell for bored pile. Photograph courtesy of Cementation Piling and Foundations Ltd., Rickmansworth

for work on small sites or amongst existing buildings, or even inside buildings. Short rigs can be used in limited headroom with a slight loss in efficiency. Very little site preparation is required. Minimal soil disturbance occurs when boring through clays, and piles can be placed close to buried services. In granular soils, the possibility of removal of material from around the bore can lead to undermining of adjacent footings, and greater care is necessary. During boring for a pile the temporary casing may be advanced to considerable depths, and the single line pull of the winch is then inadequate to extract it. By a pulley system, the standard winch pull of around 20 kN may be increased to 80 kN or more. Even this may be insufficient to extract long lengths of casing driven into dense soils (especially if the casing has been in the ground for some time), and hydraulic jacks are occasionally used to initiate extraction.

In granular soils, a 'shell' is used for boring. This comprises a heavy tube with a flap valve at its base with an open top (see Figures 3.13 and 3.14). The removal of soil by a shell relies on water being present. Dropping the shell into a granular soil below the water table causes the flap at the base to open, the flap closing as the shell is raised. By working the shell up and down, the granular soil is loosened by liquefaction as a low pressure is produced under the shell as it is rapidly raised, and this material then flows into the shell as it is dropped. Spoil from the shelling action is simply tipped out at the surface by up-ending the shell. There is a danger during shelling that the soil may be over-loosened and material drawn from the sides of the bore, and in extreme cases, the shell may become buried below the casing. To prevent this over-break, the temporary casing should be advanced by driving it into the ground as the borehole advances.

With a long length of heavy casing in loose sandy gravels, it is possible for rapid shelling action to so loosen the ground that the casing falls under its own weight. Whilst rapid progress is achieved, there is always a risk of over-break by this means, and the ground is loosened around the bore to a significant degree. It is important that a positive head of water is kept in the pile bore when boring through granular deposits to minimize the loosening effect, and to prevent piping up the pile bore. Water may have to be added in order to achieve this.

Obstructions, either natural or man-made, are cleared by the use of a heavy 'I' section chisel, with a hardened flat point or blade (Figure 3.15).

Ground conditions and boring expertise vary, and the ultimate bearing capacity of a bored pile in granular soils can be affected by these factors. To take some account of this, Weltman and Healy (1978) proposed a reduction factor, R, which may be applied to SPT N values obtained in granular soils to account for variations in boring techniques and the ground conditions. This is presented in Figure 3.16.

3.4.3 *Large-diameter percussion bored piles*

Large-diameter percussion bored piling equipment is well suited to the penetration of hard strata, which may include weak sedimentary rock. Where casing is advanced as the drilling proceeds, as with the Benoto system, unstable strata are supported and may be interspersed with harder strata.

Figure 3.15   Heavy 'I' section chisel, for tripod boring rig use. Sometimes called the Californian Chisel

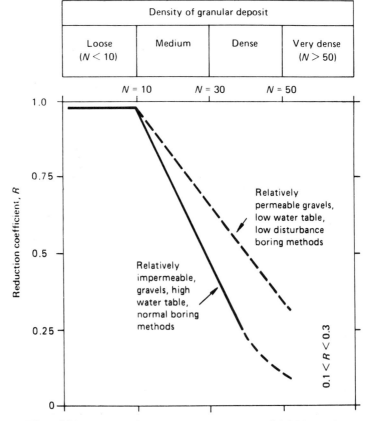

Figure 3.16   Reduction factors. Illustration courtesy of CIRIA, London

Figure 3.17 Leading dimensions (mm) of 'Benoto' piling rig. Illustration courtesy of Lilley Construction Ltd., Piling Division, London

Semi-rotary down-hole percussive hammer rigs, of the type used for well boring, are now available in sizes which approach those of large-diameter piles. This equipment does not handle casing, and a separate reaming and casing operation may be necessary using an attendant auger rig. However, the plant will bore through weak rock with little difficulty and in certain circumstances may be suitable for forming deep rock sockets for example.

The Benoto rig is illustrated in Figure 3.17. The EDF 55 boring machine is mounted on a wheeled or skid-equipped chassis and employs an oscillatory drive to achieve casing penetration. A development of the original Benoto rig by Lilley Construction Limited is known as the 'Libore', shown in Figure 3.18. The casing is sectional, but because of the semi-rotary driving action, is not screwed together, but employs a flush quick-action circumferential joint within the twin-walled casing thickness. The boring tube has a hardened cutting edge and is hydraulically clamped in a collar which transmits the semi-rotary action and vertical motion provided by hydraulic rams. Soil is removed from inside the casing by a grab in relatively weak strata. In hard ground, rock

Figure 3.18 Libore boring rig. Photograph courtesy of Lilley Construction Ltd., Piling Division, London

jaws are fitted to a hammer grab weighing approximately 1400 kg. With the jaws locked open, the grab chisels through obstructions ahead of the temporary casing. Other impact equipment can be used to break through particularly hard ground. The ability of the machine to penetrate hard ground makes it especially suitable for secant piled walls, in which each pile overlaps its neighbour by about 10% of the pile diameter. The piles are reinforced and concreted in the usual way, the casing being hydraulically jacked out of the ground during concreting. Piles up to around 40 m in length can be formed at a rake of up to 1 in 5, although in some ground conditions the full depth may not be possible. Concrete placing in the dry is relatively easy, but tremie placing can be cumbersome. For underwater placing, skips can be used as an alternative, but require to be carefully designed. Working loads are 200 to 500 tonne in suitable ground for pile diameters of between 670 and 1200 mm respectively.

3.4.4 *Rotary bored cast-in-place piles*

The majority of large-diameter piles are bored using rotary methods, generally by augers in the UK, although in other parts of the world, such as the US,

Figure 3.19 Large-diameter auger piling rig spinning off spoil from short-flight auger. Photograph courtesy of The Expanded Piling Company Ltd., Grimsby

large-diameter drilling plant is more commonly used to construct 'caisson' piles.

The augering plant is usually crane- or lorry-mounted, with a power pack to drive a ring gear within a rotary drilling table which is fixed to the crane base (Figure 3.19). The main advantage of lorry-mounted rigs is the ease of transport between sites, but this is at the expense of manoeuvrability on confined sites. The drilling table height is generally lower on lorry-mounted rigs, which restricts casing lengths unless these are installed by a separate machine (e.g. crane with casing vibrator). The crane-mounted rigs are costly to transport, but are better able to move between pile positions on uneven sites. The most powerful rigs are of this type. A number of manufacturers produce complete rotary boring rigs, such as Bachy Bauer, Soilmec, BSP or Casagrande. Other companies offer crane attachments, for fitting to a standard base machine.

In operation, the auger is driven from the ring gear at the drilling table via a sliding square or keyed circular sectioned Kelly bar, which may be telescopic for deep boreholes. The auger and Kelly bar are suspended from the crane winch rope (see Figure 3.20). An attendant crane is generally necessary for handling temporary casing, except with some specially developed rigs, usually for the smaller diameter piles, which can handle the casing from a second winch line and double cathead.

Figure 3.20   Cased auger rig (Soilmec CM39). Total weight 27 500 kg

(a)

(b)

Reversible tooth

(c)

(d)

Wisdom tooth

Super tooth

Figure 3.21 Boring tools used in the construction of rotary bored cast-in-place piles: (a) reaming tool, (b) drilling bucket, (c) short-flight clay auger, (d) heavy-duty auger

The usual crane-mounted auger will bore to a diameter of at least 3000 mm. Depths of boring vary from around 25 m from cased auger rigs to as much as 60 m for the larger crane-based auger rigs, with telescopic, triple extension Kelly bars. Typical design loads vary from 1 to 20 MN, in suitable ground conditions. Although the auger rigs can bore raked piles at up to 1:3.5 (more usually 1 in 4), it is probably better practice to take lateral and inclined loads on a suitably reinforced vertical pile of sufficient cross-section.

The augers fitted to this type of rig are usually of the 'short' flight type for large diameters. Alternatively, in ground which breaks up excessively after cutting, a bottom-opening bucket may be used. Both augers and buckets can be equipped with hardened teeth for breaking up lightly cemented deposits. A slight, if irregular, increase in bore diameter is obtained with reaming tools attached to the periphery of the auger. Various items of auger equipment are illustrated in Figure 3.21. Augered soil is brought to the surface and removed from the auger by 'spinning off'.

A special feature of auger bored piles which is sometimes used is the underream, which enables the bearing capacity of suitable strata to be exploited by providing an enlarged base. The soil has to be capable of standing open unsupported to employ this technique, and stiff to hard clays, such as the London Clay, are ideal. An underreaming tool is shown in Figures 3.22 and 3.23. In its closed position (Figure 3.22) it will fit inside the straight section of a pile shaft, and can then be expanded at the base of the pile to produce the underream as shown in Figure 3.23. Loose 'crumbs' of material from the base of the bore may have to be removed by hand. The shaft and underream of such

Figure 3.22 Underreaming tool in the closed position. Photograph courtesy of The Expanded Piling Company Ltd., Grimsby

Figure 3.23 Underreaming tool in the open or working position. Photograph courtesy of The Expanded Piling Company Ltd., Grimsby

a pile would normally be inspected before concreting is carried out. A man-carrying cage is lowered down the pile bore for this purpose, provided it has a diameter of 750 mm or more. Safety precautions in this operation will include the provision of an air line and possibly a running casing sleeve within the pile bore. Remotely-operated cameras may also be used to inspect the base of a pile, thereby obviating the need for personnel to enter the bore. It is however important that loose material does not remain at the pile toe, and a suitable means of removing this is required if remote inspection is used. Underreaming is not possible in flooded boreholes.

Although the sliding Kelly bar and fixed rotary table is a common arrangement, an hydraulically driven motor which slides in vertical guide rails can be mounted on the top of the drilling stem. Because of the weight of such a motor, this system is more popular in the smaller-diameter and continuous-flight auger rigs (Figure 3.24).

A new development is the rope-suspended rotary drill in which a submersible electric drive motor is mounted directly over a triple cutting head, as shown in Figure 3.25. There is no torque to be resisted by the rig with this arrangement, and deep holes can be drilled without any drill stem handling or extension problems. The drill is manufactured by the Tone Boring Co. Ltd. of

Figure 3.24 A continuous-flight auger drill with sliding hydraulic motor mount. Photograph courtesy of Cementation Piling and Foundations Ltd., Rickmansworth

Japan (type RRC-U). The cutting heads can be expanded and contracted by remote control to provide underreams or changes in shaft diameter at selected depths. Verticality is also under remote control, with a drill-mounted pick-up giving a continuous record of the head position. Spoil is removed by reverse circulation, using a suction pump at the surface, supplemented with an airlift for deep holes. Hole stabilization is maintained by a head of water or drilling mud. A range of five machines is produced, to cover a range of hole diameters from 800 mm to 6.56 m. The system is however specialised and is not commonly used. A somewhat similar system, using large cutting wheels, is now used for the formation of barettes and diaphragm wall panels. These machines, commonly referred to as 'mills' (Figure 3.26) are effective in cohesionless soils or in hard strata but may encounter difficulties in clay soils, especially if there are boulders present.

For large-diameter deep drilling of pile bores in weak sedimentary rocks augers may not be suitable, and a core barrel or rock roller bit is employed. High capacity drilling tables are mounted on heavy base cranes, such as the NCK Atlas or Manitowoc, and air or direct circulation water flush drilling employed.

Figure 3.25   Tone RRC-U rope suspended rotary drill for variable-diameter boreholes/pile shafts. Photograph courtesy of Tone (Japan) and Geo Publications Ltd., Brentwood

Core barrels are fitted with a calix basket to collect cuttings above the core barrel and have tungsten carbide cutting bits (see Figure 3.27). Core barrels are typically up to 2 m in diameter, and rock rollers in excess of 3 m are available, although these large sizes are rarely used.

3.4.4.1 *Pile formation with rotary boring equipment.*   The upper portion of a pile bore is likely to be in loose or weak ground, and some form of temporary support is then necessary. It is usual for a steel casing to be employed for this, even when the remainder of the bore is made under the stabilizing action of a bentonite suspension, which is described in a later section. The temporary casing also seals off water below the water table in permeable ground.

Shallow, dry, granular deposits are loosened by the auger, and will generally stand open long enough for a short length of casing to be inserted. To advance

1. POWER PACK
2. CRANE
3. MUD RETURN AFTER
   TREATMENT
4. HYDRAULIC FEED
   CYLINDER
5. DRILLING TOOL GUIDE
6. CLINOMETERS
7. PUMP BODY
8 DRILLING TOOL

Figure 3.26   Milling rig for barettes

the bore further in granular soils, the auger can be used to loosen the soil, followed by driving the casing with a drop hammer. Alternatively, the casing can be advanced with an hydraulic casing vibrator suspended from the attendant crane. The process can be made more efficient in some instances by adding bentonite powder to the granular soil, and 'mudding-in'. Water may also need to be added. The effect is to produce a column of loosened soil with a stable-walled bore through which the casing can be driven more easily to depths of about 10 m. A casing vibrator can increase the depth of insertion to around 25 m. Casagrande casing oscillators provide an alternative means of installing casing. At the level of an impermeable soil, such as a stiff clay, the casing can be forced into the ground to form a seal against entry of water by

Figure 3.27 CEZ rotary rig with 1.8 m diameter core barrel fitted with tungsten carbide teeth. Upper part of barrel comprises a 'calix basket' to collect cuttings when drilling by direct circulation water flush methods. Photograph courtesy of Pigott-Bachy Ltd., Ormskirk

Figure 3.28 Sequence of large-diameter rotary pile construction

'crowding' action. The 'crowd' force is applied via the Kelly bar by clamps gripping the casing and applying an axial thrust via retracting hydraulic rams against the reaction of the rig. A sequence of construction for large rotary bored piles is shown in Figure 3.28.

3.4.5 Boring and concreting

Concreting in 'dry' bores is usually a straightforward procedure. On completion of the borehole, the reinforcement is placed with suitable spacers to locate it centrally in the bore. During concreting, the casing is withdrawn, always maintaining a head of concrete within the casing. When pouring concrete in a dry vertical bore, a hopper and tube should be used to ensure that the mix is not directed towards the reinforcement. In raking piles, a chute may be required or a first charge of grout placed.

Any sudden drop in concrete level as the casing is withdrawn should be noted. This would probably indicate a zone of over-break; it is possible in certain circumstances as shown in Figure 7.11 for intermixing of soil, water and concrete to take place. This occurrence is sometimes accompanied by a loss or distortion of pile reinforcement, the steel cage being carried downwards and outwards at the overbreak zone. To prevent this, boring ahead of the pile casing in unstable ground should be avoided. It is often the casing-off of such zones of over-break that leads to conditions which produce contaminated or segregated concrete in the pile shaft—a feature sometimes mistakenly ascribed to the cross-flow of groundwater.

With a self-compacting mix, it is not necessary to vibrate the concrete, and a vibrator should not be used inside the pile casing. It is sometimes considered beneficial to vibrate the top 1 to 2 metres of a pile after the casing has been withdrawn, but this should not be overdone, or soil mixing can occur.

In weaker soils, such as alluvial clays, it is possible for the fluid pressure of the wet concrete in the pile bore to fail the soil, and a slump occurs with a bulge in the pile shaft. A critical undrained cohesion of around $15 kN/m^2$ has been observed to result in this effect, and in a typical alluvial deposit this may occur at 3 to 5 m depth. The possibility of soil failure from this cause can be lessened by pulling the casing at a steady rate, rather than in short rapid pulls, and limiting the head of concrete in the casing, provided this can be done without risk of the concrete slumping below the bottom of the casing in the bore.

3.4.5.1 *Boring and concreting in water-bearing ground.* In a stratigraphic succession in which water-bearing gravel overlies a stiff clay, such as is common in the London area and other areas in the UK, the temporary casing is used to support the borehole through the gravel, and is then driven a short distance into the clay in order to seal off the water. Boring then continues under dry conditions. This may not always be possible, and the base of the pile may be in a granular deposit below the water table. The tremie method of

Figure 3.29 Placing concrete by the tremie method. Note that in stage 2, any slurry should be removed by pump or air-lift prior to concreting, especially if end-bearing is significant

concreting is then employed. This is a practical method of forming a pile shaft provided certain precautions are taken. It is essential that a total collapse slump mix is used (i.e. greater than 175mm), and that the tremie pipe is always well below the water/concrete interface during concreting. The tremie pipe and hopper connections should be watertight and in clean condition to permit free flow of the concrete. A minimum internal diameter for the tremie pipe of 150 mm is suggested for use with concrete having a maximum aggregate size of 20 mm, increasing for larger aggregates. A vermiculite or expanded poly-styrene plug should be used in the tube between the water and first batch of concrete, and immediately before placing the tremie pipe in the pile bore, a check on sedimentation should be made, and any significant sediment should be air-lifted out. The procedure is illustrated in Figure 3.29. Where the pile cut-off level is above ground, concrete should overflow from the pile head on completion. The upper part of the pile is composed of weak concrete from an excess of water and may also be contaminated from boring detritus. Where there is a deep cut-off, the concrete level should be raised to allow for around 0.5 to 1.0 m for trimming off the weak concrete.

3.4.6 Concrete mixes for bored piles

The integrity of the pile shaft is of paramount importance, and the concreting mixes and methods that have evolved for bored piles are directed towards this, as opposed to the high-strength concrete necessary for pre-cast piles or structural work above ground.

This prerequisite has led to the adoption of highly workable mixes, and the total collapse mix for tremied piles has been mentioned. In order to ensure that the concrete flows between the reinforcing bars with ease, and into the interstices of the soil, a high-slump, self-compacting mix is called for. A minimum cement content of 300 kg/m^3 is generally employed, increasing to

Table 3.10 Recommended concrete slumps for cast-in-place piles (based on Federation of Piling Specialists' Specification, for cast-in-place piling)

Piling mix	Typical slump (mm)	Conditions of use
A	125	Poured into water-free unlined bore. Widely spaced reinforcement leaving ample room for free movement of the concrete between bars.
B	150	Where reinforcement is not spaced widely enough to give free movement of concrete between bars. Where cut-off level of concrete is within casing. Where pile diameter is < 600 mm.
C	> 175	Where concrete is to be placed by tremie under water or bentonite in slurry.

400 kg/m^3 at slumps greater than 150mm, with a corresponding increase in fine aggregate content to maintain the cohesion of the mix. Three mixes recommended by the Federation of Piling Specialists are given in Table 3.10.

3.4.7 Reinforcement for bored piles

For piles loaded in compression alone, it is only necessary to reinforce the shaft to a depth of 2m greater than the depth of temporary casing, to prevent any tendency for concrete lifting when pulling the casing. Piles subject to tension or lateral forces and eccentric loading (possibly by being out of position or out of plumb) require suitable reinforcement designed to cope with these forces. Nominal reinforcement for piles in compression only would comprise say four 12mm dia. bars for a 400mm dia. pile to five 16mm dia. bars for a 550mm dia. pile. A special cage of 5mm steel, or hoops of flat steel, are employed as lateral ties. The assembled cage should be sufficiently strong to sustain lifting and lowering into the pile bore without permanent distortion or displacement of bars. Bars should not be so densely packed that concrete aggregate cannot pass freely between them, and hoop reinforcement is not recommended at closer than 100 mm centres.

It is frequently necessary to lap bars in long piles, and it is here that trouble can arise, with aggregate blocking at the laps and preventing concrete flowing to the borehole walls. Such an occurrence results in a defective pile, but is sometimes not appreciated by designers. To overdesign the steel reinforcement in a bored pile can lead to practical difficulties in formation of the pile shaft. Where large steel cross-sections are unavoidable, consideration can be given to staggering the laps, although this can be difficult to arrange in prefabricated cages. If bar connectors are used, staggering is not such a problem as the lengths involved are shorter, but these are designed for use either in tension or alternatively in compression, and it is not generally possible to predetermine the orientation of the cage in the pile bore. Mild steel bars can be butt joined by welding. Provided high-quality welds are used, this can be an effective solution. Finally, provided the cage can be oriented, maximum steel need only be placed over that part of the pile subjected to maximum stress (if this is known), and a reduced density can be used in the plane of the neutral axis.

3.4.8 Excavation using a bentonite suspension

A method of borehole wall support developed from slurry trench techniques is the use of a bentonite suspension. Where pile bores are required to penetrate considerable depths of unstable soil, the installation of long lengths of temporary casing is a time-consuming process.

By employing a bentonite suspension of around 6% by weight bentonite, the borehole walls may be effectively supported, provided there is a positive head of bentonite slurry above the groundwater table of at least 1.5m. In permeable

Figure 3.30   The bentonite method of pile bore formation

granular soils, there is an initial flow of slurry into the borehole wall, where a 'filter cake' rapidly builds up which then prevents further flow, and the excess hydrostatic pressure in the borehole supports the ground. An additional benefit is the holding in suspension of fine detritus from the boring operations. Before concreting, the bentonite in the pile bore is checked for contamination and if necessary is exchanged for fresh bentonite. Bentonite slurry is returned to the surface for settling and desanding. Cleaned slurry may be re-used, with adjustment of bentonite content by the addition of new bentonite or mains water as necessary. After some re-uses, the bentonite slurry may become too contaminated for effective cleaning, and it is then discarded.

Concreting is carried out by the tremie method, using concrete with a slump in excess of 175mm. With clean bentonite slurry in the bore, the suspension is effectively scoured and displaced by the fluid concrete. No decrease in pile adhesion has been observed in properly constructed bentonite piles, whether the shafts are in granular soils or cohesive clays. A full description of the constructional method, and the results of many pile tests on bentonite piles, are given by Fleming and Sliwinski (1977). The process is illustrated in Figure 3.30.

3.4.9 Diaphragm piers (barettes)

A form of non-displacement piling suitable for exceptionally large working loads is the diaphragm pier or 'barette' which comprises a short section of diaphragm wall, of rectangular or other suitable cross-section. Excavation of barettes is as for diaphragm walls, i.e. grabbing methods, with trench support by bentonite slurry. In practice, an excavation is made to say 1.5 to 2m depth and of slightly larger plan size than the required cross-section of the barette. Guide walls are then formed by back-shuttering methods, generally to a size to suit the grab to be used. Excavation to final depth continues as for a normal diaphragm wall, with concreting by identical techniques. The advantages of the large cross-section of the units is best exploited in end-bearing, but support from the soil may be derived from a mixture of shaft friction and end-bearing, as for other piles. This type of pile is capable of supporting large lateral loads.

3.5 Continuous-flight augered piles with grout or concrete intrusion

Continuous-flight auger (CFA) piles are now widely available in sizes generally up 750 mm diameter, and exceptionally 900 mm diameter. The commonly used diameters are 300, 450, 600, and 750 mm. These piles offer considerable environmental advantages during construction. Vibration is minimal, and noise outputs are low. In permeable soils with a high water table, their use removes the need for concreting by the tremie method, and temporary support for the borehole walls using casing or bentonite slurry is not necessary. The method of pile construction is suitable in sands, gravels and clays. A typical CFA rig is illustrated in Figure 3.24.

The piles are constructed by screwing the continuous-flight auger into the ground to the required depth, which is restricted to approximately 25 m. At the full depth, grout (or highly workable concrete mix in the larger pile diameters) is injected down the hollow auger stem to the head of the auger. The auger is lifted out of the ground as the grout continues to be intruded. In early attempts at using this method of construction, there was difficulty in controlling the relative rates of auger extraction and grout intrusion. An important feature of present day CFA piles is the use of a reliable flow meter, developed by Cementation, such that many machines carry instrumentation to monitor and record the whole process of construction of this type of pile. The items recorded are usually penetration/uplift per revolution, auger depth, concrete supply per increment of auger uplift during placing, and injection pressure at the auger head. The data can be printed on site and stored for analysis and record purposes. Depths of reinforcement can be up to 20 m in wet soils, and insertion may be assisted by the attachment of a small vibrator to the top of the cage. It is essential that cages are welded for this purpose, and that excessive links are not used. They may be placed at 1 m centres or more in regions of the pile where shear forces are small and away from the pile head.

Although concrete with a slump of about 150 mm is now more commonly used than grout, a typical grout mix is sand:cement in a ratio of 1 to $1\frac{1}{2}$ or 1 to 2 (depending on the quality of the sand) by weight with possibly an additive to improve workability. Crushing strength may be between 21 and 30 N/mm^2 at 28 days. Typical pile diameters and pile loads are given below.

Pile diameter	Design load (approx.)
mm	kN
300	350
400	500
450	500 to 750
500	750 to 1000

Piles of this type are constructed by a growing number of contractors. The construction process for the Cementation 'Cemcore' pile, for example, is shown in Figure 3.31.

This process has been extended to allow construction of piles of up to 750 mm diameter (for loads of up to 2600 kN), but in this case, the intruded material is a pumpable concrete. Reinforcement may be placed after completion of the concrete pour, but to depths typically not exceeding about 15 m.

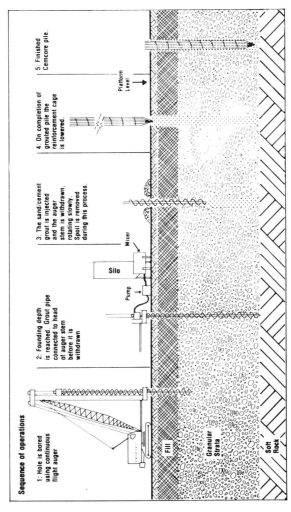

Figure 3.31 Construction process for 'Cemcore' piles. Illustration courtesy of Cementation Piling and Foundations Ltd., Rickmansworth

3.6 Micro piles*

The term 'micro pile' is commonly used to denote piles less than approximately 250mm in diameter, although sometimes the term is restricted to those less than 150mm diameter, and the term 'mini' piles is used to cover the range 150 to 250mm diameter.

Several decades ago, the Building Research Establishment described the use of 'short bored piles' to reduce settlements of lightly-loaded structures on weak soil. The micro pile performs a similar role and may be an economical alternative to deep trench footings or reinforced slab foundations for domestic and light industrial buildings in soils likely to produce excessive settlement or heave if traditional shallow footings are used. A number of other uses also exist, and are described in the following section. Their various methods of installation are also discussed.

3.6.1 Uses of micro piles

Three main categories of use can be identified, viz. as bearing piles, for soil reinforcement and as tension piles. The applications under these categories are given below.

Bearing piles.
  (i) For light industrial or domestic dwellings in (a) weak soils, (b) swelling soils, (c) shrinkable soils (see Figure 3.32), (d) to reduce differential settlement between old and new construction.

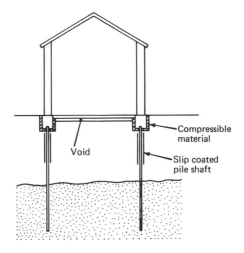

Figure 3.32 Construction in shrinkable clay using micro piled ring beam and suspended floor slab. The beam is isolated from the soil by compressible material. Illustration courtesy of *Ground Engineering*, Foundation Publications Ltd., Brentwood

*Section 3.6 by S.M. Herbert, M.Sc. F.G.S.

Figure 3.33 Retaining wall strengthened against settlement and overturning. Load is transferred to the micro pile by grouting into the structure. The structure is reinforced and no pile cap is required. Illustration courtesy of *Ground Engineering*, Foundation Publications Ltd., Brentwood

Figure 3.34 Support of isolated machine base using micro piles. Illustration courtesy of *Ground Engineering*, Foundation Publications Ltd., Brentwood

(ii) For underpinning often combined with strengthening of the structure (see Figure 3.33).
(iii) For support of isolated machine bases by design or later necessity (see Figure 3.34).
(iv) For support of floor slabs in industrial buildings with conventional piling for the structure. Often a remedial measure (see Figure 3.35).
(v) As 'settlement reducers'.

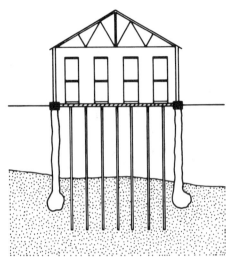

Figure 3.35 Support of floor slab with uniform load from storage racks, with micro piles. Illustration courtesy of *Ground Engineering*, Foundation Publications Ltd., Brentwood

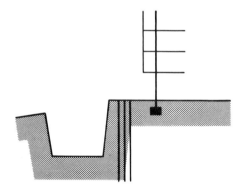

Figure 3.36 Micro pile wall to provide resistance against soil movement adjacent to an excavation. Illustration courtesy of *Ground Engineering*, Foundation Publications Ltd., Brentwood

Soil retaining and reinforcement
  (i) To form a reinforced soil 'wall' (see Figure 3.36).
 (ii) Stabilization of slopes (see Figure 3.37).
(iii) Bored micro pile wall (see Figure 3.38).

In tension
  (i) Below buoyant structures.
 (ii) Below towers and masts.

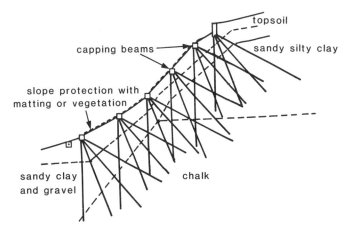

Figure 3.37 Fondedile 'Pali radice' reticulated structure to provide resistance in unstable ground. Illustration courtesy of *New Civil Engineer*.

Figure 3.38 Bored micro pile wall (Drillstress Ltd., photo R.L. Wyman.)

3.6.2 *Installation methods*

As for larger piles, there are two basic techniques of installation: drilling/boring cast-in-place methods or displacement methods.

(i) *Drilled or bored cast-in place methods* vary according to the type of soils

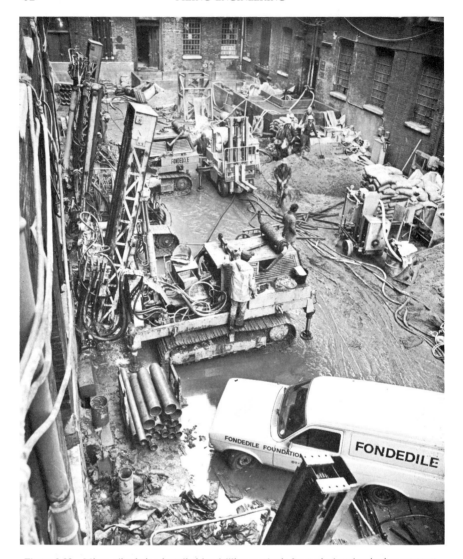

Figure 3.39 Micro piles being installed by drilling methods for typical underpinning contract. Photograph courtesy of Fondedile Foundations Ltd., Yiewsley

encountered. In non-cohesive soils, the method employed by Fondedile Foundations Ltd. is typical (see Figure 3.39). Casing of diameter 75 to 250mm fitted with a tungsten bit is drilled into the ground to the full depth of the pile. The drilling fluid, which is either water or bentonite in suspension, passes down the inside of the casing and carries away cuttings and fines from the surrounding soil, via the annular space between the drill casing and the soil,

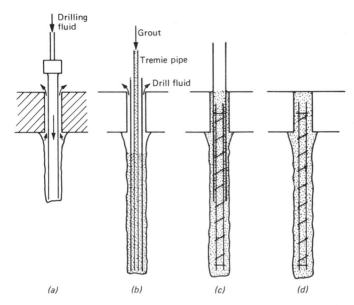

Figure 3.40 Method of installing typical drilled cast-in-place micro piles. (a) Drilling with external flushing; (b) casing drilled to full depth, grout being introduced via a tremie pipe and displacing drilling fluid; (c) reinforcement placed, and casing being withdrawn; (d) completed pile. Illustration courtesy of *Ground Engineering*, Thomas Telford Ltd., London

thus producing an irregular, rough surfaced bore of larger diameter than the casing employed. After the bore has been formed, sand-cement grout is pumped into the borehole via a tremie pipe or injection tube from the base of the bore. When clear grout issues from both inside and outside the casing, the reinforcing cage is placed and the casing gradually withdrawn. One important advantage of this method is the facility to apply pressure to the grout during withdrawal of the casing. This causes grout to penetrate into the surrounding ground, thus enhancing the frictional resistance of the pile. Similar methods of installation are used by a number of companies. The method is illustrated in Figure 3.40.

In cohesive soils, a clay cutting bit or continuous-flight auger is used. Casing may be reduced in diameter with depth, if necessary, or omitted apart from a top casing in stiff clays, where stable unlined bores can be formed.

Motive power for the drilling or boring is hydraulic, electric or pneumatic, and boreholes can be successfully formed though materials such as concrete, steel, timber or natural obstructions, and in areas of confined space or limited headroom.

(ii) *Displacement methods* of micro pile installations are generally similar to those encountered in conventional piling, with a range of rigs being offered according to pile size and available working space. Two alternative approaches are also employed—systems which use a jacking force to drive the pile into the ground and methods involving the use of mandrels. The former have the advantages of being almost vibration-free and quiet, and can provide load-testing of the piles as they are installed. The mandrel system employed by Menard Techniques Ltd. involves the creation of a hole to a maximum depth of 8 m by vibratory driving of a tapered steel mandrel. This is then filled with a dry concrete mix and the mandrel redriven. Some expansion of the pile occurs against the soil during this operation and a pile 'skin' develops. The pile is completed by filling with concrete and inserting the steel reinforcement.

Pre-cast or preformed piles are not suitable for use in ground with significant obstructions, and where used in underpinning by passing through existing foundations, the hole in the footing has to be drilled as a separate process.

3.6.3 *Sizes of micro piles and load capacities*

Many types of proprietory micro and 'mini' piles are available, in a range of sizes, from approximately 75 mm to 300 mm in diameter. Design load capacities are clearly difficult to state with any certainty, but generally fall in the range of 20 to 500 kN. A full range of installation processes are employed by the various contractors in this field. Both drilled and grouted (cast-in-place) piles, and displacement methods using hammered or jacked casing/pile installation methods are used over the range of available pile diameters. However, the systems are frequently specialized with respect to a particular contractor, and each system requires careful consideration before use. Some are tailored to particular applications, such as underpinning, and may not be suitable in other situations.

3.6.4 *Special considerations relating to micro piles*

Micro piles do not have large end-bearing areas although they may be installed to depths comparable with piles of larger diameters. As a consequence, they work primarily by shaft friction. However, should the shaft friction be fully developed, there will be relatively little base resistance to mobilize and unless a composite action between a reinforced slab foundation and the piles is deliberately planned, settlements can be excessive. In non-cohesive soils or weak rock, higher than normal unit shaft resistance has been noted in the case of drilled grouted-in-place micro piles, whilst displacement types may not develop resistances comparable with conventional piles as the volume displacement is less. It is however, usual practice to adopt empirical methods of design for granular soils, modified in the light of experience by the specialist contractor. Settlements are generally found to be similar to, or slightly less than, those associated with traditional friction piles. In cohesive soils, for small diameter piles designed on undrained strength, the 'α' method has been employed. The magnitudes of strains and displacements associated with small-diameter piles in London Clay are reported by Cooke and Price (1973).

The costs of load testing to check design assumptions can be prohibitive for a small number of micro piles and some systems of integrity testing may be impractical because of their small diameter. It is therefore particularly important that grouting is carefully supervised in bored or drilled micro piles, especially since tremieing in small-diameter holes is more difficult than in larger-diameter bores.

A further important factor is that of corrosion, which can more rapidly lead to serious effects than in conventional piles in view of the small sizes of micro piles. That which may be a surface effect in a normal-sized pile can form a significant proportion of the cross-section of a micro pile. Pre-cast or preformed piles have some advantages here, in that there is usually no problem with the quality or integrity of the shaft and coatings can be applied to combat corrosion (or to reduce down-drag or heave effects).

3.7 Corrosion

3.7.1 Corrosion of steel piles

The corrosion of steel piles is most likely to be a serious problem in two situations:

(a)   piles driven into disturbed ground (i.e. fill)
(b)   piles in a marine environment.

Steel piles and the reinforcement in reinforced concrete piles installed in undisturbed ground are not generally subject to significant corrosion below the water table. Above the water table, or in a zone of alternate wetting and drying from tidal action, corrosion rates can be increased, depending on temperature, pH and chemistry of the aqueous environment.

Corrosion rates of steel in air in industrial areas of 8.5 to 9.0 mm per 100 years are quoted by Chandler (1976), representing the most severe conditions. The following general rates are quoted overall:

in air   2.5 to 10.0 mm/100 years
soil   0.5 to 7.5 mm/100 years
water   5.0 to 25.0 mm/100 years.

Recent case studies by the British Steel Corporation have revealed a maximum corrosion rate in the atmosphere of up to 20 mm/100 years in industrial environments, but rates not exceeding 10 mm/100 years are expected to be more typical. In soil, a maximum corrosion rate of 3 mm/100 years was found in extracted piles after up to 25 years' service. Average rates of corrosion in soil, however, were found to be 1 mm/100 years. In fresh water, the BSC studies suggest that over a pH range of 4 to 9, corrosion rates may be up to 5 mm/100 years but rates may well be less for hard water, at around 1 mm/100 years. Considerably greater rates of corrosion, of up to 34 mm/100 years were recorded for steel piles at, or just below the water line.

In sea-water, rates of corrosion are rapid at first then decrease. The BSC study quotes a thickness loss rate of about 8 mm/100 years on average for totally submerged sections, and for exposed steel in the intertidal zone. In the splash zone considerably higher corrosion rates were recorded in exposed steel, of up to 25 mm/100 years, and typical UK rates of 15 mm/100 years for steel exposed to sea-water on one side only.

In a marine atmospheric zone, maximum rates of thickness loss of 10 to 25 mm/100 years were found in the BSC study in the splash zone, falling to 10 to 5 mm/100 years some little distance above the splash zone.

In a marine environment Chandler quotes higher rates, as illustrated in Figure 3.41. As can be seen, in the splash zone above high tide level, a corrosion rate of around 60 mm/100 years is almost twice as great as the rate of corrosion in that part of the pile below low tide which is always submerged.

There are a number of measures which may be taken to combat corrosion in steel piles.

(1) The use of copper-bearing steel—effective against atmospheric corrosion only.
(2) The use of high-yield steel (B.S. 4360, Grade 50B) in a structure designed for mild steel (Grade 43A stresses). This permits a greater loss of metal before stresses become critical.
(3) Increasing the steel thickness by using the next heavier section in the range ('H' section and tubular piles).
(4) The use of a protective coating, which may be restricted to the exposed

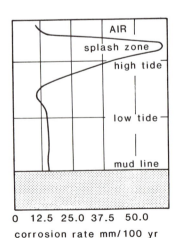

Figure 3.41   Corrosion rates of steel in sea-water

section of the pile. The coating may be of metallic type (e.g. metal spraying or galvanizing) or a paint system.

(5) Cathodic protection of piles in soil below the water table or marine piles. This type of protection is complex and tends to be costly.

(6) Concrete encasement of steel piles above the waterline. The inherent impact absorbing properties of the pile are lost by this method.

Experience in the use of low-alloy steels, containing 0.2 to 0.35% copper, is not yet well established, but the method could be effective in reducing splash zone rates of thickness loss. The danger in this method is that corrosion does not take place evenly, particularly after long periods of time, and deep corrosion pits can occur, locally reducing the strength of the pile.

If it is considered necessary to use coatings to provide protection below ground level these should be able to resist abrasion during driving. Coal tar epoxies are now considered to be most durable in marine conditions, and the current view of the BSC is that isocyanate-cured pitch epoxy paints are most effective. When such systems eventually fail, however, they cannot be replaced. The shaft friction of piles coated in this way may well be different to that of untreated steel, and comparative pile tests should be carried out to investigate this possibility. Strict control of the coating procedure may be necessary, and it may be better that this is not performed on site.

If fill is to be placed around or against steel piles, its pH value can be increased by mixing in lime or using limestone if available. In permeable, low pH fill, corrosion can be rapid, and some form of protection would be necessary.

3.7.2 Attack on concrete piles

Attack on both pre-cast and cast-in-place concrete piles can occur, in soils and groundwater having high sulphate and chloride concentrations and low pH. The degree of concrete attack is however a function of the availability of harmful chemicals in the groundwater. The objective should be to provide a dense concrete for all piles in this environment.

3.7.3 Sulphate attack

Attack on concrete by sulphates is well documented. Reaction between sulphate ions in groundwater and portland cement results in a chemical product of greater volume, and hence cracking occurs, leading to spalling and an even greater rate of attack. In order to cause corrosion, sulphates must be in solution, and therefore the solubility of the particular sulphate in question has a considerable bearing on the severity of attack. Groundwater mobility (and therefore soil permeability) is also a relevant factor. The sulphates commonly encountered are those of calcium, magnesium and sodium.

Various standards are available for assessing the aggressiveness of the groundwater, based on its sulphate concentration. The standard most commonly used in the UK is that of BS 8110. This is for general application to buried concrete. A scheme designed specifically for concrete in piles is given by Tomlinson (1977). For the especially severe conditions that may obtain in the Middle East, a conservative classification is given by Fookes and Collis (1975). The classifications are compared in Figure 3.42 (Bartholomew, 1979). The onset of moderate attack is generally at sulphate concentrations of around 200 to 400 parts per million (ppm). At these levels, portland cement, sulphate-resisting cement, or in some cases 25% pozzolanic material cement replacement can be used, the former rendered safe from attack by increasing the cement content and decreasing the water/cement ratio. At levels of around 1500 ppm (British Standard and Tomlinson classification), sulphate-resisting cement is considered necessary, and at concentrations greater than approximately 5000 ppm, some exclusion of groundwater from the concrete is considered necessary. The classification given by Fookes and Collis invokes protective measures at lower sulphate concentrations.

3.7.4 Chloride attack

Chlorides are encountered in particular in Middle Eastern soil conditions but could well also be present in industrial waste. Concrete itself is not usually subject to attack by chlorides in solution, but at even very low concentrations, chlorides can cause accelerated corrosion of steel reinforcement with consequential bursting of the concrete as the products of corrosion are formed. The corrosion can then progress at an enhanced rate. Protection of the reinforcement is therefore important, and increasing the cover of dense concrete is one way of improving matters. Stainless steel, galvanized or epoxy-coated reinforcement are alternatives. A further measure is the use of the

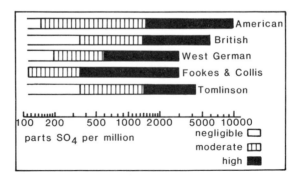

Figure 3.42   Comparison of classifications of aggressive agencies. After Bartholomew (1979)

protective membranes around the pile shaft. Bartholomew (1979) considers permanent sleeving of pile shafts is necessary under tropical conditions for chloride concentrations exceeding that of sea-water (3% salt by mass).

3.7.5 Acidic groundwater

This condition is most frequently met in peaty soils, but can also be present because of industrial contamination. In the former situation, the condition may not be severe, but in the latter situation, conditions can be extreme and localized, perhaps affecting only a small number of piles. By using dense concrete, at as low a water/cement ratio as possible, perhaps with workability being achieved using plasticizer, the permeability of the concrete may be reduced, and some protection achieved. Where industrial contamination is involved, the maximum likely concentrations should be determined, and an assessment made of the distribution and extent of the contamination. External protective measures to the pile shaft may be necessary. The *BRE Digest* ranks pH values of less than 6.0 as moderately aggressive, and less than 3.5 as very highly aggressive.

3.7.6 Miscellaneous aggressive agencies

Barry (1982) lists the following possible sources of attack on concrete.

Agent	Attack
(i) Inorganic acids (e.g. sulphuric, nitric, hydrochloric acid)	Can dissolve components of concrete; severity depends on concentration.
(ii) Organic acids (e.g. acetic, lactic, formic, humic acid)	Can slowly dissolve concrete. The attack is generally slow.
(iii) Alkalis (e.g. sodium and potassium hydroxide)	Alkalis such as these will dissolve concrete when highly concentrated.
(iv) Plant and animal fats (e.g. olive oil, fish oil, linseed oil)	Bond strength may be reduced.
(v) Mineral oils and coal tar distillates (e.g. light and heavy oils, paraffin)	If of low viscosity, these agents can degrade concrete. Phenols and creosols corrode.
(vi) Organic matter (e.g. landfill refuse).	Organic matter can degrade concrete if hydrolysis results in lime removal.

3.7.7 Protection of concrete piles

Bartholomew (1979) classifies the aggressiveness of the soil conditions, and provides a scheme of suitable protective measures. The scheme is presented in Table 3.11. The type of pile is a factor, precast piles having the most dense concrete, and tremied piles potentially the least, these respectively therefore being least susceptible and most susceptible to attack. Typical cement contents and water-cement ratios are shown in Table 3.12 (after Bartholomew).

Table 3.11 Precautionary measures to protect concrete piles. After Bartholomew (1979)

Requirement or measure	Normal conditions	Moderately aggressive conditions	Highly aggressive conditions	Very highly aggressive conditions
(1) Dense impermeable concrete	Yes	Yes	Yes	Yes
(2) Ordinary portland cement	Yes	Yes	No	No
(3) Sulphate resisting portland cement *or*	No	Yes	Yes	Yes
(4) Other special cements or pozzolanas	No	Yes	Yes	Yes
(5) Minimum cement content and maximum water cement ratio specified	Yes	Yes	Yes	Yes
(6) Special aggregates	No	e.g. Limestone aggregate sometimes used in acidic conditions		
(7) Sacrificial layer	No	Rarely relied upon by itself		No
(8) Minimum cover specified	Yes	Yes	Yes	Yes
(9) Use of permanent linear				
—Steel	No	No	No	Yes
—Rigid PVC	No	No	No	Yes
—Flexible PVC	No	No	No	Yes
(10) Use of coating e.g. epoxy, resin, bitumen on precast concrete shaft or steel casing	No	No	No	Yes

Table 3.12 Range of cement contents and water cement ratios in general use for concrete mixes. After Bartholomew (1979)

Pile type	Normal conditions	Moderately aggressive conditions	Highly and very highly aggressive conditions
Precast	450–475 kg/m³ 0.4–0.5	450–475 kg/m³ 0.4	450–475 kg/m³ 0.4
Driven cast-in-situ*	280–370 kg/m³ 0.25–0.6	330–450 kg/m³ 0.3–0.55	370–500 kg/m³ 0.3–0.45
Bored or augered	300–450 kg/m³ 0.5–0.55	350–450 kg/m³ 0.475–0.5	380–500 kg/m³ 0.45–0.5
Tremied under bentonite	350–450 kg/m³ 0.5–0.6	350–450 kg/m³ 0.475–0.5	400–500 kg/m³ 0.43–0.45

*Includes both standard and vibrated shaft Franki piles

Sleeving cast-in-place piles is a procedure sometimes carried out to combat corrosion. The sleeving may be flexible or rigid (usually uPVC). The sleeve may be placed inside the temporary casing which has been advanced into the ground and may be fixed to the reinforcing cage. Alternatively, a flexible sleeve may be attached to the outside of the temporary casing, and remains behind when the casing is withdrawn. In granular soils, it is possible for the sleeve to tear, and a test drive of the system should be carried out.

4 Design of single piles

Although entitled 'Design of single piles', this chapter deals mainly with the analytical methods used in design, rather than with the actual design process. Practical and economic factors play an essential role in choosing the type and relative size of pile for a given application. For each pile considered, analysis is then required in order to assess whether that pile will satisfactorily fulfil its function as a foundation element. This analysis will generally involve calculation of the ultimate capacity of the pile under working load conditions, and an assessment of any other potential problems, for example, whether the pile can be installed with a given size of pile driving rig.

Analytical methods in pile design have advanced very rapidly over the last 15 years. This advance has enabled many of the traditional empirical approaches to be substantiated, or in some cases replaced by more soundly based theory, particularly as regards estimating the likely deformations—vertical and horizontal displacements, and rotation—that the pile group may undergo. Economic pressures have led to more efficient design of pile groups, with a greater emphasis on limiting deformations (particularly settlement) to an acceptable level. The aim of the authors has been to present some of these newer methods of analysis, in a practical manner, with less emphasis on more traditional methods that are adequately covered elsewhere.

The main sections of the chapter deal sequentially with methods of estimating the axial capacity and deformation of single piles, followed by lateral capacity and deformation. In the later part of the chapter, specific problems that involve both axial and lateral response of the pile are considered. These include the use of raking piles, problems associated with piles in soil which is consolidating, and consideration of the buckling potential of piles, both during driving and under static loading. Some discussion of how construction techniques may affect the choice of design parameters is included at appropriate points in the body of the chapter.

4.1 Axial capacity of piles

A pile subjected to load parallel to its axis will carry the load partly by shear generated along the shaft, and partly by normal stresses generated at the base of the pile (see Figure 4.1). The ultimate capacity, Q, of the pile under axial load is equal to the sum of the base capacity Q_b, and the shaft capacity Q_s. Thus

$$Q = Q_b + Q_s = A_b q_b + A_s \bar{\tau}_s \qquad (4.1)$$

where A_b = area of the pile base,
         q_b = end-bearing pressure,
         A_s = area of the pile shaft, and
         $\bar{\tau}_s$ = average limiting shear stress down the pile shaft.

The relative magnitude of the shaft and base capacities will depend on the geometry of the pile and the soil profile. Piles which penetrate a relatively soft layer of soil to found on a firmer stratum are referred to as 'end-bearing' piles and will derive most of their capacity from the base capacity, Q_b. Where no particularly firm stratum is available to found the piles on, the piles are known as 'friction' or 'floating' piles. In cohesive soil, the shaft capacity is generally paramount, while in non-cohesive soil or where the pile base is underreamed, the overall capacity will be more evenly divided between shaft and base. In most circumstances, it is generally assumed that the shaft capacity in uplift is similar to that under downward loading (though see section 4.1.1).

The shaft capacity of a pile is mobilized at much smaller displacements of the pile (typically 0.5 to 2% of the pile diameter) than is the base capacity.

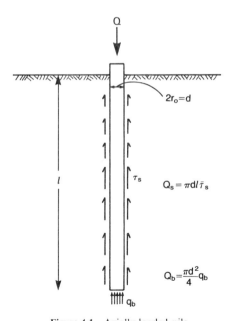

Figure 4.1   Axially loaded pile

The latter may require displacements as large as 5 to 10% of the pile base diameter (even larger for low-displacement piles in granular soil) in order to be fully mobilized. This difference between the load deformation characteristics of the pile shaft and that of the pile base is important in determining the settlement response of a pile, and the sharing of load between shaft and base, under working conditions. To illustrate this by a simple example, based on results from instrumented pile load tests (Whitaker and Cooke, 1966), consider a bored pile of 0.6 m diameter and 10 m long, installed in stiff clay. The shaft capacity will be taken as 600 kN, mobilized at a displacement of 3 mm, while the base capacity is 400 kN, mobilized at a displacement of 30 mm. Treating the load settlement curves as linear up to failure, the complete load settlement response will be as shown in Figure 4.2.

The ultimate capacity of the pile is 1000 kN, of which 60% is provided by the shaft. However, at a working load of say 400 kN, Figure 4.2 shows 94% of the load being carried by the pile shaft, and only 6% reaching the base. Choice of factor of safety for such a pile must be made in the light of the different response of pile shaft and base. Once the pile shaft capacity has been fully mobilized (at a load of 640 kN), the 'stiffness' of the pile–soil system reduces to that of the pile base, and the displacements start to increase rapidly. This type of consideration is of particular importance in the design of underreamed or partially end-bearing piles, where the base capacity is large. Settlements under working load conditions will be larger than for comparable straight-shafted friction piles.

The above example illustrates an important approach to the design of piles. In the first instance, unit values of 'skin friction' (the limiting value of shear stress between pile and soil) and of end-bearing pressure must be estimated in order to calculate the shaft and base capacities. The same overall capacity may be achieved with a variety of combinations of pile diameter and pile length. However, before the final geometry of the pile is determined,

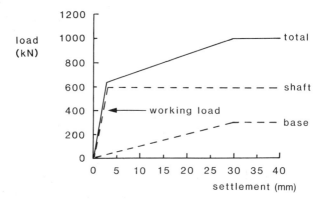

Figure 4.2   Idealized load settlement response

some consideration must be given to the load settlement characteristics of the pile—and also, of course, to other factors such as the economics of installing piles of one size or another.

In general, long slender piles may be shown to be more 'efficient' than short stubby piles, both as regards their capacity per unit volume installed, and also in terms of the stiffness of the pile (Randolph, 1983). Limitations on the slenderness of a pile come from economic reasons (since, once a large boring or pile driving rig is mobilized, the rate of pile installation becomes largely independent of the pile diameter), and also from considerations of its stability during driving (Burgess, 1976) and of the axial compressibility of the pile. In the extreme case, piles may be so long that no load at all reaches the base under working conditions. Working settlements will probably be sufficient to mobilize full skin friction in the upper part of the pile shaft (Randolph, 1983), affecting the load settlement characteristics and bringing the possibility of degradation of the shear transfer between pile and soil.

The following sections discuss methods of estimating the ultimate capacity of piles in different soil types. In all cases, the capacity may be assumed to be the sum of shaft and base capacities (see equation (4.1)). Methods of estimating values of the end-bearing pressure, q_b, and the skin friction, τ_s, will be broadly divided into two categories, those based on fundamental soil properties such as shear strength or angle of friction, and those based directly on in-situ measurements such as Dutch cone or standard penetration tests.

It should be emphasized that estimation of pile capacity is still largely based on empirical methods, derived from correlations of measured pile capacity with soil data of variable quality. There is generally a wide scatter in the correlations, and different approaches suit different soil types better than others, albeit still with a significant margin of error. Many soils do not fit neatly into the categories discussed in the following sections of non-cohesive (implying free-draining, granular), or cohesive (implying less permeable, fine-grained) soil, and of rock. Pile design must thus be seen as a compromise, adopting the most appropriate method available for the particular soil types and site conditions prevailing. Where a pile penetrates layers of differing soil types, the shaft capacity in each layer is cumulative, with minimal 'interaction' effect between adjacent layers.

4.1.1 *Capacity in non-cohesive soils*

End-bearing pressure. It might be expected that the end-bearing pressure beneath a pile in a uniform deposit of non-cohesive soil would be directly proportional to the local vertical effective stress, and thus would increase approximately proportionally with depth. However, extensive research by Vesic (1977) has shown that the end-bearing pressure appears to reach a limiting value, beyond which no further increase is observed with further penetration of the pile. This limit is often taken, conservatively, as 11 MN/m²

(Tomlinson, 1977), although Coyle and Castello (1981) quote values of end-bearing pressure for piles in clean dense sand as high as 15 MN/m².

At one time, this limiting behaviour was attributed to some form of arching effect. However, a more rational explanation lies in the variation of rigidity index (ratio of stiffness to strength) and friction angle, ϕ', with confining pressure. The effect of rigidity index has been discussed by Kulhawy (1984). Here, attention will be focused on the variation of friction angle, which has a dominant effect except in very loose soils.

Bolton (1986) discusses the strength and dilatancy characteristics of sand and shows that, when dense sand is sheared at high confining stresses, the peak friction angle will be that associated with shearing at constant volume. As will be shown in this section, the implication of this for bearing capacity is that, as the embedded length of pile increases, values of end-bearing pressure increase at a gradually decreasing rate, and, for practical pile lengths, appear to asymptote towards values in the range 10 to 15 MN/m².

The end-bearing pressure q_b, may be expressed in terms of the effective vertical stress, σ_v', and bearing capacity factor, N_q as

$$q_b = N_q \sigma_v' \tag{4.2}$$

Values of N_q quoted in the literature vary considerably, but those derived by Berezantzev et al. (1961) are used most widely for the design of deep circular foundations. Figure 4.3 shows the variation of N_q with friction angle ϕ'. It is then necessary to choose an appropriate value of ϕ' consistent with the type of non-cohesive material, its relative density, and the average stress

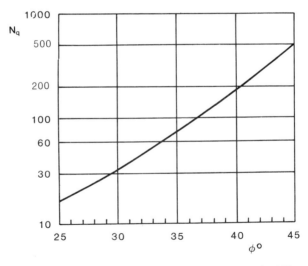

Figure 4.3   Variation of N_q with ϕ' (Berezantzev et al., 1961)

level at failure. Following Bolton (1986), ϕ' may be related to the relative density of the sand, corrected for the mean stress level, p', and a critical state angle of friction, ϕ'_{cv}, which relates to conditions where the soil shears with zero dilation (i.e. at constant volume). The corrected relative density, I_R, is given by

$$I_R = I_D[5.4 - \ln(p'/p_a)] - 1 \qquad (4.3)$$

where I_D is the uncorrected relative density, and p_a is atmospheric pressure $(100\,\text{kN/m}^2)$. A lower limit of zero may be taken for I_R at very high stress levels, while values of I_R greater than 4 should be treated with caution. In a subsequent discussion, Bolton (1986) suggests restricting equation (4.3) to mean effective stress levels in excess of $150\,\text{kN/m}^2$, below which the corrected relative density is taken as $I_R = 5I_D - 1$.

The appropriate value of ϕ' may than be calculated from

$$\phi' = \phi'_{cv} + 3I_R \text{ degrees} \qquad (4.4)$$

The average mean effective stress at failure may be taken approximately as the geometric mean of the end-bearing pressure and the ambient vertical effective stress, i.e.

$$p' \approx \sqrt{N_q \sigma'_v} \qquad (4.5)$$

The end-bearing pressure q_b, may now be calculated, for given values of ϕ'_{cv}, I_D, and σ'_v, by iterating between equations (4.3) to (4.5) and the chart for N_q shown in Figure 4.3. As an example, consider the case of $\phi'_{cv} = 30°$, $I_D = 0.75$ and $\sigma'_v = 100\,\text{kN/m}^2$. Assuming an initial value of N_q of 50, equation (4.5) gives $p' = 707\,\text{kN/m}^2$ from which ϕ' may be calculated as 35°. For this value of ϕ', Figure 4.3 yields a new value of N_q of 75. Further iteration yields a final value of $N_q = 66$, and an end-bearing pressure of $q_b = 6.6\,\text{MN/m}^2$.

Figure 4.4 presents charts of end-bearing pressure against ambient vertical effective stress, for different values of ϕ'_{cv} and I_D. The charts are plotted on logarithmic axes, which obscures the natural variation of q_b with depth. This variation is non-linear, showing a gradually reducing rate of increase with depth, similar to that found from pile load tests. Only at great depths and for dense deposits of sand, do end-bearing pressures exceed $20\,\text{MN/m}^2$. Bolton quotes values of ϕ'_{cv} for different types of sand ranging from 25° for mica up to 40° for feldspar. In practice, presence of silt particles will mean that ϕ'_{cv} for most deposits will rarely lie much above 30°.

Strong experimental support for the approach outlined above has been reported by Neely (1988), from measurements on expanded base piles. In a database of 47 pile test results, the mean ratio of predicted to measured capacity was found to be 0.92, with a coefficient of variation of 0.32.

Lower values of end-bearing pressure should be adopted for calcareous or very siliceous sand deposits, due to the more compressible nature of these

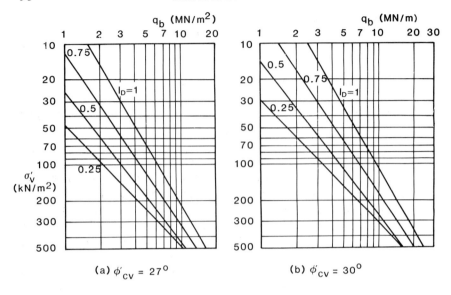

(a) $\phi'_{cv} = 27^\circ$ (b) $\phi'_{cv} = 30^\circ$

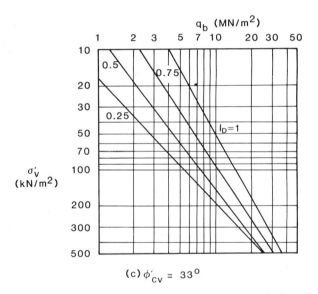

(c) $\phi'_{cv} = 33^\circ$

Figure 4.4   End-bearing pressures in non-cohesive soil

soils. It will usually be preferable to estimate end-bearing pressures for such
soils directly from field tests such as cone penetration or pressuremeter tests.
Where such test results are not available, the correlation from laboratory
test results reported by Houlsby *et al.* (1988) may be used, where the
end-bearing pressure in uncemented calcareous sand was given in terms of

the mean effective stress, p', as

$$\frac{q_b}{p_a} = 38\left(\frac{p'}{p_a}\right)^{0.6} \tag{4.6}$$

where p_a is atmospheric pressure ($100\,\text{kN/m}^2$).

Skin friction. The starting point for calculating values of skin friction τ_s, for piles in non-cohesive soil is the expression

$$\tau_s = \sigma'_n \tan\delta = K\sigma'_v \tan\delta \tag{4.7}$$

where σ'_n is the normal effective stress acting round the pile shaft after installation, and δ is the angle of friction between pile and soil. The normal stress may be taken as some ratio K of the vertical effective stress σ'_v, resulting in the second form of the expression. The appropriate value of K will depend on (a) the in-situ earth pressure coefficient, K_0, (b) the method of installation of the pile and (c) the initial density of the sand.

Piles driven into loose to medium deposits of sand provide considerable compactive effect. As will be discussed in Chapter 5, the capacity of individual piles in sand tends to increase as neighbouring piles are driven in a group, due to this compactive effect. For dense deposits, the greater tendency of the sand to dilate will limit the amount of densification close to the pile shaft. Again, though, driving further piles close by will have a beneficial effect on the shaft capacity of a pile, due to vibratory compaction.

From pile test results, a similar trend has been observed for the skin friction as for the end-bearing pressure (Vesic, 1977); the rate of increase of skin friction with depth gradually reduces and there is a tendency towards some limiting value. Vesic quotes values of the ratio, q_b/τ_s, which lie between 50 and 200. It seems reasonable to expect the value of K to vary in a similar fashion to N_q. For full displacement, driven piles, K may be estimated from

$$K \approx N_q/50 \tag{4.8}$$

This expression will give values of K varying between about 1.5 at low stress levels, down to unity or even lower at greater depths, with an average value typically around 1.2.

The value of δ in equation (4.7) will depend on the roughness of the pile surface. Kishida and Uesugi (1987) have reported a detailed study of the effects of surface roughness, and shown how the interface friction angle may be related to the friction angle of the soil in terms of a normalized roughness coefficient, defined as the maximum roughness of the pile surface (over a gauge length of D_{50} for the soil) normalized by the value of D_{50}. For typical pile surfaces (oxidized mild steel or concrete), the normalized roughness coefficient will exceed 0.05, and the coefficient of friction at the interface will lie in the range 0.75 to 1 times that for the soil itself.

Reducing the coefficient of friction for the sand by, say 20%, or reducing the friction angle by a few degrees, may be achieved in a consistent fashion by taking the interface friction angle as $\delta = \phi'_{cv}$, the critical state angle of friction. This may be justified on the basis that no dilation is to be expected between the sand and the wall of the pile. The skin friction may then be calculated as

$$\tau_s = \frac{N_q}{50} \sigma'_v \tan \phi'_{cv} \tag{4.9}$$

The ratio q_b/τ_s implied by equation (4.9) is $50/\tan \phi'_{cv}$, which is in broad agreement with the range quoted by Vesic.

On the basis of instrumented pile tests, some authors have recommended that skin friction for piles loaded in tension should be taken as half that for piles loaded in compression. However, many such pile tests took insufficient account of the residual stresses which exist after pile installation, and considerably underestimated the end-bearing capacity of the pile, thus overestimating the skin friction in compression. Current opinion is that there is no systematic difference in the value of skin friction which may be mobilized by a pile loaded either in tension or compression, except for relatively slender piles. For such piles, the tendency for the pile to contract radially under tensile axial load may lead to lower radial stresses, and hence lower values of skin friction. This effect may be readily quantified provided an estimate may be made of the shear modulus of the soil.

From analysis of pressuremeter tests, the relationship between change in radial stress, $\Delta\sigma_r$ and circumferential strain, ε_θ, is

$$\varepsilon_\theta = \frac{\Delta\sigma_r}{2G} \tag{4.10}$$

The circumferential strain at the pile edge may be written in terms of the axial stress in the pile, $P/(AE)_p$ (where $(AE)_p$ is the cross-sectional rigidity of the pile and P is the axial load) as

$$\varepsilon_\theta = -v_p \frac{P}{(AE)_p} \tag{4.11}$$

where v_p is the Poisson's ratio for the pile. These expressions may then be combined to give a local change in skin friction of

$$\Delta\tau_s = -2G \frac{v_p P}{(AE)_p} \tan \delta \tag{4.12}$$

For typical values of pile and soil parameters, it may be shown that this change in skin friction may amount to 10 to 20%. The effect starts to become significant where the pile slenderness ratio, l/d, is greater than about 5% of the stiffness ratio, E_p/G (taking E_p as the Young's modulus of an equivalent solid pile).

For partial displacement piles, the normal effective stress acting on the pile shaft will be lower than for full displacement piles and the value of skin friction should be reduced by 20%. For bored piles, or driven cast-in-situ piles where the casing is removed, loosening of the surrounding soil may occur, followed by some increase in the stress level as the concrete is placed. The appropriate value of K in equation (4.7) will depend on the details of the construction method.

For conventional bored piles it is common practice to adopt values of K of 0.7, while for piles bored with a continuous-flight auger, values of 0.9 (for sandy soils) down to 0.6 (for silts and silty sands) are taken. In general, the roughness of the hole will ensure that the full soil friction is mobilized at the pile edge. However, potential loosening of the soil during the installation process may be allowed for by taking the friction angle δ between ϕ' and ϕ'_{cv}.

For driven cast-in-situ piles, while the value of K may be 1.2 or higher outside the casing, some reduction in normal stress may occur during extraction of the casing. If wet concrete is placed, a value of K of 1.0 may be taken, while values up to 1.2 are appropriate where dry concrete is rammed into the pile shaft.

For continuous-flight auger piles, experience indicates that a value of $K = 0.9$ is appropriate in sands and gravels. Rather lower values, as low as 0.5 to 0.6, have been found with silts and silty sands, possibly due to entraining material from the sides of the pile into the spoil during augering, thus reducing the horizontal stresses in the ground. Close control is needed during construction in order to monitor the amount of material removed and the rates of concrete delivery and auger extraction as the pile is formed.

It should be emphasized that, particularly for relatively short piles, the shaft capacity of bored piles may be less than half that of driven piles of the same nominal size. This factor, together with potential difficulties in construction, renders bored piles unattractive in soils with low cohesion.

Correlations with field tests. Where cone penetration testing has been carried out, the end-bearing pressure for piles may be obtained directly from the cone pressure, q_c. Nordlund (1963) recommends taking an average value of q_c over a depth range of 3 pile diameters above the pile base down to 2 diameters below the pile base (see Figure 4.5). The end-bearing pressure is then taken as this average value of q_c. Fleming and Thorburn (1983) recommend more detailed schemes for averaging the cone readings, in order to put more weight on the minimum values. The range over which the average is taken is extended up to 8 pile diameters above the level of the pile base. Thus, in homogeneous sand, end-bearing pressure is estimated as (see Figure 4.5)

$$q_b = (q_{c1} + q_{c2} + 2q_{c3})/4 \qquad (4.13)$$

where $\quad q_{c1} =$ average cone resistance over 2 diameters below pile base,

Figure 4.5 Averaging of cone data to give end-bearing pressure

q_{c2} = minimum cone resistance over 2 diameters below pile base,
q_{c3} = average of minimum values lower than q_{c1} over 8 diameters above pile base.

Correlations have also been developed for estimating q_b directly from the results of standard penetration testing. The ratio q_b/N (or q_c/N) where N is the number of blows per 0.3 m penetration, varies with the composition of noncohesive soil. As has been discussed in Chapter 2 (see Figure 2.7), there is an approximately linear relationship between q_c/N and average particle size of the soil, with the ratio varying from as high as $0.8\,MN/m^2$ for medium-size gravel, down to $0.25\,MN/m^2$ for silt.

Values of end-bearing pressure estimated from results of standard penetration tests, or even from the results of cone tests, should be assessed by means of equation (4.2), adopting estimated values of the various soil parameters where necessary, before being used in design. Correlations of results from standard penetration tests, with the relative density of the sand and effective overburden stress, may be used for this purpose, as indicated in Chapter 2 (Figure 2.9).

Modern cone penetration techniques, where a friction sleeve is used immediately behind the cone, enable the skin friction to be estimated directly for full displacement piles, taking the skin friction, τ_s, as equal to the sleeve friction, f_c, measured behind the advancing cone. Piles will generally have a somewhat rougher surface texture than the friction sleeve. However, differences due to this will be balanced by scale effects, and the presence of rather higher normal stresses on the sleeve just behind the advancing cone, than will be present around the pile shaft. Where no friction sleeve results are available, the skin friction may be estimated approximately from the cone resistance, using the empirical formula (Vesic, 1977)

$$\tau_s = 0.11 \, e^{-\tan \phi'} q_c \qquad (4.14)$$

Results from standard penetration tests may also be used to provide an approximate guide to the shaft capacity of driven piles. Thorburn and McVicar (1979) suggest

$$\tau_s = 2N \text{ kN/m}^2 \qquad (4.15)$$

Comparison of this expression with the relationship between q_b and N shown in Figure 2.7 indicates a ratio q_b/τ_s of 100 to 200, towards the upper end of the range found by Vesic and slightly higher than that implied by equation (4.9). This may reflect conservatism in the estimate of skin friction given by equation (4.15).

The above results should again be modified for cases of partial displacement piles (reducing τ_s by 20%) and for bored piles (reducing τ_s by a factor of 2 to 3). Where tapered piles are used, the values of skin friction may be increased, following the guidelines suggested by Nordlund (1963). As an approximate guide, the skin friction mobilized by gradually tapering piles (not step-tapered piles) may be taken as a factor of $(1 + t)$ greater than for straight shafted piles, where t is the taper in %. Values of skin friction should be limited to $K_p \tan \delta$, where K_p is the passive earth pressure coefficient of $(1 + \sin \phi')/(1 - \sin \phi')$.

The expressions given in this section for calculating the shaft capacity of driven piles should be used with caution in certain types of cemented sands where the in-situ voids ratio is high, or in sands containing crushable particles. Breaking down of the cementation and crushing of the soil particles during pile installation can cause liquefaction of the sand. This can lead to very low effective normal stresses acting on the pile shaft, and values of skin friction which are less than 20% of those normally associated with silica sands possessing similar values of ϕ'. McClelland (1970) has recommended a maximum skin friction of 20 kN/m^2 in such sands. Detailed discussions of pile capacity in calcareous soil have been presented by Murff (1987) and Randolph (1988), who emphasize not only the low values of skin friction, but also the susceptibility of those values to cyclic loading. In general, bored (drilled and grouted) piles or post-grouted driven piles are to be preferred to driven piles in calcareous soils.

4.1.2 *Capacity in cohesive soils*

End-bearing pressure. The long-term, drained, end-bearing capacity of a pile in clay will be considerably larger than the undrained capacity. However, the settlements required to mobilize the drained capacity would be far too large to be tolerated by most structures. In addition, the pile must have sufficient immediate load-carrying capacity to prevent a short-term failure. For these reasons, it is customary to calculate the base capacity of piles in clay in terms of the undrained shear strength of the clay, c_u, and a bearing capacity factor, N_c. Thus the end-bearing pressure is

$$q_b = N_c c_u \qquad (4.16)$$

For depths relevant for piles, the appropriate value of N_c is 9 (Skempton, 1951), although due allowance should be made where the pile tip penetrates a stiff layer (underlying soft soil) by only a small amount. A linear interpolation should be made between a value of $N_c = 6$ for the case of the pile tip just reaching the bearing stratum, up to $N_c = 9$ where the pile tip penetrates the bearing stratum by 3 diameters or more.

Skin friction. Most piles in clay develop a high proportion of their overall capacity along the shaft. Whereas the ratio of end-bearing pressure to skin friction may be typically 50 to 100 for piles in sand, the corresponding range is 10 to 20 for piles in clay. As such, considerably more effort has been devoted to developing reliable methods for estimating values of skin friction for piles in clay, than has been the case for sand. Historically, the skin friction

Figure 4.6 Relationship between skin friction and shear strength

around a pile shaft has been estimated in terms of the undrained shear strength of the soil, by means of an empirical factor, α, (Tomlinson, 1957) giving

$$\tau_s = \alpha c_u \qquad (4.17)$$

The value of α deduced from pile load tests appears to reduce from unity or more for piles in clay of low strength, down to 0.5 or below for clay of strength above about $100 \, \text{kN/m}^2$. Figure 4.6 shows measured values of skin friction from driven piles, plotted against the shear strength of the soil, illustrating this. This large amount of scatter apparent from the figure is a major cause for concern, and has prompted a more scientific approach.

Driven piles. In 1968, Chandler described an alternative approach, considering the bond between pile and soil as purely frictional in nature, with the resulting skin friction a function of the normal effective stress, σ'_n and an interface friction angle, δ, in much the same way as for piles in free-draining soils. The normal stress was related to the effective overburden stress, σ'_v, by a factor, K, to give

$$\tau_s = \sigma'_n \tan \delta = K\sigma'_v \tan \delta = \beta\sigma_v \qquad (4.18)$$

where $\beta = K \tan \delta$. The value of K will vary depending on the type of pile (driven or bored) and the past stress history of the soil. For piles in soft, normally or lightly overconsolidated clay, Burland (1973) and Parry and Swain (1977) have suggested values of K lying between $(1 - \sin \phi')$ and $\cos^2 \phi'/(1 + \sin^2 \phi')$. Neither of these suggestions takes due account of the stress changes which occur during and after pile installation.

Meyerhof (1976) deduced, empirically, an expression for the ratio $\beta = \tau_s/\sigma'_v$, from the results of a number of pile tests in clay of varying overconsolidation ratio. Assuming that the full angle of internal friction, ϕ', was mobilized at the shaft of the pile, he showed that the value of K consistent with the measured shaft capacities varied between 1 and 2 times the in-situ horizontal stress, with an average ratio of 1.5. Francescon (1982) showed from instrumented model pile tests, where the normal effective stress acting on the pile was measured directly, that this stress was indeed close to 1.5 times the original horizontal effective stress in the soil. The angle of friction between soil and pile was consistent with that measured in simple shear tests, being somewhat lower than that measured in triaxial compression tests. Denoting this angle by ϕ'_{ss}, the skin friction was thus given by

$$\tau_s = 1.5 \, K_0 \sigma'_v \tan \phi'_{ss} \qquad (4.19)$$

Attempts to follow, theoretically, the full stress history of soil elements in the vicinity of a driven pile have shown that the appropriate value of skin friction will depend not only on the shear strength of the soil, but also on its past stress history and overconsolidation ratio (Randolph *et al.*, 1979;

Randolph and Wroth, 1981; Kraft, 1982). This is best reflected by the strength ratio, c_u/σ'_v, of the soil (Randolph and Wroth, 1983). Thus either of the parameters, $\alpha = \tau_s/c_u$, or $\beta = \tau_s/\sigma'_v$, should show a consistent variation with the strength ratio for different soils. Figure 4.7 shows α, deduced from load tests on driven piles (Randolph and Murphy, 1985), plotted against the average in-situ strength ratio. The data fall in a band with a gradient of -0.5 at low values of c_u/σ'_v, the gradient reducing to about -0.25 for c_u/σ'_v greater than unity. Semple and Rigden (1984) have undertaken a similar exercise, and arrive at a slightly different variation of α with the strength ratio, although with the same general pattern (see later, Figure 4.26).

If it is assumed that the value of α should be unity for normally consolidated clay, then the value of α may be given for the general case by the two expressions

$$\alpha = (c_u/\sigma'_v)_{nc}^{0.5}(c_u/\sigma'_v)^{-0.5} \tag{4.20}$$

for c_u/σ'_v less than or equal to unity, and

$$\alpha = (c_u/\sigma'_v)_{nc}^{0.5}(c_u/\sigma'_v)^{-0.25} \tag{4.21}$$

for c_u/σ'_v greater than unity. The subscript nc in these expressions refers to the normally consolidated state of the soil. Values of skin friction may be deduced directly from similar expressions:

$$\tau_s = (c_u/\sigma'_v)_{nc}^{0.5}(c_u\sigma'_v)^{0.5} \tag{4.22}$$

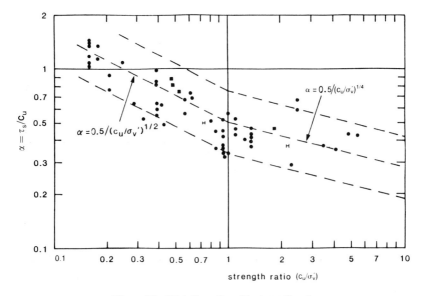

Figure 4.7   Variation of α with strength ratio

for c_u/σ_v' less than or equal to unity, and

$$\tau_s = (c_u/\sigma_v')_{nc}^{0.5} \, c_u^{0.75} \, \sigma_v'^{0.25} \tag{4.23}$$

for c_u/σ_v' greater than unity.

The above expressions have been used to predict the capacities of the piles referred to in Figures 4.6 and 4.7. Figure 4.8 shows a comparison between the predicted, Q_c and measured, Q_m, capacities, showing very good agreement. Equations (4.22) and (4.23) offer a number of advantages over alternative methods for estimating skin friction. The equations contain no artificial parameters other than the assumption that τ_s equals c_u for normally consolidated clay. Since pile installation will remould the soil, the strength ratio for normally consolidated clay should be that for remoulded material. This may be estimated with reasonable accuracy from a knowledge of the friction angle ϕ' for the soil, using concepts of critical state soil mechanics (Schofield and Wroth, 1968). To within a few per cent, the strength ratio for normally consolidated, remoulded clay with friction angle ϕ', may be

Figure 4.8 Comparison of calculated and measured pile capacities

estimated as $\phi'/100$ (where ϕ' is given in degrees). The square root sign means that the variation will be small; for most soils $(c_u/\sigma'_v)^{0.5}$ falls within the range 0.45 to 0.55. The value of undrained shear strength, c_u, occurs to the power of 0.5 or 0.75 in equations (4.22) and (4.23), thus reducing the sensitivity of the approach to accurate determination of this property.

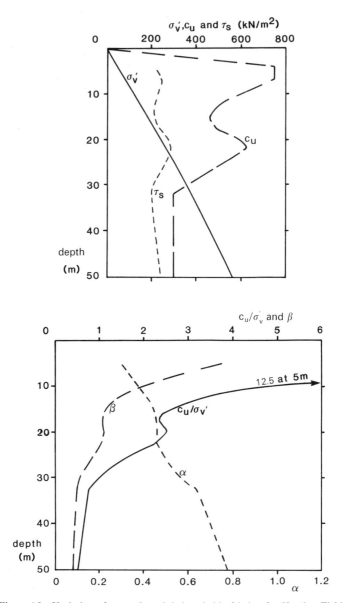

Figure 4.9   Variation of strength and deduced skin friction for Heather Field

In applying the above expressions, design profiles of c_u and σ'_v with depth should be determined, and then a profile of skin friction with depth deduced. For most soils, apart from at shallow depths, the value of c_u will be less than σ'_v and equation (4.22) will apply, giving values of skin friction which are proportional to the geometric mean of the shear strength and the effective overburden stress.

As an example of how the approach may be applied in practice, the capacity of a driven pile in stiff overconsolidated clay will be determined. Soil conditions and pile properties are taken from a group of six papers published in *Ground Engineering* in November, 1979, and January and March, 1980, in particular Parry (1980), describing the design of foundations for an offshore platform situated in the Heather Field in the North Sea. The piles were 1.524 m in diameter, 63.5 mm wall thickness, driven with a driving shoe of internal diameter 1.346 m to a final penetration of 45 m. The piles were to carry maximum working loads of 29.5 MN in compression, with a cyclic component of ± 21.2 MN. For a factor of safety of 1.5 (a typical figure for piles where the amount of settlement is not critical), the required pile capacity should be 44.3 MN.

Profiles of σ'_v, c_u and c_u/σ'_v are shown in Figure 4.9. The site showed strong sandy clay deposits, with shear strengths ranging up to $750 \, \text{kN/m}^2$ in the upper 10 m, which was very heavily overconsolidated. Parry (1980) has discussed the magnitude of these strengths, and a probable profile of the overconsolidation ratio. He quotes values of ϕ' for the soil of about $32°$ for depths between 5 and 50 m. In deducing a profile of skin friction, the upper 5 m of soil (3.3 pile diameters) has been ignored due to the likelihood of loss of skin friction due to lateral loading (and resulting post-holing) of the pile. Below this depth, the relationships given above have been used to obtain the profiles of α, β, and thus of the skin friction, τ_s, which are shown in Figure 4.9.

Table 4.1 Shaft capacity with depth

Depth (m)	Cumulative shaft capacity (MN)	
	Present approach	Parry (1980)
5	0	1.3
10	5.6	4.9
15	10.8	10.7
20	16.3	17.6
25	22.9	24.0
30	28.4	29.7
35	33.1	35.0
40	38.0	40.2
45	43.2	45.6
50	48.6	51.2

It is interesting to compare the deduced cumulative shaft capacity of the pile with that estimated by Parry (1980), using a different effective stress approach (see Table 4.1). At the design penetration of 45 m, the two approaches agree to within 5%. In addition to the shaft capacity, the base capacity of the pile may be estimated in terms of an end-bearing pressure of 9 times the shear strength of 300 kN/m^2, giving a total of 4.9 MN (assuming the pile is fully plugged under static loading conditions). Thus the overall predicted capacity is at least 48.1 MN.

Bored piles. Turning to bored piles, assuming that no chemical bond develops between the pile and soil, an effective stress approach is again more attractive than the empirical approach based purely on the undrained shear strength of the soil. Provided the pile is formed promptly after excavation of the shaft, little change in the in-situ effective stress state in the soil should occur, and equation (4.18) may be applied with $K = K_0$. In heavily overconsolidated clay, where the value of K_0 is large, some allowance for stress relaxation should be made, reducing the value of K by 20%. Alternatively, the mean stress between the in-situ horizontal stress and that due to the concrete poured into the pile shaft may be taken, replacing K by $(1 + K_0)/2$. It must be emphasized, however, that delays in pouring the concrete may reduce the in-situ horizontal effective stress considerably. Back analysis of tests of bored piles in heavily overconsolidated London Clay show deduced values of K below unity, where delays have occurred between augering and forming the piles (Fleming and Thorburn, 1983).

The question of what friction angle should be used in equation (4.18) is still subject to debate. The surface of most bored piles will be sufficiently rough to ensure that failure takes place in the soil immediately around the pile, rather than actually on the interface. However, there have been suggestions (notably Burland and Twine, 1988) that a residual angle of friction is appropriate, at least for bored piles in heavily overconsolidated clay. The friction angle mobilized on the (vertical) failure surface at peak shear stress will depend on the complete stress state. Randolph and Wroth (1981) have argued that, where the horizontal effective stress exceeds the vertical effective stress, the friction angle will be significantly lower than that obtained from a triaxial compression test, and will be similar to that mobilized on horizontal surfaces at failure in a simple shear test, conducted under appropriate normal stress conditions.

Adoption of the residual angle of friction in equation (4.8) (based on arguments of slip-planes formed during the augering process) does not appear consistent with the assumption of failure *within* the clay (due to the roughness of the pile shaft). However, residual friction angles may be appropriate where the augered shaft is relatively smooth, and failure occurs at the concrete–soil interface. Overall, a reasonable design approach seems to be to take the horizontal effective stress as $0.5\,(1 + K_0)\sigma'_v$, and the friction angle as that mobilized at failure in a simple shear test. Where the horizontal effective stress is greater than the vertical effective stress (i.e. in heavily overconsolidated clays), the friction angle will lie in the range 15 to 20° (Randolph and Wroth, 1981).

For the design of bored piles in soil deposits where the profile of K_0 with depth is not known, either the approach for driven piles may be used with an appropriate factor to allow for stress relief on excavation of the pile shaft, or a direct 'total stress' approach may be used. Assuming that little change in the in-situ effective stress state takes place during construction of the pile, equation (4.19) would indicate that the skin friction for a bored pile may be estimated as about 70% of that for a driven displacement pile. This figure may be conservative for piles in soil with low in-situ values of K_0, since some increase in lateral stress may occur during the process of casting the pile.

For bored piles in stiff, heavily overconsolidated clay, it has been customary to assume an α value of 0.45 in order to estimate the shaft capacity (Skempton, 1959). This approach is based largely on experience in London Clay, and care should be taken when applying the same value of α to clays where the strength ratio, c_u/σ'_v, is higher than for London Clay. An effective stress approach may well indicate substantially lower values of skin friction than the total stress approach based on $\alpha = 0.45$.

Weltman and Healy (1978) have analysed a number of pile tests in boulder clays and other glacial tills, and suggest values of α varying with the undrained shear strength of the soil as shown in Figure 4.10. The ratio of values for bored and driven piles is just over 80%, but both curves are derived from data showing considerable scatter. The figure also shows a separate curve for very short, stubby piles driven into till overlain by soft clay. For such piles, the ratio of c_u/σ'_v will be high and the resulting low α values are consistent with the effective stress approach outlined earlier.

Correlations with field tests. Comparisons of measured pile capacities with results of cone penetration or standard penetration tests show a much greater

Figure 4.10 Variation of α with shear strength of glacial till (Weltman and Healy, 1978)

scatter than for non-cohesive soils. The drainage conditions during a penetration test will be rather different from those existing during a pile test, and this may account for some of the variation. The end-bearing pressure mobilized by a pile should, in principle, be comparable with that measured during a cone penetration test. However, effects of scale and of different rates of testing combine to yield variations in the ratio q_b/q_c ranging from under 0.5 to over 2. Unless there is particular evidence to the contrary, for a given soil, the end-bearing pressure should be taken equal to that measured in a cone penetration test.

Estimates of the skin friction for driven piles are often made from friction sleeve measurements, taking $\tau_s = f_c$. However, caution should be exercised using this approach, owing to the sensitivity of the value of friction measured to the degree of consolidation in the soil (a) around the pile and (b) around the friction sleeve. Pile tests show that the value of skin friction measured during installation (a similar process to the continuous penetration of the cone and friction sleeve) may be as little as 20% of the long-term skin friction (see section 4.1.5). Estimates of skin friction directly from the cone resistance, q_c, offer a better guide. However, recommended correlations range from $\tau_s = q_c/10$ (Fleming and Thorburn, 1983) down to $\tau_s = q_c/40$ (Thorburn and McVicar, 1979). It is probably more reliable to deduce profiles of undrained shear strength from the cone penetration tests and then to use the expressions given earlier, relating skin friction to the undrained shear strength and to the effective overburden stress.

As discussed in Chapter 2, results from standard penetration tests in medium to firm clays and soft rocks may be used to estimate the shear strength of the soil, taking a ratio for c_u/N of 4 to $5 \, \text{kN/m}^2$ for clays of medium plasticity, rising to 6 to $7 \, \text{kN/m}^2$ for plasticity indices less than 20 (see Figure 2.4). The values of shear strength may then be used to provide estimates of end-bearing pressure and of skin friction in the normal way.

4.1.3 Capacity of piles in rock

Piles driven to refusal on firm or intact rock are often designed as purely end-bearing piles ignoring any shaft capacity the pile might have, although, in practice, considerable shear transfer may occur in the upper layers of soil, at working loads. In many cases, the maximum design load for such a pile will be determined by the stresses in the pile material itself, rather than the permitted bearing pressure on the rock.

Great care should be exercised where steel piles are driven to bedrock, to ensure that good sound rock contact is achieved. Pile driving may shatter the rock, and indeed it is good practice to ensure that the piles penetrate a small distance into the rock. This will help to prevent the piles heaving and coming away from the bearing stratum during driving of subsequent piles. Alternatively, where access permits, each pile should be re-tamped at the end of installing the neighbouring piles, to ensure good contact with the bearing stratum.

In weathered rock, careful monitoring of driving records is essential owing to the wide variability in strength exhibited by soft rocks such as chalk and marl in a weathered state. Refusal of the piles may not be achieved and the contribution of the pile shaft to its overall capacity may need to be relied upon. Accurate estimation of the skin friction around a pile driven into soft rock is difficult owing to the disruption of the structure of the rock as the pile penetrates. The skin friction will depend on the magnitude of the normal effective stress, which is liable to vary in a haphazard fashion. Recorded values of skin friction for piles driven into low-grade chalk vary from as low as $8 \, \text{kN/m}^2$ (Hodges and Pink, 1971) up to $26 \, \text{kN/m}^2$ (Hobbs and Robins, 1976) for similar pile penetrations.

End-bearing pressure. It has been customary for design codes to stipulate allowable bearing pressures on rocks of different types. Tomlinson (1986) presents an extensive list of allowable pressures, in terms of type, quality and joint spacing of the rock. In view of the great variability in strength (and allowable end-bearing pressure) of individual rock types, it appears best to express allowable bearing pressures for pile design in terms of the unconfined compressive strength of the rock. Theoretical (and experimental, where failure has actually been achieved) values of ultimate end-bearing pressure are generally in excess of $10 \, q_u$ (where q_u is the unconfined compressive strength). Large settlements would be required to mobilize such values of end-bearing, and, in practice, a rather lower 'ultimate' end-bearing pressure of about $3 \, q_u$ is often adopted (Rowe and Armitage, 1987).

A rational design approach is to limit allowable bearing stresses to below that which would initiate crushing of the rock beneath the pile. Even where the joints are relatively open, this value will not fall below the unconfined compressive strength of the rock. Thus, allowable end-bearing stress, q_{ba}, may be expressed as (Rowe and Armitage, 1987)

$$q_{ba} = q_u \tag{4.24}$$

The above value is rather higher than that of $0.3 \, q_u$ suggested by, for example, Poulos and Davis (1980). However, the higher value is supported by experimental evidence and is consistent with a factor of safety of 3 applied to the (conservative) ultimate bearing pressure of $3 \, q_u$ discussed above. Lower values, down to $0.3 \, q_u$, may be adopted where there are solution cavities or open rock joints close beneath the pile tip, or where the pile penetrates the rock by less than one diameter.

Skin friction. Shear transfer along the shaft of a rock socket is a complex phenomenon that depends on the frictional characteristics of the interface, the degree of roughness of the socket and the strength properties of the host

Figure 4.11 Correlations of skin friction and end bearing pressures with SPT value, for piles in chalk (Hobbs and Healy, 1979). (Top) Ultimate shaft resistance against N value for all piles. (Bottom) Ultimate end bearing resistance for some driven piles. *A*, Driven cast-in-place piles (fresh concrete in contact with the chalk). *B*, displacement piles (including preformed and cast-in-place types).

rock. Values of peak friction will depend strongly on the normal stress along the shaft, and the amount of dilation that occurs during shearing (Johnston and Lam, 1989). In practice, it has been customary to correlate measured values of skin friction with the unconfined compressive strength of the rock (Horvath *et al.*, 1980; Williams and Pells, 1981; Rowe and Armitage, 1987). While there are minor differences, most of these correlations show a trend of skin friction which is proportional to the square root of the compressive strength, and a relationship of the form

$$\frac{\tau_s}{p_a} = 1.3\left(\frac{q_u}{p_a}\right)^{0.5} \tag{4.25}$$

appears to offer a reasonably conservative design approach (where p_a is atmospheric pressure, $100\,kN/m^2$). The above relationship should be restricted to values of unconfined compressive strength in excess of $500\,kN/m^2$, for bored piles where the shaft is sufficiently rough to ensure full keying of the concrete and host rock. A check should be made in any design, that the assumed skin friction is not greater than 5% of the concrete strength.

Correlations with SPT in chalk. Hobbs and Healy (1979) have presented an extensive review of the performance of piles in chalk, and have correlated skin friction and end-bearing pressures for different pile types with the SPT value of the chalk. For large displacement driven piles, they recommend using an effective stress approach for estimating the shaft capacity, taking a value for K of 0.06 N, and interface friction angles of 20° (for steel piles) to 24° (for concrete piles). Higher values of K (up to 0.15 N) were recommended for driven cast-in-situ piles where high-slump concrete was compacted against the chalk, while lower values of K (0.03 N) were suggested for low displacement piles. End-bearing pressures were shown to correlate with N values with ratios of q_b/N lying between 0.2 and 0.25 MN/m². Figure 4.11 shows summaries of measured values of skin friction and end-bearing pressure, plotted against SPT values, taken from Hobbs and Healy (1979).

   For bored piles in chalk, where interlocking and possible chemical bonding occurs between the pile and the chalk, Hobbs and Healy found that the skin friction correlated well with the compressive strength of the chalk, according to the relationship given in equation (4.25).

4.1.4 Driving formulae and dynamic estimates of pile capacity

In many instances, particularly in heavily stratified soil or rock of varying degree of weathering, the capacity of each pile installed needs to be checked from the driving record. The capacity of the pile at the time of installation should represent a conservative value. With time after installation, dissipation of excess pore pressures generated during the driving process leads to an increase in effective stresses around the pile, and a resulting increase in the pile capacity. For piles driven into clay, this increase may be as much as a factor of 5 to 6 (Vesic, 1977). Thus pile load tests aimed at assessing the long-term capacity of a working pile should allow sufficient time for this capacity to be achieved—often a matter of 3 to 4 weeks (see section 4.1.5).

Pile driving formulae. The most widespread method of estimating the dynamic capacity of piles is the use of some form of pile driving 'formula', relating the measured permanent displacement (or 'set') of the pile at each blow of the hammer, to the pile capacity. Driving formulae are based on an energy balance between the (dynamic) input energy of the hammer, and the (static) work required to move the pile permanently a small distance, s. As such, they offer a compromise (often rather a poor one) between the static and dynamic capacities of a pile.

There is a variety of pile driving formulae in common use, many of which have been reviewed by Whitaker (1975), who comments that, in some situations, pile capacities predicted by the different formulae may differ by a factor of 3. Rather than describe all the different approaches here, the basic approach will be outlined and some of the main variables discussed.

A simplified picture of the driving process, and variation of pile resistance with displacement is shown in Figure 4.12. This picture gives rise to the

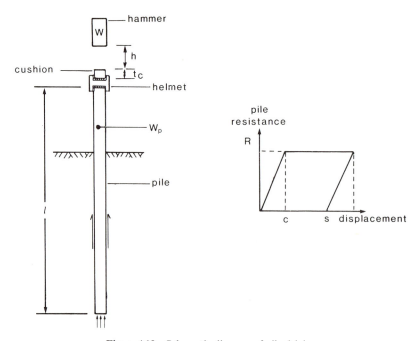

Figure 4.12 Schematic diagram of pile driving

fundamental pile driving formula

$$R = \frac{\eta W h}{(s + c/2)} \qquad (4.26)$$

where: R is the pile resistance,
       η is the efficiency of the hammer (allowing for energy loss on impact),
       W is the weight of the hammer,
       h is the drop height,
       s is the permanent set of the pile, and
       c is the elastic, or recoverable, movement of the pile.

This expression is widely used in the American continent, adopting a value of 0.8 to 1 for η, and 0.1 in. (2.5 mm) for c (giving the so-called *Engineering News* formula). In Europe, both η and c are chosen with regard to the type of hammer, the material used as a cushion at the head of the pile, and the physical properties of the pile. Probably the best known variant is that due to Hiley (1925), where the overall hammer efficiency is given by

$$\eta = \frac{k(W + e^2 W_p)}{W + W_p} \qquad (4.27)$$

where k is the output efficiency of the hammer (ratio of power delivered at the cushion, to rated power),
      W_p is the weight of the pile, and
      e is the coefficient of restitution between the hammer and the pile cushion (or top of pile if there is no cushion).

The recoverable movement, c, is taken as the sum of the cushion compression, $c_c = R t_c/(AE)_c$ (where t_c is the thickness and $(AE)_c$ the cross-sectional rigidity of the cushion), the elastic shortening of the pile (considered as a column), $c_p = Rl/(AE)_p$, and the recoverable movement (or

Table 4.2 Values of parameters for pile driving formulae

Type of hammer	Power efficiency, k
Drop hammer (triggered fall)	1
Steam or compressed air hammer	0.9
Drop hammer (winch operated)	0.8
Diesel hammer	0.6–0.8

Cushion type	Coefficient of restitution, e	Young's modulus, E_c (MN/m^2)
Micarta plastic	0.8	3×10^3
Greenheart oak	0.5	3×10^2
Other timber	0.3	2×10^2

'quake') of the ground, c_g. The last quantity may be taken as 0.5% of the pile diameter. Table 4.2 shows commonly adopted values for the quantities k, e and E depending on hammer type and cushion material.

It should be emphasized that the form of equation (4.26) is inherently inaccurate for assessing, by means of the measured set, s, whether a required pile resistance, R, has been achieved. Uncertainties in the values of k and the various components of c can lead to wide variations in the deduced pile resistance. Over a period of a few hours driving, the characteristics of both the hammer and the cushion may alter significantly; thus piles driven to a given set at the start and end of the period may have widely varying capacities.

Dynamic analysis of pile driving. A more rigorous assessment of the driving characteristics of a pile may be achieved by means of a proper dynamic analysis of the pile–soil system. The inertial effects of the soil around the pile, and the viscous nature of the soil, may both be taken into account by appropriate numerical techniques, such as the finite element method (Smith and Chow, 1982). A common simplification of the soil is as a massless medium providing frictional resistance alone, while the pile is modelled as a discrete assembly of mass elements, interconnected by springs. This 'one-dimensional' model is discussed in detail in section 9.3 (Dynamic testing of piles).

At the design stage, the main objective of dynamic analysis of pile driving is to assess the 'drivability' of a pile in given soil conditions, and with a particular type and size of hammer. This is commonly accomplished using one-dimensional wave equation programs, based on the method described in section 9.3. Probably the most widely used program is the WEAP program developed originally by Goble and Rausche (1976) for the Federal Highway Administration in North America. The program has recently been updated (Goble and Rausche, 1986; Rausche *et al.*, 1988).

The main outcome of a drivability study is a series of curves that give the penetration rate (or blow count) as a function of the assumed 'static' resistance of the pile. These curves may be used to choose an appropriate hammer, to assess the time and cost of installation of each pile, and to provide a quality control on pile installation (in terms of required 'set' per blow for a given capacity). In addition, a drivability study will include assessment of the maximum tensile and compressive stresses in the pile during driving, maximum acceleration levels, and range of required (or permitted) hammer stroke. Limiting the range of hammer stroke may be particularly necessary during the early driving of concrete piles, in order to avoid excessive tensile stresses in the pile.

4.1.5 *Increase in capacity of piles with time*

Discussion of the axial capacity of piles would not be complete without some comment on the effects of time and also of loading rates. The design methods that have been described are based on observed performance of piles which have been load tested over a relatively short period (of tens of minutes or hours), usually a matter of a few days or weeks since installation. Except in particular circumstances involving dynamic loading, piles must generally withstand their design loads for very long periods of time, and the question arises whether creep, or stress relaxation, in the soil around the pile, may lead to reduced capacity. In soils which are known to creep, some allowance may need to be made for these effects, although it should be borne in mind that the stress level in the soil will generally be relatively low. It is unlikely that creep would lead to more than 5 to 10% reduction in the capacity of a pile.

The question of the effect of time since installation is of much more consequence, particularly for displacement piles installed in soils of low permeability. Piles driven into cohesive soil generate high excess pore pressures close to the pile. The pore pressures arise partly as a result of a decrease in effective stress, as the soil is sheared and remoulded, and partly due to the increase in total stress, as the pile forces soil out of its path. Typically, the excess pore pressures may be as high as the effective overburden stress, and may extend out over a zone up to 10 times the diameter of the pile. Randolph *et al.* (1979) suggest that the excess pore pressure, Δu, close to a driven pile may be estimated from the expression

$$\Delta u = 4c_u - \Delta p' \tag{4.28}$$

where $\Delta p'$ is the change in mean effective stress due to shearing and remoulding the soil. In normally or lightly overconsolidated clay, $\Delta p'$ will be negative and may be as high as 2 to $3c_u$ for sensitive clays. In more heavily overconsolidated clay, $\Delta p'$ will become positive as the clay attempts to dilate on shearing. Figure 4.13 shows typical field measurements of excess pore pressures due to pile installation. The values reflect the trend implied by equation (4.28).

After the pile is installed, the excess pore pressures will dissipate, primarily by radial flow of pore water away from the pile, and the soil will consolidate.

Figure 4.13 Field measurements of excess pore pressure

During this process, the water content of the soil will decrease, and its shear strength increase. Seed and Reese (1955) report a decrease in water content of 7% close to a pile driven into San Francisco Bay mud. This was accompanied by a threefold increase in the remoulded strength of the soil (measured from samples taken from close to the pile). A similar figure for the decrease in water content adjacent to a displacement pile has been reported by Francescon (1982), from tests on model piles installed in soil of varying overconsolidation ratio.

The axial capacity of a driven pile will vary with time after installation, because of this consolidation process. Seed and Reese report a sixfold increase in pile capacity, over a period of 30 days. Similar increases in capacity have been discussed by Vesic (1977) and by Thorburn and Rigden (1980). Analytical studies of the stress changes caused by pile installation, and of the consolidation process, have shown that the shear strength of the soil may increase by between 30 and 100% close to the pile, over a time period which is governed by the dimensionless group $c_h t/r_0^2$, where c_h is the coefficient of consolidation in a horizontal plane, t is the time since installation of the pile, and r_0 is the radius of the pile (Randolph *et al.*, 1979; Randolph and Wroth, 1979).

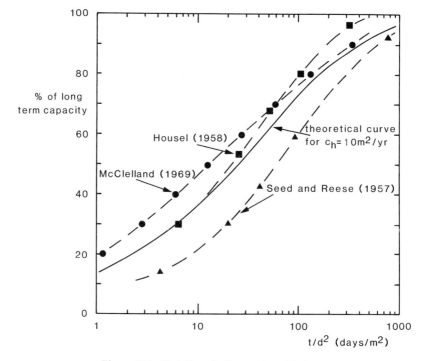

Figure 4.14   Variation of pile capacity with time

Figure 4.14 shows the variation of pile capacity with time, for three piles driven into soft silty clay. The time scale has been normalized by the square of the pile diameter. Other factors which will affect the rate of dissipation of excess pore pressures are (a) the consolidation coefficient, c_h, and (b) the extent of the zone of excess pore pressures. The latter will be a function of the rigidity index, G/c_u, of the soil (where G is the shear modulus), and the sensitivity of the soil. Also shown on Figure 4.14 is a theoretical curve of the change in excess pore pressure, normalized by the maximum value, with time. The curve is taken from Randolph and Wroth (1979), assuming that the excess pore pressures extend out to a distance of 10 pile radii from the axis of the pile. A value of $10\,m^2/year$ has been taken for c_h, which is reasonable for silty clays under conditions of horizontal drainage. The measured variations of pile capacity follow the theoretical curve reasonably

(a) Piles driven into sensitive clay close to an existing building

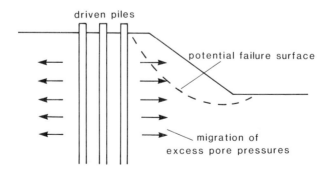

(b) Piles driven close to an existing slope

Figure 4.15 Potential problems due to pile driving

well. When piles are driven in a group, the magnitude of the excess pore pressures generated will be similar to that for a single pile. However, the zone over which they extend will be considerably larger, and the resulting consolidation process will be more protracted.

Pile installation may temporarily reduce the stability of nearby buildings or slopes. The remoulding of soil during pile driving may decrease the strength of a sensitive clay sufficiently to cause a bearing capacity failure or slope instability (see Figure 4.15(a)). Dissipation of the excess pore pressures generated during pile driving may allow pore pressures to rise in the vicinity of a slope, thus leading to failure of the slope, even where the soil has not been remoulded (see Figure 4.15(b)). In general, the effects of pile installation will not extend beyond 10 to 15 pile diameters. Where there is danger of instability, consideration should be given to the use of non-displacement piling techniques, or to preboring a slightly undersize hole before driving piles—a technique commonly adopted in Scandinavia.

4.2 Axial deformation of piles

The load deformation response of piles under axial load has been examined extensively, using numerical methods, particularly integral equation or boundary element methods (Poulos and Davis, 1968; Poulos and Mattes, 1970; Butterfield and Banerjee, 1971a, b; Banerjee and Davies, 1977). Such methods enable charts to be developed showing how the settlement of a pile depends on the various parameters of pile geometry and stiffness, and soil stiffness. Poulos and Davis (1980) have compiled an extensive collection of such charts, enabling the load settlement response of any given pile to be estimated readily. As an aid to understanding the manner in which a pile transfers load to the soil under working conditions, an approximate solution is developed here for the load settlement response of a pile in elastic soil. The solution leads to an expression for the pile stiffness (that is applied load divided by settlement) in closed form, which may be used as an alternative to charts.

In developing this solution, the manner in which load is transferred to the soil from the pile shaft, and from the pile base, will be examined separately, before combining the two to give the response of the complete pile. In the first instance, the soil will be treated as an elastic material, characterized by an appropriate secant value of elastic modulus, varying in some fashion over the depth of interest. Choice of values of elastic modulus for different soil types is discussed in Chapter 5.

Some discussion is appropriate concerning the geometric modelling of real piles. Analytical solutions for the response of piles have tended to treat the pile as a circular cylinder of diameter d (or radius r_0) and length l. In practice, noncircular piles, such as square or hexagonal precast concrete piles or H-piles, are common. Such piles may be adequately represented as circular

Figure 4.16   Idealization of H section pile

piles by taking a pile of equivalent (gross) cross-sectional area. Thus for an H-section pile of outer dimensions $b \times w$ (see Figure 4.16), a circular pile of radius r_0 may be taken, where $\pi r_0^2 = bw$. It is also convenient to consider the pile as solid, with an appropriate Young's modulus E_p, such that the rigidity is equivalent to that of the original pile. Thus, when considering axial deformation, the equivalent pile modulus is

$$E_p = \frac{(EA)_p}{\pi r_0^2} \qquad (4.29)$$

where $(EA)_p$ is the cross-sectional rigidity of the actual pile.

4.2.1   Basic solution for a rigid pile

Pile shaft. Finite element and boundary element analyses of the response of friction piles (Frank, 1974; Randolph, 1977) have shown that load is transferred from the pile shaft by shear stresses generated in the soil on vertical and horizontal planes, with little change in the vertical normal stress (except near the base of the pile). Schematically, a pile may be considered as surrounded by concentric cylinders of soil, with shear stresses on each cylinder. For vertical equilibrium, the magnitude of the shear stress on each cylinder must decrease inversely with the surface area of the cylinder (Cooke, 1974; Frank, 1974). Writing the shear stress on the pile shaft as τ_0, the shear stress at radius r is given by

$$\tau = \frac{\tau_0 r_0}{r} \qquad (4.30)$$

Since the mode of deformation around a pile is primarily one of shear, it is more natural to develop the solution in terms of the shear modulus G and Poisson's ratio, v, rather than Young's modulus, E (noting that the shear modulus may be related to Young's modulus by $G = E/2(1 + v)$). The shear strain in the soil, γ, will thus be given by $\gamma = \tau/G$. Since the main deformation in the soil will be vertical, the shear strain may be written approximately as

$$\gamma \sim \frac{dw}{dr} \qquad (4.31)$$

where w is the vertical deflection.

These relationships may be assembled and integrated to give

$$w = \int_r^{r_m} \frac{\tau_0 r_0}{Gr} dr = \frac{\tau_0 r_0}{G} \ln(r_m/r) \qquad (4.32)$$

A maximum radius r_m has been introduced at which the deflections in the soil are assumed to become vanishingly small. Empirically, this radius has been found to be of the order of the length of the pile (Randolph and Wroth, 1978). The deflection of the pile shaft, w_s, is given by

$$w_s = \zeta \frac{\tau_0 r_0}{G} \quad \text{where } \zeta = \ln(r_m/r_0) \qquad (4.33)$$

This approximate analytical derivation of the deflection of the pile shaft in terms of the average applied shear stress τ_0 shows a number of important features of the response of a pile to axial load.

(i)   Stress changes induced in the soil are primarily shear stresses, decreasing inversely with distance from the pile axis; thus, only soil very close to the pile is ever highly stressed.

(ii) The resulting deflections decrease with the logarithm of distance from the pile axis; thus, significant deflections extend some distance away from the pile (up to about one pile length).

(iii) The deflection of the pile shaft, w_s, normalized by the pile radius r_0, is ζ times the local shear strain, $\gamma_0 = \tau_0/G$, in the soil: the parameter ζ is found to vary between 3 and 5, with an average value of about 4 (Baguelin and Frank, 1979).

Equation (4.33) is of particular importance in deducing the magnitude of deflection necessary to mobilize full shaft friction (see section 4.2.4).

The overall load taken by the pile shaft is $P_s = 2\pi r_0 l \bar{\tau}_0$, where $\bar{\tau}_0$ is the average shear stress mobilized at the pile shaft. Thus the load settlement ratio (or stiffness of the pile–soil system) is

$$\frac{P_s}{w_s} = \frac{2\pi l \bar{G}}{\zeta} \qquad (4.34)$$

where $\bar{G}$ is the average shear modulus of the soil over the depth of penetration of the pile.

Pile base. At the pile base, it is sufficient to ignore the pile shaft and surrounding soil, and treat the base as a rigid punch acting at the surface of a soil medium (that, in reality, starts at a depth, $z = l$). The deflection is obtained from the standard solution (Timoshenko and Goodier, 1970) as

$$w_b = \frac{P_b}{r_b G_b} \frac{(1 - v)}{4}$$ (4.35)

where the subscript b refers to the pile base.

Combining shaft and base. For a stiff pile, the base settlement and shaft settlement will be similar to the settlement of the pile head, w_t. The total load, P_t, may thus be written as

$$P_t = P_b + P_s = w_t \left(\frac{P_b}{w_b} + \frac{P_s}{w_s} \right)$$ (4.36)

In developing a general solution for the axial response of a pile, it is convenient to introduce a dimensionless load settlement ratio for the pile. The pile stiffness is P_t/w_t and this may be made dimensionless by dividing by the radius of the pile and an appropriate soil modulus. It has been customary to use the value of the soil modulus at the level of the pile base for this purpose, written as G_l. Thus equation (4.36) becomes (making use of equations (4.34) and (4.35))

$$\frac{P_t}{w_t r_0 G_l} = \frac{4 r_b G_b}{(1 - v) r_0 G_l} + \frac{2 \pi \bar{G}}{G_l} \frac{l}{r_0}$$ (4.37)

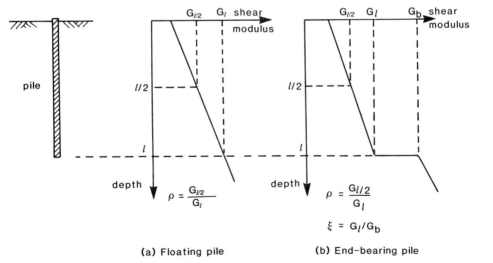

(a) Floating pile   (b) End-bearing pile

Figure 4.17 Assumed variation of soil shear modulus with depth

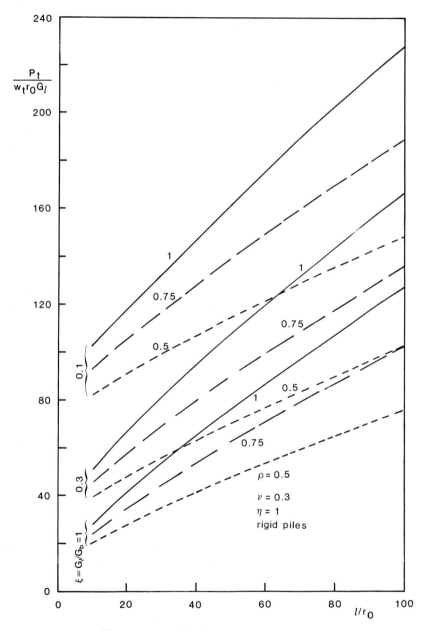

Figure 4.18   Load settlement ratios for rigid piles

In order to obtain a feel for the relative importance of the terms in equation (4.37), the shear modulus variation with depth may be idealized as linear, according to $G = G_0 + mz$ (where z is the depth), with the possibility of a sharp rise to G_b below the level of the pile base (see Figure 4.17). Defining parameters $\rho = \bar{G}/G_l$ and $\xi = G_l/G_b$, the constant ζ has been found to fit the expressions

$$\zeta = \ln\{[0.25 + (2.5\rho(1 - v) - 0.25)\xi]l/r_0\} \tag{4.38}$$

$$\zeta = \ln[2.5\rho(1 - v)l/r_0] \text{ for } \xi = 1 \tag{4.39}$$

Figure 4.18 shows how the load settlement ratio varies with slenderness ratio l/r_0 for a range of values of ρ and ξ, taking $r_b/r_0 = 1$. It should be noted that the dependence on v is small and the results in Figure 4.18 are all for $v = 0.3$. It may be seen that, as might be expected, the stiffness of the pile increases nearly linearly with pile length, holding other factors constant. However, it must be borne in mind that this result is only true of piles that may be considered rigid.

4.2.2 Pile compression

Most piles exhibit some shaft compression at working loads, and this should be allowed for in estimating pile deflection. The axial strain at any level down the pile is

$$\varepsilon_z = -\frac{dw}{dz} = \frac{P}{\pi r_0^2 E_p} \tag{4.40}$$

taking E_p as the Young's modulus of an equivalent solid pile (see equation 4.29). The load P will vary down the pile as load is shed into the surrounding soil, so that

$$\frac{dP}{dz} = -2\pi r_0 \tau_0 \tag{4.41}$$

Finally, w_s and τ_0 may be related by equation (4.33) to give

$$\frac{d^2w}{dz^2} = \frac{2G}{\zeta E_p r_0^2} w \tag{4.42}$$

This differential equation may be solved to yield w_s in terms of hyperbolic cosine and sine functions of z. Substituting in the appropriate boundary conditions at the pile base yields an expression for the load settlement ratio of the pile head of

$$\frac{P_t}{G_l r_0 w_t} = \frac{\dfrac{4\eta}{(1-v)\xi} + \dfrac{2\pi\rho}{\zeta}\dfrac{\tanh(\mu l)}{\mu l}\dfrac{l}{r_0}}{1 + \dfrac{4\eta}{\pi\lambda(1-v)\xi}\dfrac{\tanh(\mu l)}{\mu l}\dfrac{l}{r_0}} \tag{4.43}$$

where, summarizing the various dimensionless parameters,

$\eta = r_b/r_0$ (ratio of underream for underreamed piles)
$\xi = G_l/G_b$ (ratio of end-bearing for end-bearing piles)
$\rho = \bar{G}/G_l$ (variation of soil modulus with depth)
$\lambda = E_p/G_l$ (pile–soil stiffness ratio)
$\zeta = \ln(r_m/r_0)$ (measure of radius of influence of pile)
$\mu l = \sqrt{2/\zeta\lambda}(l/r_0)$ (measure of pile compressibility)

Figure 4.19 shows the variation of the load settlement ratio with slenderness ratio, l/r_0, for $\eta = \xi = 1, v = 0.3$.

The load at the pile base is given by

$$\frac{P_b}{P_t} = \frac{\dfrac{4\eta}{(1-v)\xi}\dfrac{1}{\cosh(\mu l)}}{\dfrac{4\eta}{(1-v)\xi} + \dfrac{2\pi\rho}{\zeta}\dfrac{\tanh(\mu l)}{\mu l}\dfrac{l}{r_0}} \tag{4.44}$$

Figure 4.20 shows this ratio as a function of l/r_0 for a range of values of ρ and λ (again for $\eta = \xi = 1, v = 0.3$). It is clear that, under working conditions, very little load reaches the base of long piles. The situation would be different for underreamed or for end-bearing piles and also for groups of piles, where additional load is transferred to the base of the piles.

From Figures 4.19 and 4.20, it may be seen that there are combinations of slenderness ratio, l/r_0, and stiffness ratio, λ, beyond which very little load is transmitted to the pile base. Further increase in pile length yields no corresponding increase in the load settlement ratio of the pile. This limiting behaviour is the converse of a stiff, rigid pile, and corresponds to the case where the pile starts behaving as if it were infinitely long, with no load reaching the lower region.

The two limits may be quantified. Piles may be taken as essentially rigid where l/r_0 is less than about $0.5\sqrt{E_p/G_l}$, and equation (4.43) then reduces to equation (4.37). At the other extreme, for piles where l/r_0 is greater than about $3\sqrt{E_p/G_l}$, $\tanh(\mu l)$ approaches unity and equation (4.43) reduces approximately (exactly for $\rho = 1$) to

$$\frac{P_t}{w_t r_0 G_l} = \pi\rho\sqrt{2\lambda/\zeta} \tag{4.45}$$

As expected, the load settlement ratio is now independent of the length of the pile (since no load reaches the lower end). The modulus G_l should be interpreted as the soil shear modulus at the bottom of the active part of the pile, that is at a depth that corresponds to $z/r_0 = 3\sqrt{E_p/G_l}$, rather than at $z = l$.

In deriving the solution for the load settlement ratio of a single pile, various simplifying assumptions had to be made. It is therefore essential to compare

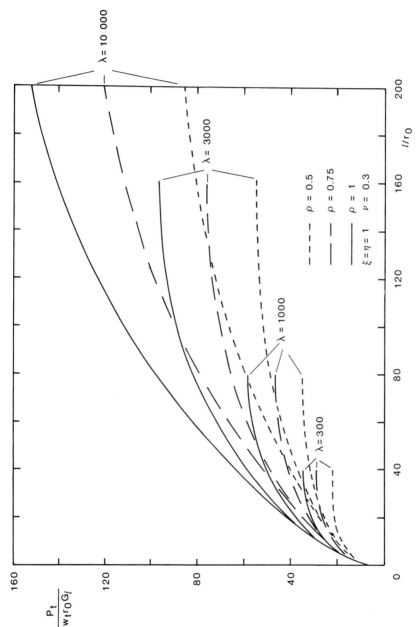

Figure 4.19 Load settlement ratios for compressible piles

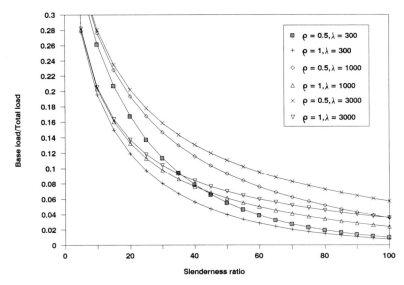

Figure 4.20   Proportion of load taken up by pile base

values of pile stiffness obtained from equation (4.43) with those by the alternative methods described by Poulos and Davis (1980). Poulos and Davis write the load settlement ratio of the pile in terms of a coefficient, I, where

$$\frac{P_t}{w_t r_0 G} = \frac{4(1+v)}{I} \tag{4.46}$$

The coefficient I is obtained from the product of a number of other coefficients which reflect features such as the pile–soil stiffness ratio, the value of Poisson's ratio for the soil and so forth. Thus

$$I = I_0 R_k R_h R_v \tag{4.47}$$

where I_0 is the value of I for a rigid pile in an infinitely deep homogeneous deposit of incompressible soil ($v = 0.5$),

   R_k allows for the compressibility of the pile,

   R_h allows for the finite depth of a layer of soil: and

   R_v allows for values of Poisson's ratio less than 0.5.

For comparison, values of the pile load settlement ratio computed by this means, using charts from Poulos and Davis (1980), are plotted in Figure 4.21. The agreement between the two different approaches is reasonably good, considering, on the one hand, the simplifying assumptions adopted in the solution developed here, and, on the other hand, the possible scope for error when using a number of multiplicative factors taken from charts. For long compressible piles, the results from Poulos and Davis give higher values of pile stiffness than obtained using equation (4.43). The higher values are partly

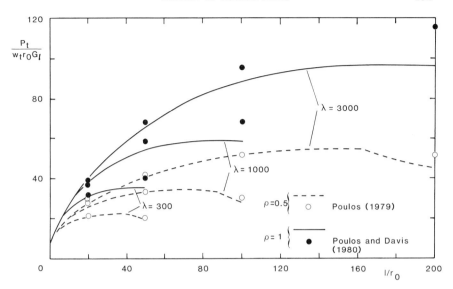

Figure 4.21 Comparison of load settlement ratios calculated by present method with that of Poulos and Davis (1980)

due to relatively coarse discretization of the very long piles, leading to numerical inaccuracies.

At this stage, it may help to illustrate the approach described above with an example or two.

Example 1

Estimation of the working settlement of bored pile, 460 mm in diameter, founded at a depth of 16 m in stiff clay; the upper 2 m of soil is made ground, with no shear transfer assumed. The strength of the clay increases uniformly from $70 \, \text{kN/m}^2$ just below the made ground up to $200 \, \text{kN/m}^2$ at the pile base, with an average value of $135 \, \text{kN/m}^2$. The design load for the pile is 800 kN. Taking a shear modulus for the soil of $G = 150c_u$ (see section 5.2.4), with $v = 0.2$, and assuming the Young's modulus for the pile is $E_p = 25\,000 \, \text{MN/m}^2$, we obtain

$$G_l = 30 \, \text{MN/m}^2, \rho = 0.675, \lambda = 833$$

For the lower 14 m of pile, $l/r_0 = 60.9$, thus

$$\zeta = \ln(r_m/r_0) = 4.4; \quad \mu l = \left(\frac{2}{\zeta \lambda}\right)^{0.5} \frac{l}{r_0} = 1.42$$

$$\frac{\tanh(\mu l)}{\mu l} \frac{l}{r_0} = 38.2$$

Thus $\qquad P_t/w_t = G_l r_0 \times \dfrac{\dfrac{4}{0.8} + \dfrac{2\pi \times 0.675}{4.4} \times 38.2}{1 + \dfrac{1}{\pi\lambda}\dfrac{4}{0.8} \times 38.2} = 269\,\text{kN/mm}$

The deflection at depth of 2 m, for the working load of 800 kN, is thus $800/269 = 3.0\,\text{mm}$. In addition, 0.4 mm of compression will occur in the upper 2 m of the pile, giving an overall working settlement of 3.4 mm.

Example 2

Estimation of the working settlement of a driven cast-in-situ pile, 510 mm in diameter and driven through low-grade chalk to bear on grade I/II chalk at a depth of 12 m.

Back analyses of earlier pile tests have indicated that the shear modulus for the upper chalk may be taken as increasing at a rate of $1\,\text{MN/m}^2$ for every metre of depth, taking $v = 0.24$. The shear modulus of the bearing chalk, allowing for disturbance due to pile installation, may be taken as $G = 36\,\text{MN/m}^2$ (Lord, 1976). Young's modulus of the piles is $E_p = 28\,000\,\text{MN/m}^2$.

The various parameters may be calculated as

$$G_l = 12\,\text{MN/m}^2, \rho = 0.5, \xi = G_l/G_b = 0.33, \lambda = E_p/G_l = 2333,$$
$$l/r_0 = 47, \zeta = 3.1 \text{ (from equation (4.38))}, \mu l = 0.78;$$

$$\frac{\tanh(\mu l)}{\mu l}\frac{l}{r_0} = 39.3$$

Thus: $\qquad P_t/w_t = G_l r_0 \times \dfrac{\dfrac{4}{0.76} \times \dfrac{1}{0.33} + \dfrac{2\pi \times 0.5}{3.1} \times 39.3}{1 + \dfrac{1}{\pi\lambda}\dfrac{4}{0.76} \times \dfrac{1}{0.33} \times 39.3} = 157\,\text{kN/mm}.$

For a working load of 1000 kN, the settlement will therefore be $1000/157 = 6.4\,\text{mm}$.

4.2.3 *Layered soil profiles*

The aim of the solution developed above has been to retain flexibility so that it may be extended or adapted to different conditions. The solution was originally based on separate consideration of the response of the pile shaft and of the pile base. The complete solution may be extended to deal with step-tapered piles or piles which penetrate more than one soil type (with widely differing soil stiffnesses) by treating each section of the pile shaft independently, retaining compatibility of displacement of the pile shaft between each section (see Figure 4.22). Effectively, the load settlement ratio for the pile base (the term $4/(1-v)$ in equation (4.43)) is replaced by the load settlement ratio for the section of pile below that which is currently being considered.

An example calculation will illustrate this. Consider a straight-shafted bored pile, 0.6 m in diameter and 20 m long, with Young's modulus of $25 \times 10^3 \text{ MN/m}^2$, which penetrates

(i)   20 m of stiff clay, with shear modulus reasonably constant at 40 MN/m^2;
(ii) the same stiff clay, but now overlaid by 8 m of a softer deposit with G varying from 5 MN/m^2 at the surface up to 15 MN/m^2 at 8 m depth.

In the first case, straightforward substitution of the pile and soil parameters into equation (4.43) results in $\eta = \xi = \rho = 1$, $E_p/G_l = 625$, $\zeta = 4.76$ (taking $v = 0.3$). Thus the overall load settlement ratio is $P_t/(w_t r_0 G_l) = 48.4$, or $P_t/w_t = 581 \text{ kN/mm}$. In the second case, the lower 12 m of pile may be treated in the normal way to give $P_t/w_t = 540 \text{ kN/mm}$ (surprisingly high considering how much shorter this section of pile is than before). Now, the top 8 m of pile may be analysed, substituting the appropriate value of the load settlement ratio at the lower end instead of the terms $4/(1-v)$. In this example,

$$P_b/(w_b r_0 G_l) = 540/(0.3 \times 15) = 120 \qquad (4.48)$$

and, taking $\rho = 0.67$, $\zeta = 3.44$, the calculated stiffness is $P_t/w_t = 431 \text{ kN/mm}$.

It is interesting to note that, in the second case, the average shear modulus

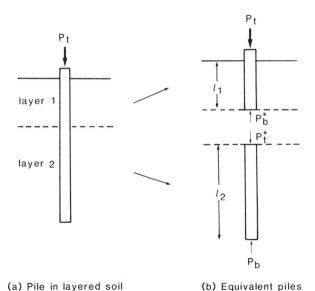

(a) Pile in layered soil   (b) Equivalent piles

Figure 4.22   Modelling of pile in layered soil

over the full depth of penetration of the pile is $28 \, \text{MN/m}^2$. If the pile were to be analysed in a single step, then the parameter, ρ, would be $28/40 = 0.7$ and the stiffness from equation (4.43) for the full 20 m pile is calculated as $P_t w_t = 444 \, \text{kN/mm}$, only 3% higher than that given in the more precise calculation.

4.2.4 Behaviour of long slender piles

It was shown earlier that as the length of a pile is increased, the stiffness of the pile response approaches a limiting value. If piles are being used to reduce settlements, it would clearly be illogical to design piles of such a geometry that this limit was reached or even approached. However, there are many cases where piles are used mainly to provide sufficient bearing capacity, for example, to support storage tanks on soft ground, or for offshore oil production platforms. In these situations, it may prove economically advantageous to use very long slender piles, and, in many cases, compression of the pile may be sufficient to cause relative slip between pile and soil in the upper part of the pile, under working conditions.

The load level at which slip starts to occur between pile and soil close to the ground surface may be estimated by combining equations (4.33) and (4.45). Thus, the local movement of the pile to mobilize full skin friction is obtained by substituting τ_s for τ_0 in equation (4.33) to give

$$(w_s)_{\text{slip}} = \zeta r_0 \tau_s / G \tag{4.49}$$

The ratio G/τ_s varies from between 100 to 400 for piles in clay, to over 1000 for piles in sand. Thus the movement for slip to occur will be 1 to 4% of the pile radius, in clay, and as low as 0.25% of the radius, in sand. Substitution of equation (4.49) into equation (4.45) leads to an expression for the pile load, P_{slip}, at which slip starts to occur at ground level (where the pile movement is greatest), given by

$$P_{\text{slip}}/Q_s = 1/\mu l = \sqrt{\zeta E_p / 2G} (r_0 / l) \tag{4.50}$$

where Q_s is the ultimate shaft capacity of the pile, which equals $2\pi r_0 l \bar{\tau}_s$. It is not uncommon for this ratio to be as low as 0.2, particularly for piles used offshore, in which case considerable slip will occur at working loads, where P/Q_s may be of the order of 0.5 to 0.6.

It is clear that an elastic solution for the pile load settlement ratio becomes inappropriate at load levels greater than that given by the expression above. Estimation of the pile settlement must then allow for the region of slip between pile and soil. This may be accomplished in the same manner as for a pile in a layered soil profile, essentially treating the upper part of the pile (where full skin friction has been mobilized) independently from the lower part. Some iteration will be required, since, for a given applied load at the pile head, the transition point between the two regions must be estimated initially. This procedure will be illustrated by an example, taken from pile load tests reported by Thorburn *et al.* (1983).

The two piles in question were 250 mm square, of precast concrete, driven into silty clay to penetrations of 29 m and 32 m. The shear strength varied with depth according to

$$c_u = 6 + 1.8z \text{ kN/m}^2 \text{ (with } z \text{ measured in m).} \qquad (4.51)$$

Taking $\alpha = 0.9$ for the clay, which was lightly overconsolidated, the shaft capacities of the two piles may be estimated as 0.84 MN and 1.00 MN respectively. In calculating the load settlement ratio for the piles, a cylindrical pile of radius 141 mm (giving the same cross-sectional area) will be taken. The short-term Young's modulus of the piles was $E_p = 26\,500 \text{ MN/m}^2$.

The response of the piles to constant rate of penetration tests may be matched by adopting a shear modulus of $G = 250c_u = 1.5 + 0.45z \text{ MN/m}^2$, taking Poisson's ratio equal to 0.5. Equation (4.45) then gives the pile stiffness as $P_t/w_t = 90 \text{ kN/mm}$. The local movement to mobilize full skin friction may be estimated from equation (4.49) as 2.5 mm. Thus, slip between pile and soil will start at the top of the pile when the applied load reaches $90 \times 2.5 = 225 \text{ kN}$. Figure 4.23 gives the measured pile responses, showing both piles having the

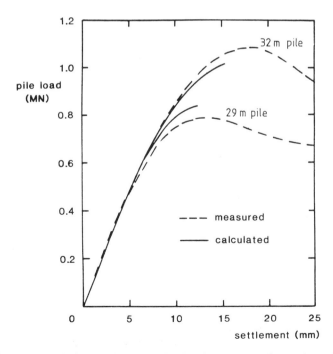

Figure 4.23 Calculated and measured load settlement curves for two long piles

same initial stiffness, with an essentially linear response up to loads of about 500 kN.

The complete load settlement curve may be estimated by calculating the load and corresponding settlement for a given depth of slip between pile and soil. Thus, consider the stage at which full skin friction has been mobilized over the top 10 m of the pile. The load shed over the top 10 m may be calculated from the given profile of shear strength, as 150 kN. For the bottom part of the pile, the load settlement ratio may be calculated from equation (4.43) as $P/w = 112 \, \text{kN/mm}$. Thus the load level at the transition point (where the settlement is 2.5 mm) is 280 kN. The overall load at the pile head is then $280 + 150 = 430 \, \text{kN}$. The pile compression over the top 10 m of the pile may now be calculated as 2.3 mm, giving an overall settlement at the pile head of 4.8 mm.

The above process may be repeated for different depths of slip between pile and soil to give the two estimated load settlement curves shown in Figure 4.23. In practice, adequate estimates of these curves may be obtained merely from the initial slopes of the load settlement curves, and estimates of the pile settlements when full shaft capacity has been mobilized. Thus for the shorter pile, the compression of the pile for the shaft capacity of 838 kN is 9.3 mm. Together with 2.5 mm of settlement at the bottom of the pile (in order to mobilize full skin friction there), the pile head movement is 11.8 mm. This corresponds reasonably well with the measured movement to peak load.

It should be noted that, for simplification, no allowance has been made in the above calculations for the small amount of load taken by the pile base. In addition, the measured pile responses show clear evidence of strain softening, which will affect the distribution of skin friction down the pile at the point of failure of the pile (see following discussion). The working loads for these piles were 440 kN and 550 kN respectively for the short and long piles. Maintained load tests at these load levels showed some creep movement, with final settlements of about 8 mm, of which probably about half reflected slip between the upper part of the pile and the soil.

As may be seen from the pile tests just discussed, large relative movement between the pile shaft and the soil may lead to a reduction in shear transfer from a peak value of skin friction to a residual value. Such strain softening may be significant (a) in loose compressible sands (particularly calcareous sediments) and (b) in soft, lightly overconsolidated clays. Model pile tests in soft clay reported by Francescon (1982) and by Chandler and Martins (1982) indicate that the residual value, $(\tau_s)_{\text{res}}$, may be estimated from

$$(\tau_s)_{\text{res}} = K_0 \sigma_v' \tan \phi_{\text{res}}' \qquad (4.52)$$

where ϕ_{res}' is the residual angle of friction for the soil. The model pile tests indicate that shear transfer from the pile shaft may reduce by a factor of two or more after as little as 30 mm of movement of the pile, particularly for cyclic relative displacements. Similar findings have been reported from ring shear tests (Bishop *et al.*, 1971; Lupini *et al.*, 1981). The consequences of this in the design of long slender piles to anchor offshore platforms have been discussed by Randolph (1983).

The strain softening behaviour can also have consequences for the ultimate capacity of long piles, where a form of progressive failure becomes possible. At failure, the displacements at the head of a long pile may have been sufficient to reduce the skin friction to a residual value (see Figure 4.24). Thus the measured capacity will be lower than would be calculated from the profile of peak values of skin friction. In order to quantify this effect, a flexibility ratio, K, may be introduced (Randolph, 1983) given by

$$K = \frac{\pi d \tau_s l^2}{(AE)_p \Delta w_{res}}$$

(4.53)

where Δw_{res} is the relative movement needed between pile and soil for the value of skin friction to reduce from peak to residual. The flexibility ratio is

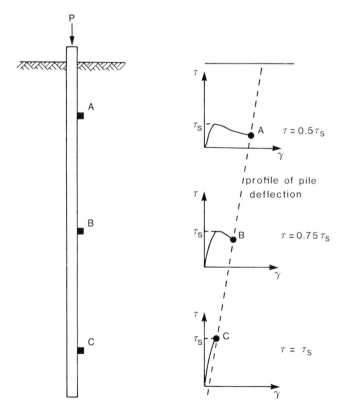

Figure 4.24 Progressive failure of a long pile

similar to the parameter π_3, introduced by Murff (1980), but based on Δw_{res} rather than the movement w_{slip} to mobilize peak skin friction initially.

Figure 4.25 shows values of reduction factor, R_f, which should be applied to pile shaft capacities based on peak values of skin friction, in order to allow for progressive failure of the pile. The different curves reflect different ratios of residual to peak skin friction. In general, land-based piles will have values of K which are greater than unity, and the effect of strain softening will be small. However, piles used offshore frequently have values of K in the range 2 to 4, and the effect may be significant. For a particular situation, estimation of the reduction factor, R_f, will require an estimate of the ratio of $(\tau_s)_{res}$ to $(\tau_s)_{peak}$, and also of the movement, Δw_{res} to cause the degradation. Ring shear, or direct shear box tests on the soil in question, will enable the value of ϕ'_{res} to be assessed, while Δw_{res} may be taken, conservatively, as 30 mm. This value is likely to be largely independent of pile size, since the degradation is concentrated in a thin band close to the pile shaft, with re-orientation of soil particles responsible for most of the decrease in the angle of friction.

The most important quantity in the ratio, K, given above, is the aspect ratio, l/d, of the pile. Semple and Rigden (1984) have studied the results of load tests on long slender piles driven into clay and have introduced a correction factor, F_l, to allow for progressive failure, which is based purely on the overall aspect ratio of the pile. The correction factor is to be used in

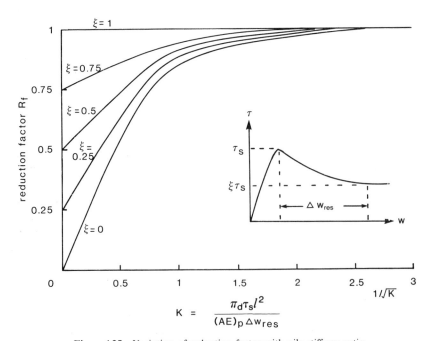

$$K = \frac{\pi_d \tau_s l^2}{(AE)_p \Delta w_{res}}$$

Figure 4.25   Variation of reduction factor with pile stiffness ratio

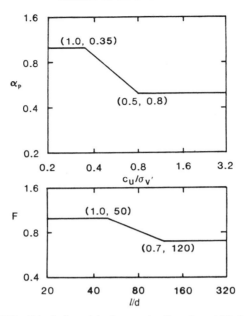

Figure 4.26 Criteria for axial pile capacity (Semple and Rigden, 1984)

conjunction with estimates of peak values of α which differ somewhat from those suggested in section 4.1. Figure 4.26 shows the suggested variations of α and the correction factor F_l. It may be noted that a minimum value of $\alpha = 0.5$ is assumed at high values of the strength ratio, and also that a minimum correction factor of $F_l = 0.7$ is taken, even for very long slender piles.

4.2.5 Rock sockets

The axial deformation of rock sockets deserves special consideration. Although rock sockets are generally relatively short, with typical length-to-diameter ratios of less than 10, the high stiffness of the host rock compared with the pile stiffness entails that the rock socket will behave in a compressible fashion. As an example, consider a rock socket of length 10 m and diameter 1.4 m, embedded in soft rock with a shear modulus of 700 MPa. Assuming a Young's modulus for the concrete of 25 000 MPa, the quantity that controls the compressibility, μl, may be calculated as 1.9 (see equation (4.43)). The initial stiffness may be estimated approximately from equation (4.45) as $P/w = 7.3$ MN/mm.

The relatively compressible nature of rock sockets, together with high base capacity, entails that partial (or complete) slip along the pile shaft will generally occur under working conditions. Analytical and numerical

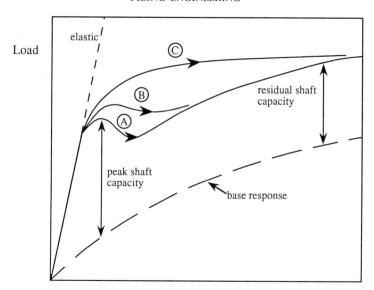

Figure 4.27   Different types of rock socket responses

solutions for estimating settlement and load transfer in rock sockets under conditions of partial slip have been presented by Rowe and Armitage (1987), and Carter and Kulhawy (1988). As slip occurs, the skin friction will often decrease, as the cohesive or dilational component of the skin friction will reduce to zero (Johnston and Lam, 1989). Since the shaft capacity is mobilized at much smaller displacements than the base capacity, the overall load–displacement response of the rock socket may then show a decrease in capacity at intermediate displacements (due to strain softening of the shaft) prior to increasing again at larger displacements as the base capacity is mobilized. Figure 4.27 shows schematically the different types of rock socket response, depending on the rapidity of strain softening and the rate of mobilization of the base capacity.

4.3 Lateral capacity of piles

Almost all piled foundations are subjected to at least some degree of horizontal loading. The magnitude of the loads in relation to the applied vertical loading will generally be small, and no additional design calculations would normally be necessary. For example, the piled foundation of a tall building will easily withstand the small wind shear loading that the building will be subjected to. In other cases, such as a piled earth retaining wall or a jetty, horizontal loading may prove critical in the design. Traditionally, piles have been installed at an angle to the vertical in such cases, providing sufficient horizontal resistance by virtue of the component of the axial capacity of the pile which acts horizontally. However, it is very conservative to ignore the capacity of a pile to withstand lateral loading (i.e. loading applied normal to the pile axis). This may typically be an order of magnitude less than the axial capacity of a pile, but may well be sufficient to obviate the need for so-called 'raking' or 'battered' piles, which are more expensive to install.

Lateral loading of a pile may be divided broadly into the two categories of 'active' loading, where external loads are applied to the pile, with the soil resisting the load, and 'passive' loading, where movement of the soil subjects the pile to bending stresses. This, and the subsequent section, are concerned primarily with the first category of loading, although some of the features discussed (for example, the limiting pressure between pile and soil) are applicable to both categories. Passive loading of piles by consolidating soil is discussed further in section 4.5.

As a pile is loaded laterally, normal stresses will increase in front of, and decrease behind the pile. Displacements in the soil will tend to be radially away from the pile in front of the pile, and radially towards the pile behind it. At some stage, near the ground surface, a gap will probably open up between the back of the pile and the soil, with soil in front of the pile failing in a wedge type of mechanism, as shown in Figure 4.28. Further down the pile shaft, soil will eventually fail by flowing around the pile, with no gap present. In calculating the distribution of limiting pressure which may be mobilized by the pile, these different types of failure mechanism in the soil must be taken into account.

Collapse of the pile itself may occur in one of two modes. For short piles, or piles with a large plastic moment, the pile will rotate essentially as a rigid body. The resulting profile of resistance offered by the soil will be as shown

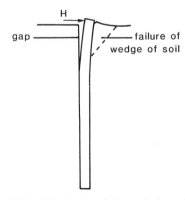

Figure 4.28 Deformation of pile under lateral load

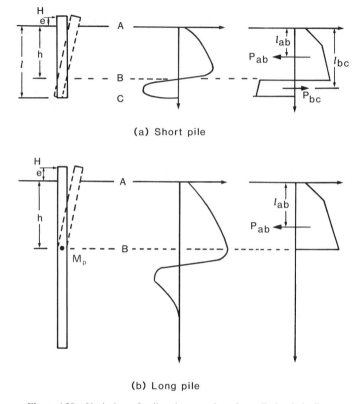

(a) Short pile

(b) Long pile

Figure 4.29   Variation of soil resistance along laterally loaded piles

schematically in Figure 4.29(a). Above the centre of rotation, passive pressures will develop in front of the pile, while below the centre of rotation, passive pressures will develop behind the pile. For longer piles, a plastic hinge will develop at some depth down the pile, and only the upper part of the pile will undergo significant displacement (Figure 4.29(b)). Although the soil will be subjected to loading below the plastic hinge, calculation of the failure load will only require knowledge of the limiting pressures acting over the upper part of the pile, above the plastic hinge.

Where a pile is embedded in a pile cap which is restrained from rotation, three different collapse modes are possible, as shown in Figure 4.30. Short piles will translate as a rigid body with the pile cap, while progressively longer piles will first form a plastic hinge at the level of the pile cap, and then a plastic hinge at some distance down the pile. In practice, the majority of piles encountered behave as long piles as regards lateral capacity. Failure will thus be by the mode shown in Figure 4.30(c).

In calculating failure loads, it is customary to treat the soil as a rigid

(a) No hinge   (b) 1 hinge   (c) 2 hinges

Figure 4.30   Failure modes for laterally loaded piles

plastic material, idealizing the profiles of limiting pressure as shown in Figure 4.29, with sharp transitions from full limiting pressure acting in front of the pile just above the hinge point, down to zero at the hinge point and, for short piles, full limiting pressure acting behind the pile just below the hinge point. The relevant equivalent forces, P, and their appropriate lever arms, l, are shown in Figure 4.29 for an unrestrained pile. For short piles, the moment at point B is $P_{bc}(l_{bc} - h)$, where h is the depth of the hinge point, and this must be less than the plastic moment for the pile section. Horizontal equilibrium then gives

$$H_f = P_{ab} - P_{bc} \qquad (4.54)$$

while moment equilibrium entails that

$$H_f(e + h) = P_{ab}(h - l_{ab}) + P_{bc}(l_{bc} - h) \qquad (4.55)$$

For a given profile of limiting pressure acting on the pile, P_{ab} and P_{bc} may be evaluated and these two equations solved to find H_f.

If the pile fails by forming a plastic hinge, this will occur at the point of maximum bending moment, and thus of zero shear force. In Figure 4.29(b), the moment at B may be set equal to the plastic moment, M_p, for the pile section, and the pressures acting on the pile below point B may be ignored. For zero shear force at the hinge point, H_f must equal P_{ab}. Taking moments about B then gives

$$H_f[e + h - (h - l_{ab})] = H_f(e + l_{ab}) = M_p \qquad (4.56)$$

For the case of a pile where the head is restrained from rotating, the above equations may be used where one or two plastic hinges form, merely adding an extra term M_p to the right-hand sides of equations (4.55) and (4.56) to allow for the hinge at the top of the pile. For short piles, where the whole pile translates laterally (Figure 4.30(a)), the failure load may be calculated in a straightforward manner, by integrating the limiting pressure over the length of the pile. It should be noted that the bending moments at the top of piles built in to a pile cap must be taken into account in the design of the pile cap itself.

The following sections discuss values of limiting pressure appropriate for different soil types.

4.3.1 Limiting lateral pressures in non-cohesive soils

At shallow depths (less than one diameter) a circular or square pile will appear to soil elements in front of the pile as a long retaining wall. At failure, a wedge of soil will be pushed ahead of the pile, and the limiting pressure acting on the pile (for non-cohesive soil) will be close to K_p times the local vertical effective stress, where K_p is the passive earth pressure coefficient, equal to $(1 + \sin \phi')/(1 - \sin \phi')$. At greater depths, however, much larger limiting pressures may develop, owing to the three-dimensional nature of the problem. In front of the pile, the soil will be deformed much as in a pressuremeter test, while, behind the pile, the contact stresses will reduce until, in the limit, the pile breaks away from the soil. From the results of pressuremeter tests in sand (Hughes et al., 1977; Fahey and Randolph, 1984), pressures greater than 10 times the effective overburden stress are typically required to increase the radius of the pressuremeter by 10%. No limiting pressure is normally reached, but rather the pressure increases proportionally with the cavity strain to some exponent. The value of the exponent depends on the friction angle and rate of dilation of the soil (Hughes et al., 1977).

In practice, of course, the pile will fail at some stage, normally by the formation of a plastic hinge at some point down the pile. Lateral movement of the pile to cause such failure will generally be in excess of 10% of the pile diameter, at which stage the pressuremeter analogy may no longer be accurate. Although no alternative analytical solution has been proposed, Broms (1964) has suggested a net limiting force per unit length of pile, P_u, given by

$$P_u = 3K_p \sigma'_v d \qquad (4.57)$$

where d is the diameter or width of the pile. This expression furnishes average limiting pressures which will be slightly greater than the figure of $10\sigma'_v$ mentioned above, for practical values of ϕ'. Using equation (4.57) in an analysis of the type described in the previous section, Broms has prepared charts in non-dimensional form giving the lateral capacity of piles in terms of the plastic moment and geometry of the pile. Comparisons with field test results show a tendency for the measured capacities to be underestimated by about 30% (Poulos and Davis, 1980).

Reese et al. (1974) suggest a more complex variation of limiting pressure with depth, taking due account of the wedge type of failure near the ground surface. The variation of P_u, with depth is initially proportional to K_p, but at greater depths becomes proportional to K_p^3. The proposed variation of limiting pressure with depth is not amenable to the production of charts of the form given by Broms (1964).

An intermediate variation of limiting pressure with depth is to take P_u proportional to the square of the passive earth pressure coefficient, giving

$$P_u = K_p^2 \sigma_v' d \qquad (4.58)$$

For almost all naturally occurring sand, K_p will be greater than 3, so that this expression will give a better fit to results of load tests than did equation (4.57). Barton (1982) has computed the limiting lateral force per unit length on model piles loaded laterally in a centrifuge. The piles, strain-gauged internally to record the distribution of bending moment, were embedded in uniform dense sand with friction angle $\phi' = 43°$ (giving $K_p = 5.3$). Figure 4.31 shows the calculated values of average limiting force per unit

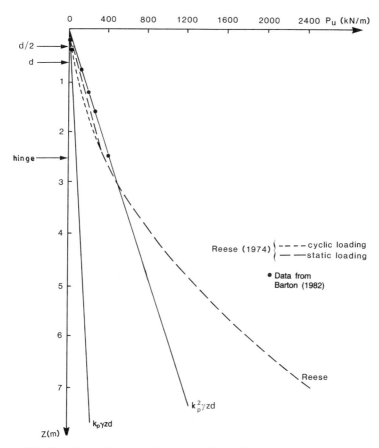

Figure 4.31   Variation with depth of measured ultimate force down laterally loaded pile

length, P_u, with depth, compared with the three variations discussed above. As anticipated, close to the soil surface, the limiting force is about $K_p \sigma'_v d$. Below about 1.5 diameters, however, the limiting pressures closely follow the variation given by equation (4.58).

Adopting this expression for the limiting force per unit length acting on the pile, the results derived in the previous section may be evaluated and charts may be prepared giving the lateral capacity of short and long piles embedded in uniform soil. Assuming that the vertical effective stress, σ'_v is equal to γz, where γ is the appropriate unit weight for the sand (i.e. dry weight above the water table, and buoyant weight below), the lateral capacity and the plastic moment of the pile may be non-dimensionalized by the quantities $K_p^2 \gamma d^3$ and $K_p^2 \gamma d^4$ respectively. The charts are shown in Figure 4.32(a), for short piles, and Figure 4.32(b) for long piles. Barton (1982) comments that values of ultimate lateral capacity obtained from these charts tended to underestimate the actual measured failure loads by about 6%.

4.3.2 *Limiting lateral pressures in cohesive soils*

For piles in cohesive soil, the limiting pressure exerted on the front of the pile by the wedge of soil near the ground surface may be taken as $2c_u$, where c_u is the undrained shear strength of the soil. Reese (1958) and Matlock (1970) have suggested taking a net limiting pressure at shallow depths, given by

$$p_u = \frac{P_u}{d} = 2c_u + \sigma'_v + \alpha c_u \frac{z}{d} \tag{4.59}$$

where the factor α lies between 0.5 and 3. Randolph and Houlsby (1984) suggest taking $\alpha = 1.5$. (Note that σ_v should be replaced by σ'_v where water is free to flow in behind the pile.)

At greater depths, two modes of failure in the soil may be considered. The first of these assumes flow of the soil around the pile in a plane normal to the pile axis. Idealizing the soil as a rigid plastic material, with shear strength c_u, classical plasticity theory allows the limiting force per unit length acting on the pile to be deduced. The second approach uses the pressuremeter analogy discussed in the previous section, where soil is assumed to move radially outward, in front of the pile, and a gap opens up behind the pile.

Plasticity solution. Randolph and Houlsby (1984) describe an analytical solution for the flow of soil around a cylindrical object. The solution is exact, within the confines of the rigid plastic model of the soil, and allows for limiting friction, f_s, at the pile–soil interface which is less than or equal to the shear strength of the soil. Defining Δ by

$$\sin \Delta = f_s/c_u \tag{4.60}$$

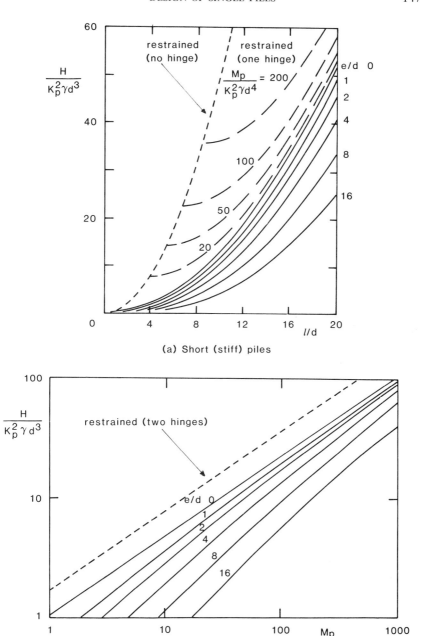

(a) Short (stiff) piles

(b) Long (flexible) piles

Figure 4.32 Design chart for laterally loaded piles in non-cohesive soil the limiting force per unit length of pile is given by

$$P_u = c_u d [\pi + 2\Delta + 4\cos\{(\pi - \Delta)/4\}[\sqrt{2} + \sin\{(\pi - \Delta)/4\}]] \quad (4.61)$$

The variation of the limiting pressure with friction ratio f_s/c_u is shown in Figure 4.33. The value of P_u varies from $9.14c_u d$ for a perfectly smooth pile, up to $11.94c_u d$ for a perfectly rough pile, with a mean value of about $10.5c_u d$. Figure 4.34 shows the failure mechanism for the case of $f_s/c_u = 0.5$

Pressuremeter analogy. Experimental evidence for the usefulness of a pressuremeter analogy comes from tests conducted by Swain (1976), on deeply-embedded strip anchors. The geometry of the strip anchor is similar to that of a pile loaded laterally. The experimental results indicated a pattern of outward radial movement of the soil in front of the strip anchor, and inward radial movement behind the anchor (see Figure 4.35). Only at loads near failure were there indications of flow around the strip anchor. For this pattern of deformation, it seems reasonable to assume that the pressure exerted by soil in front of a laterally-loaded pile will approach the limiting pressure measured from a pressuremeter test. Behind the pile, the lowest

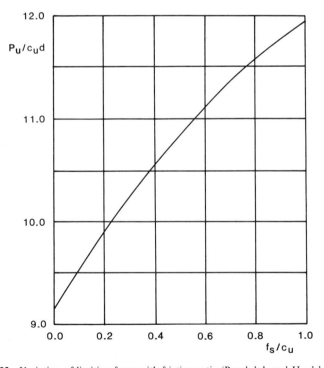

Figure 4.33 Variation of limiting force with friction ratio (Randolph and Houlsby, 1984)

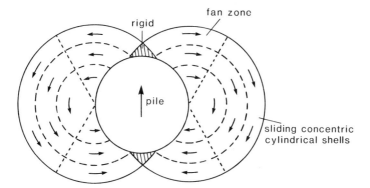

Figure 4.34 Flow mechanism for soil around laterally loaded pile

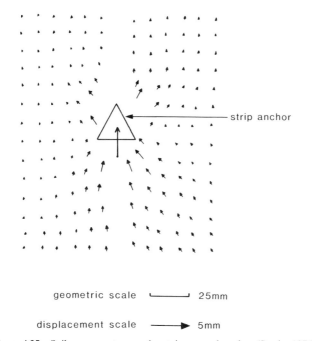

geometric scale ⌞————⌟ 25mm

displacement scale ———▶ 5mm

Figure 4.35 Soil movement around a strip ground anchor (Swain, 1976)

normal stress attainable will be a suction of the order of $100 \, kN/m^2$. If a gap forms, then this stress will increase to zero (in a dry hole) or to the ambient head of water if free water is available (such as for offshore piles). In addition, at the sides of the pile, some friction may be present between the pile and the soil. Typically, this friction may be estimated as $c_u d$, per unit length of pile.

The limit pressure, p_l, obtained from a pressuremeter tests may be written

$$p_l = \sigma_h + c_u[\ln(G/c_u) + 1] \qquad (4.62)$$

where σ_h is the (total) in-situ horizontal stress in the soil, and G is the shear modulus of the soil. In practice, for typical values of G/c_u, this expression may be approximated (Marsland and Randolph, 1977) by

$$p_l = \sigma_h + 6c_u \qquad (4.63)$$

Thus the net force per unit length acting on the pile lies between the limits

$$(\sigma_h' + 7c_u)d \leqslant P_u \leqslant (\sigma_h + p_a + 7c_u)d \qquad (4.64)$$

where the lower limit corresponds to the case of the ambient head of water acting behind the pile, and the upper limit is where a suction develops behind the pile (p_a denoting atmospheric pressure).

For normally or lightly overconsolidated clay, σ_h'/c_u will be about 2, and the lower limit in equation (4.64) corresponds to the result for a smooth pile obtained from plasticity theory. For short-term loading, some suction will probably develop behind the pile, and the plasticity solution will give the lowest estimate of limiting force per unit length acting on the pile. For stiff, overconsolidated clay, the ratio σ_h'/c_u will be considerably lower and may be as low as 0.5 at shallow depths. Even if suctions develop behind the pile, the limiting force per unit length may be below $9c_u d$.

General profile of limiting pressure. The discussion above may be useful in choosing particular profiles of limiting pressure for laterally loaded piles. In many cases, however, the suggested values for P_u lead to a profile which increases from $2c_u d$ at the ground surface up to a value of about $9c_u d$ at a depth of 3 pile diameters, much as was originally suggested by Broms (1964). (Broms conservatively took P_u equal to zero at the ground surface, to allow for any gap which might be present around the head of the pile.) This profile of P_u is shown in Figure 4.36. The limit of $9c_u d$ at depth is likely to be conservative, except for the case of piles in stiff overconsoldiated clay,

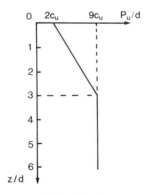

Figure 4.36 Idealized variation of limiting resistance with depth

(a) Short (stiff) piles

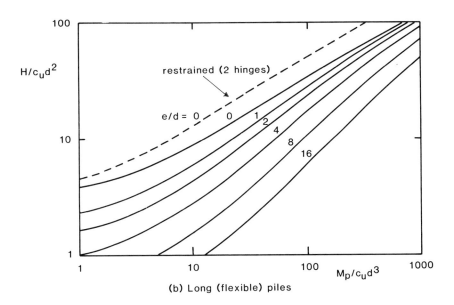

(b) Long (flexible) piles

Figure 4.37 Design chart for laterally loaded piles in cohesive soil particularly if there is free water readily available to flow into any gap behind the pile.

Adopting the profile of limiting force per unit length shown in Figure 4.36, charts may be prepared giving the lateral capacity of a pile in terms of its geometry and limiting plastic moment. The charts, shown in Figure 4.37, are similar to those presented by Broms (1964), giving marginally higher failure loads as a result of taking P_u equal to $2c_u d$ at the ground surface.

For other calculated variations of P_u with depth, the expressions derived in section 4.3 may be used to produce similar sets of charts to those shown in Figure 4.37. For layered profiles, the procedure may involve some iteration choosing a value for the depth of plastic hinge, following the principles described in section 4.3 to calculate a value of H_f, and then checking that the position of the hinge corresponds to the point of zero shear force in the pile.

4.4 Deformation of single piles under lateral loading

The deformation of a pile under lateral load is generally confined to the upper part of the pile, seldom extending beyond about 10 pile diameters below the ground surface. This has led to a common—though imprecise— idealization of laterally loaded piles in terms of equivalent cantilevers. The pile is replaced by a cantilever, fixed at some depth determined by folklore and ignoring the soil support above that depth. The limitations of this idealization are discussed further in section 4.4.3. More accurate modelling of the response of a pile to lateral load is obtained by allowing for the soil support over the upper part of the pile. Two alternative approaches are possible, one where the soil is represented by discrete springs down the length of the pile, and one where a more accurate, continuum model of the soil is adopted. The two approaches are discussed in the following two sections.

4.4.1 Subgrade reaction approach

Early analysis of the response of piles to lateral loading treated the soil as a series of springs down the length of the pile—the so-called Winkler idealization of soil (see Figure 4.38). The spring stiffness, k, giving the load per unit length of pile induced for unit lateral deflection of the pile, is generally referred to as the coefficient of subgrade reaction. If k is assumed constant down the length of the pile, then analytical solutions are possible, giving the deflected shape of the pile, and the shear force and bending moment distribution down the pile (Matlock and Reese, 1960). For a pile of a given bending rigidity $(EI)_p$, in soil with a coefficient of subgrade reaction k, there is a critical length beyond which the pile behaves as if it were infinitely long.

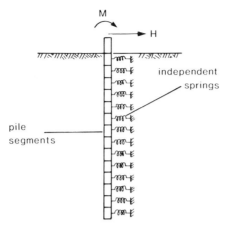

Figure 4.38 Subgrade reaction model of soil around pile

This critical length is given by

$$l_c = 4[(EI)_p/k]^{1/4} \qquad (4.65)$$

Thus, as for the case of long slender piles subjected to axial loading, the effects of load applied at the top of the pile die out at some depth down the pile. In fact the large majority of piles encountered in practice behave as 'flexible' piles (i.e. longer than their critical length). For such piles, the deflection, u, and rotation, θ, at ground level due to applied load, H, and moment, M, are given by

$$u = \sqrt{2}\frac{H}{k}\left(\frac{l_c}{4}\right)^{-1} + \frac{M}{k}\left(\frac{l_c}{4}\right)^{-2} \qquad (4.66a)$$

$$\theta = \frac{H}{k}\left(\frac{l_c}{4}\right)^{-2} + \sqrt{2}\frac{M}{k}\left(\frac{l_c}{4}\right)^{-3} \qquad (4.66b)$$

Similar expressions for the critical length and for the load deformation response may be obtained for cases where the value of the coefficient of subgrade reaction varies with depth, in particular where it is proportional with depth (Reese and Matlock, 1956). For the coefficient of subgrade reaction varying proportionally with depth according to $k = nz$, the critical length may be estimated from

$$l_c = 4[(EI)_p/n]^{1/5} \qquad (4.67)$$

and the corresponding ground level deformations are given by

$$u = 2.43\frac{H}{n}\left(\frac{l_c}{4}\right)^{-2} + 1.62\frac{M}{n}\left(\frac{l_c}{4}\right)^{-3} \qquad (4.68a)$$

$$\theta = 1.62\frac{H}{n}\left(\frac{l_c}{4}\right)^{-3} + 1.73\frac{M}{n}\left(\frac{l_c}{4}\right)^{-4} \tag{4.68b}$$

Modern computer techniques have permitted the extension of this approach to include non-linear springs for representing the soil. In place of a coefficient of subgrade reaction, a complete load transfer (or p (load per unit length of pile)–y (deflection)) curve is specified. Finite difference modelling of the bending of the pile then allows a complete load deflection curve to be obtained for the pile (Reese, 1977). Typical forms for the p–y curves have been presented by Reese et al. (1974) for piles in sand and by Matlock et al. (1970) and Sullivan et al. (1979) for piles in clay. In particular situations, notably for offshore pile design, load transfer curve analysis of piles can play an essential role in design, mainly with a view to assessing the effects of cyclic loading. However, such analyses are relatively complex and time-consuming, and are rarely justifiable for onshore applications.

4.4.2 Elastic continuum approach

For most applications, a simple linear analysis of the effects of lateral loading will be adequate. This may be achieved by means of the subgrade reaction theory. However, there are several limitations to this approach. A major difficulty concerns the numerical value to be assigned to k for a particular soil. Comparisons of the subgrade reaction approach with rigorous elastic analysis have shown that the correct choice for k depends not only on the soil properties, but also on the pile stiffness and the form of loading (that is the eccentricity of the applied lateral load). In addition, there is no rational way in which interaction effects can be quantified when a group of piles is loaded laterally.

In order to overcome these limitations, solutions based on finite element and boundary element modelling of the pile and soil continuum have been presented (Poulos, 1971; Kuhlemeyer, 1979; Randolph, 1981). These solutions allow expressions similar in form to those in equations (4.65) and (4.66), to be written for the critical pile length and the ground level deformations. Dealing initially with piles that are longer than their critical length, it is first necessary to consider the average soil stiffness over the active length of the pile. In order to avoid needing different solutions for different values of Poisson's ratio, Randolph (1981) introduced a modulus G^* defined by

$$G^* = G(1 + 3v/4) \tag{4.69}$$

Referring to Figure 4.39, a characteristic modulus, G_c, is defined as the average value of G^* over the active length of the pile. In addition a parameter, ρ_c is introduced reflecting the degree of homogeneity in the soil stiffness. For the linear variation of modulus shown in Figure 4.39, ρ_c may conveniently be defined by

$$\rho_c = G^*_{l_c/4}/G^*_{l_c/2} = G^*_{l_c/4}/G_c \tag{4.70}$$

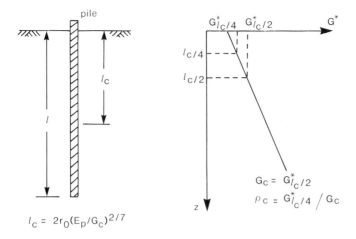

$$l_c = 2r_0(E_p/G_c)^{2/7}$$

Figure 4.39 Definition of ρ_c and G_c

It is helpful to cast the solution in terms of an equivalent solid pile of the same cross-sectional area and the same bending rigidity as the real pile. For an equivalent pile of radius r_0, the appropriate Young's modulus may be calculated as

$$E_p = (EI)_p/(\pi r_0^4/4) \tag{4.71}$$

where $(EI)_p$ is the bending rigidity of the actual pile.

The critical pile length is now defined by

$$l_c = 2r_0(E_p/G_c)^{2/7} \tag{4.72}$$

The form of this equation is similar to equation (4.65), with an exponent of 2/7 instead of 1/4. Also, since E_p for an equivalent pile is used instead of the bending rigidity $(EI)_p$, it is possible to think in terms of a critical slenderness ratio $(l/r_0)_c$ for the pile which is deduced directly from the stiffness ratio, E_p/G_c.

It may seen from Figure 4.39 that the definition of G_c requires the knowledge of the critical length, l_c, which is in turn defined in terms of G_c. Thus some iteration is required except for the extreme cases of homogeneous soil ($\rho_c = 1$) and soil where G is proportional to depth ($\rho_c = 0.5$ and the critical length reduces to $l_c = 2r_0(E_p/m^*r_0)^{2/9}$, where m^* is the rate of increase of G^* with depth). For example, suppose that the variation of shear modulus with depth, z, is approximated as

$$G = 10 + 4z \text{ MN/m}^2 \text{ (z being in m).} \tag{4.73}$$

For a pile of radius 0.3 m, with equivalent modulus $E_p = 30 \times 10^3 \text{ MN/m}^2$, a first guess for the critical pile length might be 6 m (that is 10 pile diameters). Taking Poisson's ratio as 0.3, the characteristic modulus G_c is calculated as

$$G_c = (10 + 4 \times 6/2)(1 + 3 \times 0.3/4) = 27.0 \text{ MN/m}^2. \tag{4.74}$$

This gives a revised estimate of the critical pile length of

$$l_c = 2 \times 0.3 \times (3 \times 10^4/27.0)^{2/7} = 4.45 \text{ m}. \tag{4.75}$$

Further iteration yields final values of $l_c = 4.63$ m, $G_c = 23.6$ MN/m^2.

The concept of a characteristic shear modulus G_c and a critical pile length l_c may be used to write expressions for the ground level deformation of the pile. Thus the lateral deflection, u, and rotation, θ, are given by

$$u = \frac{(E_p/G_c)^{1/7}}{\rho_c G_c}\left[0.27H\left(\frac{l_c}{2}\right)^{-1} + 0.30M\left(\frac{l_c}{2}\right)^{-2}\right] \tag{4.76a}$$

$$\theta = \frac{(E_p/G_c)^{1/7}}{\rho_c G_c}\left[0.30H\left(\frac{l_c}{2}\right)^{-2} + 0.80\sqrt{\rho_c}M\left(\frac{l_c}{4}\right)^{-3}\right] \tag{4.76b}$$

These expressions have been derived from a large number of finite element analyses (Randolph, 1981). The similarity with those obtained from the subgrade reaction approach (equations (4.66a and b)) is evident.

Generalized profiles of deflection and bending moment down the pile may be drawn for piles subjected to shear force loading H or moment loading M. These profiles are shown in Figures 4.40 and 4.41. The maximum moment for a pile under a lateral load of H, occurs at a depth between $l_c/4$ (for homogeneous soil) and $l_c/3$ (for soil with stiffness proportional to depth). The value of the maximum moment may be estimated as

$$M_{\text{max}} = (0.1/\rho_c)Hl_c \tag{4.77}$$

For piles within a group, the pile cap may prevent rotation of the head of the pile. For such 'fixed-headed' piles, equations (4.76a and b) may be used to find the fixing moment M_f. Setting $\theta = 0$, the fixing moment is given by

$$M_f = -[0.375/(\rho_c)^{1/2}]Hl_c/2 \tag{4.78}$$

The resulting deflection of the pile head may then be calculated from equation (4.76a)

$$u_f = \frac{(E_p/G_c)^{1/7}}{\rho_c G_c}[0.27 - 0.11/(\rho_c)^{1/2}]H\left(\frac{l_c}{2}\right)^{-1} \tag{4.79}$$

This is approximately half what it would have been for a free-headed pile. Figure 4.42 shows the generalized deflected profile and variation of bending moment down the pile for this case.

To illustrate the use of these solutions in design, consider the response of a steel pipe pile, 1.5 m in diameter, with 50 mm wall thickness, embedded in soft, normally consolidated clay with a shear strength which increases at a rate of 2.5 kN/m^2 per metre of depth. The equivalent modulus of the pile may be calculated as

$$E_p = E_{\text{steel}}\left[1 - \left(\frac{r_i}{r_o}\right)^4\right] = 50\,600 \text{ MN/m}^2$$

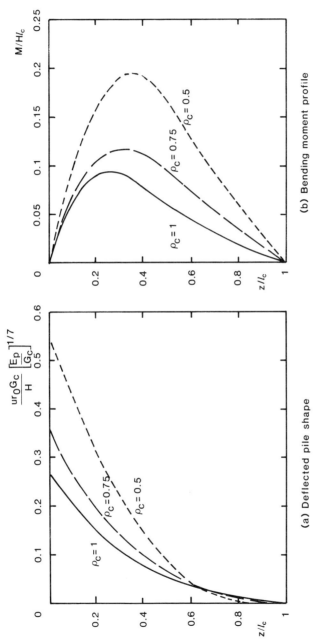

(a) Deflected pile shape

(b) Bending moment profile

Figure 4.40 Generalized curves of lateral deflection and bending moment profile for force loading

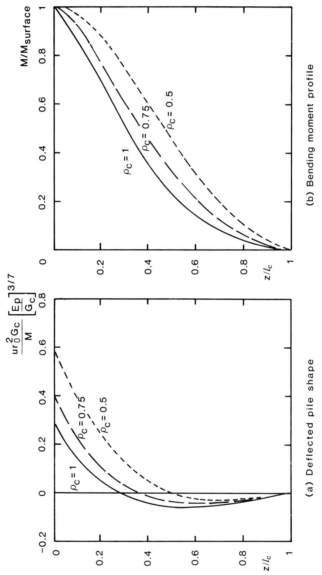

(a) Deflected pile shape

(b) Bending moment profile

Figure 4.41 Generalized curves of lateral deflection and bending moment profile for moment loading

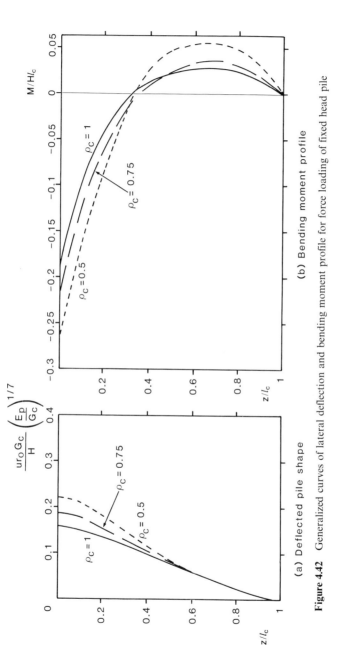

(a) Deflected pile shape

(b) Bending moment profile

Figure 4.42 Generalized curves of lateral deflection and bending moment profile for force loading of fixed head pile where r_i is the inner radius of the pile and E_{steel} has been taken as $210\,\text{GN/m}^2$. Taking a shear modulus for the soil of $G = 100c_u = 0.25z\,\text{MN/m}^2$, and Poisson's ratio as 0.3, the critical pile length may be calculated as

$$l_c = 2r_o(E_p/m^*r_o)^{2/9} = 23.1\,\text{m}$$

where $m^* = 0.25(1 + 0.75 \times 0.3) = 0.306\,\text{MN/m}^3$. The value of G_c is then $0.5 \times 23.1 \times 0.306 = 3.53\,\text{MN/m}^2$, and $\rho_c = 0.5$.

Under a lateral load of 1 MN, with no rotation allowed at ground level, the maximum bending moment and ground level deflection may be calculated from equations (4.78) and (4.79) respectively as

$$M_f = -6.1\,\text{MNm}; \quad u_f = 21.3\,\text{mm}.$$

Some comment is appropriate concerning the applicability of solutions for the response of piles to lateral load, which are based on elastic soil properties. Clearly, the large strains which occur locally around the head of the pile will reduce the relevant secant modulus to a low value. It is probable that idealization of the soil as a material with stiffness proportional to depth is a better idealization than as a homogenous material. Even with this idealization, however, choice of soil modulus will depend on the magnitude of movement expected. Centrifuge model tests on piles embedded in dense sand, reported by Barton (1982), show that the deduced stiffness of the soil decreases sharply as the head deflection of the pile increases. Taking a shear modulus for the sand which increases proportionally with depth, z, according to $G = mz$, the variation of the deduced value of $m^* = m(1 + 3v/4)$ with head deflection was as shown in Figure 4.43.

In practice, the critical design criterion will usually be the maximum bending moment down the pile rather than the head deflection. Equation (4.77) shows that this maximum moment will be relatively insensitive to the value of soil modulus adopted. Indeed, Barton's centrifuge tests showed that the elastic solution gave a reasonably good estimate of the position and magnitude of the maximum bending moment.

Short piles. For piles that are shorter than their critical length, the head deformation will be larger than given by equations (4.76a and b). The increase in deflection is small until the pile length falls below about $0.8l_c$. Carter and Kulhawy (1988) have presented a solution for essentially rigid piles in homogeneous soil ($\rho_c = 1$). For piles where $l/r_o \leqslant 0.1(E_p/G_c)^{1/2}$, the ground level deflection and rotation are given by:

$$u = 0.2\frac{H}{G_c r_o}\left(\frac{l}{r_o}\right)^{-1/3} + 0.15\frac{M}{G_c r_o^2}\left(\frac{l}{r_o}\right)^{-7/8} \tag{4.80a}$$

$$\theta = 0.15\frac{H}{G_c r_o^2}\left(\frac{l}{r_o}\right)^{-7/8} + 0.4\frac{M}{G_c r_o^3}\left(\frac{l}{r_o}\right)^{-5/3} \tag{4.80b}$$

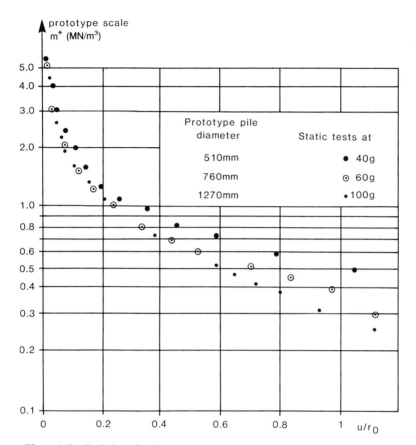

Figure 4.43   Variation of deduced soil modulus with pile deflection (Barton, 1982)

With such short piles, the design will normally hinge on the possibility of the pile failing by rigid body movement through the soil. Analysis of this type of behaviour has been discussed in section 4.3. Further charts giving the response of short piles to lateral loading may be found in Poulos and Davis (1980).

4.4.3 Cantilever idealization

The main shortcoming of the idealization of laterally loaded piles as cantilevers, fixed at some depth below ground surface, has been the lack of consideration of the role of the relative pile and soil stiffnesses in determining an appropriate depth of fixity. Equations (4.76a and b) may be rewritten in a form more familiar to engineers with a structural background, by substituting for G_c in terms of E_p and the critical length, l_c (see equation (4.72)) and then replacing E_p by $(EI)_p/(\pi r_o^4/4)$. In most cases, it will be appropriate to take ρ_c equal to 0.5, allowing for very low soil stiffness near the ground surface, where the strain level in the soil is high. For this case, equations (4.76a and b) transform to

$$u = 0.424 \frac{H}{EI}\left(\frac{l_c}{2}\right)^3 + 0.472 \frac{M}{EI}\left(\frac{l_c}{4}\right)^2 \tag{4.81a}$$

$$\theta = 0.472 \frac{H}{EI}\left(\frac{l_c}{4}\right)^2 + 0.887 \frac{M}{EI}\left(\frac{l_c}{4}\right) \tag{4.81b}$$

These equations are similar to those that would be obtained from a cantilever idealization of the pile, with a depth of fixity of $l_c/2$. In that case the three independent coefficients on the right-hand sides would be 1/3, 1/2 and 1 respectively, instead of the values of 0.424, 0.472 and 0.887. Thus the errors involved in estimating pile deformations from a cantilever idealization will be small, providing the appropriate depth of fixity of $l_c/2$ is adopted.

The cantilever idealization is not suitable for estimating the profile of induced bending moments. For example, in the simple case of a pile under force loading, H, only, equation (4.77) gives a maximum bending moment of $0.2Hl_c$ (taking $\rho_c = 0.5$), occurring at a depth of $l_c/3$. The cantilever approach would give a maximum bending moment at the base of the cantilever (a depth of $l_c/2$) of magnitude $0.5Hl_c$.

4.4.4 Layered soil profiles

The response of piles to lateral loading is greatly affected by the soil properties close to ground level. The presence of a desiccated crust or a shallow layer of granular material overlying softer cohesive soil, can have a significant beneficial effect, reducing both deflections and bending moments down the pile. General solutions for the response of piles in a layered soil profile have been presented by Davisson and Gill (1963) using a subgrade reaction approach, and by Pise (1982) for an elastic soil continuum. The two approaches yield similar results which are best presented in the form of modifications to the solution given in section 4.4.2. Thus, consider a layer of soil of thickness t_l, whose stiffness is a factor f_l greater than that of the underlying soil (see Figure 4.44). For given values of t_l and f_l, the deformation of the pile at ground level may be compared with that for $f_l = 1$ (that is, a single layer). Similarly, the maximum bending moment induced in the pile will be affected.

Figure 4.45 shows the effect of soil layering on the ground level deflection and bending moment for a fixed-headed pile subjected to lateral loading. The ratios u_f/u_{f1} and M_f/M_{f1} (where the subscript 'f' denotes fixed-head, and the subscript '1' denotes the value for $f_l = 1$) are plotted against the ratio t_l/l_c for different values of f_l. These curves may also be used with acceptable accuracy for free-headed piles. It is interesting to note that soil conditions

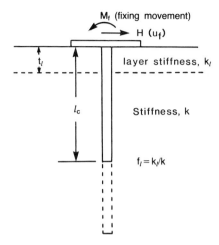

Figure 4.44   Definition of notation for pile in two-layer soil

over the upper 10% of the active length of pile are critical in determining the lateral response of piles.

An alternative approach for analysing the response of a pile in a layered soil profile may be used where ground level deflections are high (10% or more of the pile diameter). For such cases, it is sufficient to assume that the limiting resistance is mobilized in the layer. The pile deformation and bending moment distribution may then be obtained by treating the pile as a cantilever in that region, while the lower part of the pile is analysed using the expressions presented in section 4.4.2.

4.4.5   Raking piles and effect of lateral loading on axial capacity

The response of piles to inclined loading (i.e. combined axial and lateral loading) has been discussed by various authors (Broms, 1965; Meyerhof and Ranjan, 1972; Poulos and Davis, 1980). In non-cohesive soil, it has been argued that the axial capacity of a pile may be increased by small amounts of lateral load, owing to the increase in normal stress between pile and soil. This argument has only limited support from experimental data, except for very stubby, caisson-type piles. For most piles, the effects of lateral loading are confined to the upper few diameters of the pile. For small amounts of lateral load, the increase in normal stress in front of the pile will be at least partially offset by a reduction in normal stress behind the pile.

In practical terms, the axial and lateral response of a pile may be treated independently. The response of a raking pile to load applied at a particular angle may then be assessed by taking the components of load parallel and normal to the axis of the pile. The overall capacity of a pile will be determined by taking the axial and lateral components of the applied load and comparing

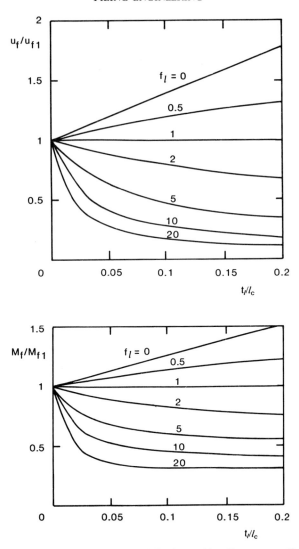

Figure 4.45 Effect of upper soil layer on pile deflection and bending moment (fixed-head piles)

them with the respective capacities. Thus for the general case shown in Figure 4.46, where a pile raking at an angle ψ to the vertical is subjected to a load inclined at θ to the vertical, axial failure will occur when

$$Q \cos(\theta - \psi) = Q_a \text{ (axial failure)} \qquad (4.82)$$

or when

$$Q \sin(\theta - \psi) = Q_l \text{ (lateral failure)} \qquad (4.83)$$

It may be seen that the maximum capacity will be achieved when both

Figure 4.46 Oblique loading of a raking pile

axial and lateral pile capacities are mobilized simultaneously. This occurs for

$$\tan(\theta - \psi) = Q_l/Q_a \tag{4.84}$$

when the overall capacity will be

$$Q = (Q_a^2 + Q_l^2)^{1/2} \tag{4.85}$$

Evangelista and Viggiani (1976) have shown that the axial and lateral deformation response of a pile is virtually independent of the angle that the pile makes with the ground surface, for angles of rake up to about 30°. Thus, as for the calculation of the pile capacity, the deformation response may be assessed by considering the components of the applied load parallel and normal to the pile axis. For vertical piles, the vertical and horizontal deformations may be written in terms of a flexibility matrix, F, and the applied load as

$$\begin{bmatrix} w \\ u \\ \theta \end{bmatrix} = \begin{bmatrix} F_{wP} & 0 & 0 \\ 0 & F_{uH} & F_{uM} \\ 0 & F_{\theta H} & F_{\theta M} \end{bmatrix} \begin{bmatrix} P \\ H \\ M \end{bmatrix} \tag{4.86}$$

For a pile battered at an angle ψ to the vertical, the matrix F must be replaced by a transformed matrix, F', where

$$F' = T^T F T \tag{4.87}$$

and

$$T = \begin{bmatrix} \cos\psi & -\sin\psi & 0 \\ \sin\psi & \cos\psi & 0 \\ 0 & 0 & 1 \end{bmatrix} \tag{4.88}$$

One aspect of the effect of lateral loading on axial capacity which deserves particular attention is the possibility of a gap opening up around the pile near ground level—a so-called 'post-holing' effect. Piles subjected to axial loading together with cyclic lateral loading generally fail at a point where the lateral component of load is zero, and the pile is then close to its neutral position. In the design of piles for offshore structures, it is generally recognized that load transfer into the soil close to the sea-floor (under axial loading) should be ignored because of the risk of post-holing. The question is: to what depth will such post-holing occur?

Some estimate of the extent of any gap forming around the pile (particularly in fine-grained soils) may be made by considering the lateral deflection at which yield in the soil close to the pile will occur. Baguelin *et al.* (1977) have considered the case of a circular disc moving sideways through elastic, perfectly plastic material, with shear modulus, G, and shear strength, c_u. They show that the lateral movement is related to the normal stress change, $\Delta\sigma_n$, in front of and behind the disc, by

$$u/r_0 \sim 2\Delta\sigma_n/G \qquad (4.89)$$

The load deflection curve for the disc was essentially linear up to a normal stress change of about $2c_u$. This would indicate that non-linear effects, and the possibility of gapping around the pile, would start to occur at a normalized deflection of $u/r_0 \sim 4c_u/G$. For a typical value of G/c_u of 100, this would be about 2% of the pile diameter.

Thus, in assessing the extent of any post-holing, and loss of skin friction, it is necessary to estimate the depth down the pile at which the cyclic lateral movements are less than about 2% of the pile diameter. Above this level, it would be imprudent to rely on significant skin friction under axial loading.

4.5 Piles in deforming soil

There are many situations in which loads are induced in a pile by virtue of the deformation of the surrounding soil. The most common example is where piles are installed into, or through, a layer of soil which is consolidating, usually due to the recent placement of surface loading. Where the downward movement of the surrounding soil exceeds the settlement of the pile, the pile resists the movement and load in excess of the external applied load will be transferred into the pile. The shear stresses on the pile shaft will act in the reverse sense by comparison with the normal situation for a pile loaded in compression; accordingly, the term 'negative friction', or 'downdrag' is used (see Figure 4.47(a)).

Piles may also be loaded laterally by soil movement, for example on a steeply sloping site, where creep of the upper layers of soil may occur. Eccentric (vertical) loading of the ground surface adjacent to a piled foundation may also lead to significant lateral pressures on the piles, as the soil moves away from the loaded area (see Figure 4.47(b)).

applied surface load

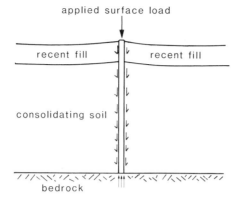

(a) Pile subjected to negative friction

(b) Piles subjected to lateral thrust from soil

Figure 4.47 Piles in consolidating soil

4.5.1 *Negative friction*

Negative friction or downdrag of piles may occur wherever piles are driven
through, or adjacent to, recently placed fill (Johanessen and Bjerrum, 1965).
Such fill may merely be to raise the existing ground level, or may be part of
an embankment, for example, for a bridge approach. The problem is normally
associated with soft, lightly overconsolidated deposits of clay. In some
sensitive clays, remoulding of the soil during driving may lead to downdrag
of the piles, even where no fill is placed (Fellenius, 1972). The settlement of
the clay is then associated not with any long-term rise in the vertical effective
stress, but rather with a reduction in voids ratio and water content as the
remoulded soil consolidates.

The starting point for any estimation of the effects of negative friction is
an assessment of the likely free-field vertical movement of the soil. Where

this movement exceeds the estimated settlement of the pile, shear stresses acting in a reverse sense will be imposed on the pile shaft. The magnitude of the shear stresses may be determined from the amount of relative movement between soil and pile, with a limiting value reached where downward slip of the soil occurs. The magnitude of relative movement necessary to cause slip has been discussed in section 4.2.4 (see equation 4.49). For relative movements greater than this limit, the mobilized shear stress may be estimated by

$$\tau_s = K\sigma'_v \tan \phi' \tag{4.90}$$

where, since the soil is consolidating, K should be taken as the appropriate value of K_0, the at-rest earth pressure coefficient corresponding to the current overconsolidation ratio. In most cases of practical interest, the consolidating soil will be normally consolidated, and K_0 may be approximated by $(1 - \sin \phi')$.

For relative movements less than that to cause slip, the mobilized shear stress (positive or negative) may be taken as a linear proportion of the limiting skin friction. Thus, for small relative movements, the local shear stress on the pile shaft is

$$\tau_0 = \frac{G}{\zeta r_0}(w_0 - w_{\text{soil}}) \tag{4.91}$$

where w_0 is the local settlement of the pile, and w_{soil} is the background consolidation settlement of the soil (see section 4.2.1).

Assessment of negative friction for a given distribution of soil settlement necessitates some iteration, since the profile of pile movement will depend on the distribution of shear stress, τ_0, down the length of the pile. The axial load at any depth, z, down the pile is given by

$$P = P_t - \int_0^z 2\pi r_0 \tau_0 dz \tag{4.92}$$

where P_t is the applied (structural) load. The axial strain in the pile may then be integrated to give the settlement of the pile at any depth, z, as

$$w_0 = w_b + \int_z^l \frac{P}{(AE)_p} dz \tag{4.93}$$

where w_b is the settlement of the pile base.

From the calculated pile settlement profile, and the estimated profile of soil settlement, the distribution of skin friction may then be calculated. This approach is similar to that described by Fellenius (1989). For end-bearing piles, the distribution of negative friction will generally increase steadily with depth, until a point close to the pile base, where the soil movement is insufficient to mobilize full skin friction (see Figure 4.48(a)). For a long friction

(a) End-bearing pile

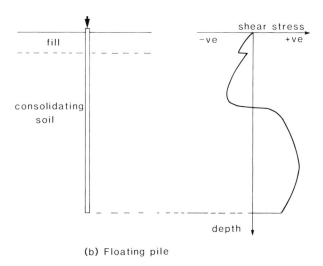

(b) Floating pile

Figure 4.48   Typical variations with depth of negative skin friction

pile, however, there will be a neutral point, generally some way above the mid-point of the pile, where pile and soil settlement are equal, and below which positive friction is mobilized to carry the load applied to the pile (see Figure 4.48(b)).

For end-bearing piles, parametric studies for single piles have been presented by Poulos and Davis (1980). The method of analysis has been extended to groups of end-bearing piles by Kuwabara and Poulos (1989). This latter work has shown that the internal piles within a group are subjected to much lower downdrag loads than the peripheral piles, and that, for a given soil settlement, the average downdrag load on the piles in a group is significantly lower than for a single isolated pile. Thus for a surface settlement varying linearly from 5% of the pile diameter at the surface, to zero at the bearing stratum, a typical single pile would attract over 90% of the maximum available load due to negative friction, while a square group of 25 piles would attract less than 50% of the potential load.

The appropriate design philosophy to be adopted in situations where negative friction may arise is a matter of some debate. If piles are designed so that they can take the full negative friction (in addition to the load applied at ground level) without yielding, then the structure founded on the piles will not settle significantly. In such cases, the soil surrounding the structure will settle and may cause as great a problem (with services, and distortion of paved areas adjacent to the building) as in the more common case of a building settling more than the surrounding soil. The problems can be particularly severe if there are adjacent, unpiled, structures. These will settle with the soil, resulting in large differential settlements between the piled and unpiled structures.

In many cases, the optimum solution is for the piled structure to settle with the surrounding soil. For this to be achieved, the piles should be designed to carry the loads applied by the structure, with a relatively low factor of safety, and with no allowance made for any negative friction. As the surrounding soil settles, negative friction will indeed drag the piles down, but, by its very nature, will not lead to bearing failure of the piles (provided they have been correctly designed for the given structural load). The low factor of safety ensures that the piles do not withstand too much negative friction before settling with the consolidating soil.

The above approach may certainly work well with floating piles, since these have a low base capacity. The shaft resistance relies on relative motion between pile and soil. Thus, if the soil settles around the piles, the pile must also settle in order to carry the applied load. The same approach is more sensitive to apply for end-bearing piles, owing to uncertainties in the load settlement response of the pile base. However, unless there is a danger of breaking through a competent stratum into softer soil beneath (which might lead to failure of the piles under the structural load alone) the same general philosophy of allowing the piles to settle with the consolidating soil should still provide a satisfactory foundation.

In situations where it is essential to restrict the absolute movement of the pile foundation, measures may be taken to relieve the piles of negative friction by coating the shafts or by sleeving them, particularly in the upper regions. Claessen and Horvat (1974) discuss the use of bitumen coatings, and recommend that, where such coatings are used, they should be omitted on the lower 10 diameters of the shaft, in order to ensure that the full end-bearing capacity may be relied on. Coatings as thin as 1.5 mm may be sufficient to eliminate nearly all negative friction (Walker and Darvall, 1973), provided that care is taken to ensure that the coating is not stripped off the pile during installation.

4.5.2 *Lateral movement of soil*

Before considering the case of vertical piles loaded laterally by horizontal soil movement, it is worth remarking that a common cause of foundation failure is due to the use of raking piles in soil which is still consolidating. Vertical settlement of the soil around a raking pile can impose lateral pressures that are sufficient to break the pile. Raking piles should thus be avoided in situations where significant consolidation settlement of the soil may occur—including the case of piles being driven into sensitive clay, as mentioned at the start of section 4.5.1.

Eccentric loading of the ground around a piled structure, for example an embankment behind a piled bridge abutment, may generate sufficient horizontal soil movement to cause distress in the piles. This form of loading can also lead to significant movement of the structure. Cole (1980) reports the case of a piled bridge abutment, founded on a deep deposit of soft silty clay, which rotated away from the bridge (that is towards the retained embankment) due to lateral soil pressure below ground level.

A simple method of estimating the lateral pressure on piles close to a surcharged area has been presented by De Beer and Wallays (1972). The method differentiates between surcharge loads (particularly embankments) which have factors of safety against collapse (without the piles) which are greater or less than 1.4. For the less stable situation, it is assumed that the full limiting pressure is mobilized against the pile (see section 4.3) throughout the depth of the soft deposit. For cases with higher factors of safety, the lateral pressure on the pile is taken as the surcharge pressure, factored by a parameter which depends on the geometry of the loaded area relative to the piles. Figure 4.49 shows how the method would be applied for typical embankment geometries. The actual embankment, of height H, is replaced by a fictitious embankment of material of unit weight $18 \, \text{kN/m}^3$, of a height H_c, such that the vertical loading is the same. The lateral pressure on the piles is then taken as

$$\Delta p_h = \Delta p_v \frac{(2\alpha - \phi')}{(\pi - \phi')} \tag{4.94}$$

where Δp_v, is the applied surcharge pressure, and ϕ' is the friction angle of the soft layer of soil. The angle α is defined in Figure 4.49.

In the above approach, no account is taken of the depth, or extent, of the soft layer, or even of its initial strength; the method is clearly very approximate. An approach based on elastic interaction of the moving soil and the pile, but with allowance made for the limiting pressure that the soil may exert on the pile, has been described by Poulos and Davis (1980). Their method relies on an initial estimate of the horizontal soil movement. A

Figure 4.49   Piles subjected to lateral thrust near an embankment (De Beer and Wallays, 1972)

boundary element analysis of the problem then allows calculation of the resulting loading in the piles. Charts are presented for some common profiles of soil movement.

Bourges and Mieussens (1979) and Bourges *et al.* (1982) have presented an approach which is similar in principle to that used by Poulos and Davis (1980), although adopting a different analytical technique. The approach rests on an estimate of the horizontal soil movement, *u*, as a function of depth, and an iterative procedure is then used to calculate the pressure loading of the pile in terms of the difference between this background movement and that of the pile, u_0, according to

$$\Delta p_h = f(u - u_0) \tag{4.95}$$

where the function f represents a load transfer curve much as would be used for analysis of a pile subjected to lateral loading applied at its head. This will depend on the stress–strain properties of the soft deposit and may be linked with experimental curves obtained from pressuremeter testing.

Carter (1982) has used the finite element method to investigate the bending moments induced in a single pile by a surface strip load extending to one side of the pile. The results, applicable to elastic conditions only, are presented for the case of a single soil layer extending to a depth of 10 pile diameters (underlain by bedrock). Maximum bending moments of $M_{max} \sim 2.7qd^3$ were obtained, where q is the surcharge load and d is the pile diameter.

An extensive series of centrifuge model tests (Springman *et al.*, 1991) has led to a more rational method of assessing the effect of surcharge loading on adjacent piles. The tests focused particularly on the common situation where a layer of soft alluvial soil overlies stiffer material, and showed that the net lateral pressure acting on the piles varied parabolically through the soft layer. The maximum pressure (at mid-depth of the soft layer) could be expressed in terms of the embankment loading, q, the thickness of the soft layer, h, the pile diameter, d, and spacing, s, as

$$\frac{p_m}{q} \sim \frac{\psi}{\psi d/s + 2d/h} \tag{4.96}$$

where the parameter ψ may be taken in the range 0.5 to 1. The maximum pressure was limited to a value of $10.5c_u$, corresponding to flow of the soft soil around the piles.

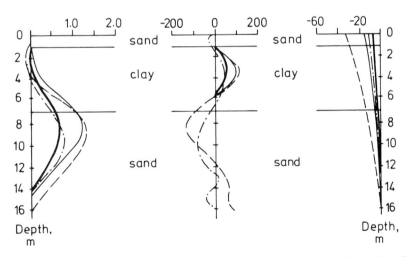

Figure 4.50 Experimental and calculated pile response due to surcharge loading: (a) bending moments (MNm), (b) lateral pressure (kPa), (c) lateral deflection (mm). —·—·—experimental (surcharge 53 kPa); ———SLAP (surcharge 53 kPa); ------- experimental (surcharge 93 kPa); ———SLAP (surcharge 93 kPa)

The bending moment and lateral deflection profiles down the piles may then be calculated by combining the effects of (a) the structural load applied to the pile head, (b) the lateral loading from the soil in the soft layer, and (c) the lateral resistance offered by the bearing stratum. The last component may be evaluated from the elastic solutions given earlier (section 4.4.2). The complete design procedure has been implemented in a spreadsheet program, SLAP (Randolph and Springman, 1991).

Figure 4.50 shows a comparison between calculated and experimental profiles of bending moment, lateral pressure and lateral deflection, for 0.635 m diameter tubular piles embedded through a 6 m thick layer of clay with average shear strength 26 kPa. The piles were spaced at 4 m, and had a bending rigidity of 5 100 MNm^2. It may be seen that the maximum bending moment occurs about 2 pile diameters into the bearing statum (the sand), and is about $7qd^3$. Maximum lateral pressure at mid-depth of the clay layer was similar to the embankment surcharge (i.e. $p_m/q \sim 1$). The lateral deflection of the piles was underpredicted at the higher surcharge pressure, owing to slight rotation of the actual piles at their tips, compared with the fully restrained condition assumed in the analysis.

In cases where surcharge loading leads to excessive bending moments or lateral deflection in nearby piles, certain construction techniques may be used to circumvent the problem. One technique, which may be relatively costly, is to shield the upper part of the piles using cylindrical caissons. An alternative solution, particularly suitable for embankment approaches to bridge abutments, is the use of piles beneath part of the embankment. This technique, known variously as 'embankment piling' or 'bridge approach support piling' (Reid and Buchanan, 1983), is shown schematically in Figure 4.51. Widely-spaced piles (up to 10 diameters apart) are used, with small pile caps that typically occupy as little as 10% of the overall area. Arching action is relied on in order to shed most of the embankment load on to the piles. Hewlett and Randolph (1988) have proposed a method of

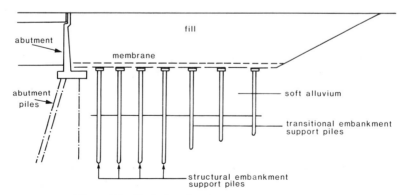

Figure 4.51   Typical section through embankment piling

analysis to assess the proporton of load carried by the piles. Field studies reported by Reid and Buchanan (1983) indicate that over 80% of the surcharge loading was shed onto pile caps spaced such as to occupy only 11% of the plan area.

The main advantages offered by embankment piling are as follows:

(a) short-term stability of the embankment will be ensured, avoiding the need for staged construction;
(b) long-term settlement of the embankment close to the bridge will be reduced, giving lower maintenance costs for the roadway;
(c) lateral pressure on the piles for the bridge abutment will be significantly reduced.

Reid and Buchanan recommend the use of a latticed fabric membrane over the tops of the pile caps, in order to prevent cavities from forming. The membrane will carry a proportion of the embankment loading, and will thus aid the degree of arching between the pile caps.

4.6 Buckling of piles

There are two main areas where the axial stability of piles may be critical in design. The first of these is during installation by driving, where instability can lead to the pile deviating significantly from its proposed alignment. The other area is under working conditions, where the axial load taken by a pile may be close to the critical buckling load for the pile. This latter case is relatively rare, but can occur for end-bearing piles driven through relatively soft alluvial deposits.

4.6.1 Instability during driving

Burgess (1976) has given a detailed analysis of the stability of piles driven or jacked into soil of uniform shear strength (see also Omar, 1978; Ly, 1980; Burgess, 1980). Two different forms of instability occur; one is a buckling phenomenon, and the other is known as 'flutter'. The latter may be viewed as directional instability at the advancing tip of the pile. Burgess' analysis for the buckling instability, which involves lateral movement of the pile, is based on a Winkler model for the response of the soil around the pile shaft. The flutter instability is independent of the lateral response of the pile.

The results of the analysis, based on the assumption that the pile is guided until just above ground level, indicate that the buckling mode of instability would rarely occur and, in any case, would always be preceded by flutter instability. The criterion for flutter to occur is shown in Figure 4.52. This figure is a dimensionless plot, essentially plotting the driving force, Q, against the proportion of that force that reaches the base of the pile, (Q_b/Q). The driving force is non-dimensionalized by the pile length, l, and the bending

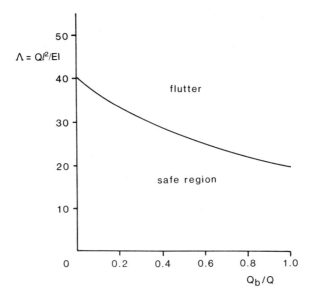

Figure 4.52   Instability due to flutter (Burgess, 1980)

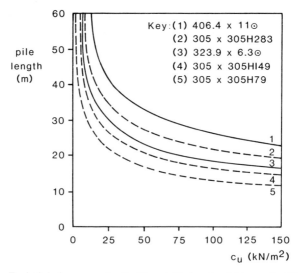

Figure 4.53   Typical design curves for limiting stable length of driven piles (Burgess, 1976)

rigidity, EI, of the pile section, in the form

$$\Lambda = Ql^2/EI \qquad (4.97)$$

Burgess (1976) gives design curves for some typical pile cross-sections (see Figure 4.53), showing how the length of pile at which flutter would occur varies with the shear strength of the soil. It has been assumed that the full shear strength of the soil is mobilized in skin friction down the pile shaft during installation. This assumption is probably inappropriate; however, the curves given in Figure 4.53 serve to illustrate that flutter may occur at relatively modest pile lengths.

4.6.2 Buckling under static loading

Buckling of piles under static loading will generally only be a problem where very soft soil overlies the founding material, giving rise to a depth in which the pile is only moderately supported laterally. The problem has been looked at by a number of workers (Davisson, 1963; Davisson and Robinson, 1965; Reddy and Valsangkar, 1970). As might have been expected intuitively, analysis shows that buckling will be confined to the critical length of the pile under lateral loading (see section 4.4). This will be a region in the upper few diameters of the pile, and, for practical purposes, it is sufficient to ignore the variation of axial load in the pile over the length of interest. This will in any case furnish a conservative estimate of the load to cause buckling.

Figure 4.54 shows a general picture of the pile, with a length l_f of pile free-standing above the ground surface. In order to calculate the buckling load, the stiffness of the embedded section of the pile under lateral force and moment loading must be known. It is sufficient to replace the embedded section of the pile by a cantilever of such a length l_e, that provides an

(a) Actual pile   (b) Equivalent cantilever

Figure 4.54 Equivalent cantilever for buckling calculation for a partially embedded pile equivalent stiffness. This cantilever idealization was discussed in section 4.4, and it was shown that the equivalent length for the cantilever should be taken as half the critical length, that is

$$l_e = l_c/2 = r_0(E_p/G_c)^{2/7} \sim 2(EI/k)^{1/4} \quad \text{or} \quad 2(EI/n)^{1/5} \qquad (4.98)$$

where the parameters have been defined in section 4.4.

With the pile now replaced by an equivalent length, $l_f + l_e$, embedded at its lower end, calculation of the buckling load may be done using the standard Euler equations. Thus, for boundary conditions at the upper end which allow lateral movement and rotation (a free end), the buckling load is

$$P_{cr} = \frac{\pi^2 EI}{4(l_f + l_e)^2} \qquad (4.99)$$

while for conditions at the upper end which prevent rotation, but still allow lateral motion, the buckling load will be a factor of 4 times greater. This approach is consistent with that of Davisson and Robinson (1965), although restricted to the practical case of piles which are longer than their critical length under lateral loading.

It is interesting to compare the numerical values of the buckling load given by equation (4.99) with the condition for flutter to occur during installation shown in Figure 4.52. For a pile with no free-standing length above the ground ($l_f = 0$), the equivalent non-dimensional form of equation (4.99) would be

$$\Lambda = 2.5(l/l_e)^2 \qquad (4.100)$$

Typically, the critical length of a pile will be about 10 to 15 diameters, giving an equivalent length of 5 to 7.5 diameters. From the example curves shown in Figure 4.53, the pile length at which flutter becomes a problem will be 50 to 100 diameters. Thus the non-dimensional parameter in equation (4.100) becomes about 250, well above that to cause flutter instability. The conclusion, then, is that buckling of long piles driven into deep layers of soft soil will be a secondary problem to that of actually installing them. However, end-bearing piles, or piles with a significant length of free-standing section, which are installed through soft deposits, should be assessed for possible buckling under static load.

5 Design of pile groups

The majority of piled foundations will consist not of a single pile, but of a group of piles, which act in the dual role of reinforcing the soil, and also of carrying the applied load down to deeper, stronger soil strata. As will be shown in this chapter, the effectiveness of a pile, in particular in respect of its stiffness, is generally reduced by the proximity of other piles. In recognition of this, current trends are towards the use of fewer, more widely-spaced piles, where the reinforcement role of piles is emphasized. Modern analytical techniques allow more realistic estimates to be made of the response of pile groups under working loads, thus giving the designer more scope to optimize the foundation layout.

In particular, two areas of progress may be singled out. The first concerns the design of pile groups to withstand horizontal loading (Elson, 1984). The traditional view of piles as pin-ended columns, incapable of withstanding loads applied normal to their axis, entailed much unnecessary use of raking piles—often with two or more rows of piles aligned parallel to the resultant loading (Hambly, 1976) (see Figure 5.1(a)). The inefficiency of such a pile group as a means of reinforcing the soil is apparent. In addition, the horizontal component of load in most pile groups will generally contain a high time-varying component. Thus there will be periods when the maximum vertical load may be applied, but the horizontal load will be substantially less than the maximum designed for. Under such conditions, the raking piles are likely to be as severely stressed in bending, as they would have been as vertical piles under the full horizontal load.

While raking piles might still be used in a modern design, in order to minimize induced bending moments, awareness of the lateral load-carrying capacity of piles allows the piles to be spread more evenly through the soil (see Figure 5.1(b)). Computer analysis of the pile group, making proper allowance for the soil support around the piles, permits estimates of the resulting bending moments and deformations to be made with some confidence.

The other area of progress concerns the use of piles in conjunction with a raft foundation, rather than as an alternative. Where reasonably competent

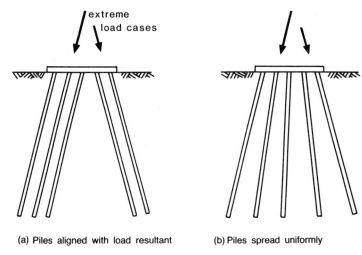

(a) Piles aligned with load resultant   (b) Piles spread uniformly

Figure 5.1   Piles as reinforcement for the soil

soil extends to the ground surface, the decision to include piles in a foundation may be made on the basis of excessive settlement of a raft foundation alone. The traditional approach has then been to design the foundation solely on the basis of the piles, ignoring any contribution of the pile cap (or raft) to the load-carrying capacity of the foundation. The illogicality of such an approach has been pointed out by Burland *et al.* (1977), who suggest the inclusion of a limited number of piles beneath a primarily raft foundation, in order to reduce settlements to an acceptable level. This approach is discussed further in section 5.3.

At the end of the chapter, there is a discussion of some of the main principles in the design of single piles and pile groups. The discussion includes comments on appropriate factors of safety to use for piles in different circumstances, and reviews the potential benefits and hazards of using raking piles.

5.1 Capacity of pile groups

A group of piles may be viewed as providing reinforcement to a particular body of soil. Failure of the group may occur either by failure of the individual reinforcing members (the piles) or as failure of the overall block of soil. When considering failure of the individual piles, it must be remembered that the capacity of each pile may be affected by the driving of subsequent piles in close proximity. This has been demonstrated experimentally by Vesic (1974) and Lo (1976), who showed that the shaft capacity of piles driven into sand could increase by a factor of 2 or more. Physically, this may be attributed to higher normal effective stresses acting on piles driven in a group, than on single piles. The beneficial effects of driving neighbouring piles may be particularly noticeable in loose sands, where compaction of the sand can lead to very low contact stresses around the pile shaft. Vibration and soil displacement due to driving of subsequent piles is likely to reduce the tendency for the soil to arch around each pile, leading to higher contact stresses, and thus shaft capacities, than for single piles.

In other situations, the capacity of a pile within a group may be reduced by comparison with a single, isolated pile. In particular, for piles driven into sensitive clays, the effective stress increase in the surrounding soil may be less for piles in a group, than for individual piles. This will result in lower shaft capacities.

Instead of failure of the individual piles in a group, block failure may occur under axial or lateral load. Soil between the piles may move with the piles, resulting in failure planes which follow the periphery of the group (or parts of the group) as shown in Figure 5.2. In general, block failure will be associated with close spacing of the piles. However, in certain cases, such as the axial failure of a pile group with a ground-contacting pile cap, block failure may be enforced at quite large pile spacings.

In the past, the difference between the capacity of a group of piles, and the sum of the individual pile capacities, has been characterized by a group 'efficiency'. This concept has led to a certain amount of confusion. As stated above, the capacity of a pile within a group may be different from that of an equivalent isolated pile, and, in that sense, the pile may be thought of as more or less efficient when acting as part of a group. However, the block capacity of a group should not be confused with the sum of the individual pile capacities by the use of an 'efficiency' factor. Independent calculations should be made of both the block capacity and the individual pile capacities, to ensure that there is an adequate factor of safety against both modes of failure.

(a) Single pile failure (b) Failure of rows of piles (c) Block failure

Figure 5.2 Block failure of pile groups

5.1.1 *Axial capacity*

The axial capacity of a pile group failing as a block may be calculated in a
similar fashion to that for an individual pile, by means of equation (4.1),
but now taking A_b as the base area of the block and A_s as the block
surface area. In general, the enclosing block as shown in Figure 5.2(c)
should be taken, but, where the pile spacing in one direction is much
greater than that in the perpendicular direction, the capacity of the
group failing as shown in Figure 5.2(b) should be assessed. It should be noted
that the settlement needed to mobilize the base capacity of the block will
usually be very large (for example, 5–10% of the width of the group).

Since the end-bearing pressure q_b is much greater than the average skin
friction τ_s, block failure only becomes more likely than the failure of individual
piles where the increase in base area, A_b, is offset by a much larger decrease
in surface area, A_s. Thus groups of a large number of long slender piles at
a particular spacing are more likely to fail as a block than groups consisting
of a few, short stubby piles at the same spacing. By a similar argument, group
failure is less likely for piles in sand (where q_b/τ_s is 50–200) than for piles in
clay (where q_b/τ_s is 10–20).

Non-cohesive soil. Calculation of the end-bearing pressure q_b for block failure
of a pile group in non-cohesive soil follows a similar procedure to that
described for single piles in section 4.1.1, except that the bearing capacity
term due to the self-weight of the soil below bearing level may now no longer
be insignificant. Thus the end-bearing pressure should be calculated from

$$q_b = N_q\sigma'_v + 0.4\gamma BN_\gamma \qquad (5.1)$$

where B is the width of the pile group (assumed approximately square).
Values of the bearing capacity factor N_γ may be obtained from standard
texts (Terzaghi and Peck, 1967).

The average frictional resistance along the sides of the block failure may
be calculated on the basis of equation (4.7), taking K as the at-rest earth
pressure coefficient, and δ equal to the angle of friction, ϕ'. Thus

$$\tau_s = K_0\sigma'_v \tan \phi' \qquad (5.2)$$

For most pile groups in sand, the calculated block capacity will be well in
excess of the sum of the individual pile capacities. For displacement piles,
the capacity of each pile in the group will itself tend to be larger than for
isolated piles, owing to the compactive effect of pile installation. This increase
in capacity has generally been interpreted in terms of a group efficiency, η_Q,
giving the ratio of group capacity to the sum of the individual pile capacities,
taken as isolated piles (Vesic, 1974; Lo, 1976). Figure 5.3 shows group
efficiencies reported by Vesic (1974). It may be seen that the base capacity

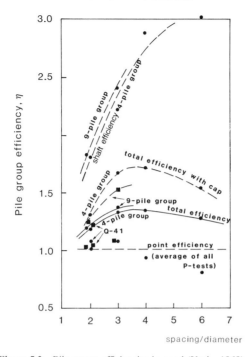

Figure 5.3 Pile group efficiencies in sand (Vesic, 1969)

of piles in a group is not affected by the neighbouring piles, but that the shaft capacities may increase by a factor of up to 3.

Accurate quantification of the increased capacity of piles in a group is difficult. As a conservative approach in design, the axial capacity of a pile group in sand is usually taken as the sum of individual pile capacities, calculated as in section 4.1.1. An exception to this is where the pile cap is cast directly on the ground, in which case the capacity of the pile group will correspond to that of the block. However, in these circumstances, the design of the pile group will probably hinge on the likely settlements. Factors of safety against overall failure will be high and accurate estimation of the group capacity will become less relevant.

Cohesive soil. For pile groups in cohesive soil, the block capacity may be calculated assuming end-bearing pressures given by equation (4.16), with an appropriate value of N_c, and assuming the full shear strength of the soil is mobilized round the periphery of the block. The factor N_c may be estimated approximately (Skempton, 1951) from the expression

$$N_c = 5(1 + 0.2B/L)[1 + (l/12B)] \qquad (5.3)$$

where B and L are the breadth and length of the pile group in plan and l is the embedded pile length; a limiting value of 1.5 should be imposed on the second (square) bracketed correction factor.

In certain types of soil, particularly sensitive clays, the capacity of individual piles within a closely spaced group may be lower than for equivalent isolated piles. However, there is a paucity of field data allowing this to be quantified, and it is rarely taken into consideration in design. The effect is not likely to be large. A more common cause for concern has been that the block capacity of the group may be less than the sum of the individual pile capacities. Evidence for this is largely from small-scale model tests. Adopting the definition of group efficiency given above, de Mello (1969) has summarized data for pile groups of up to 9×9 piles (see Figure 5.4). These results show that block failure of a pile group does not become pronounced until the pile spacing is less than 2 to 3 pile diameters.

With modern trends in pile group design for more widely spaced piles, the group capacity will rarely fall below that calculated from the sum of the individual pile capacities. Full-scale tests on a 3×3 pile group in stiff clay (O'Neill, 1982) confirm that pile group capacity may be estimated with good accuracy as the sum of the individual pile capacities. For piled foundations with a ground-contacting cap, the group capacity will correspond to that of the block enclosing the piles. However, as for groups in sand, settlement considerations will generally dominate such a design and the overall capacity will be of little relevance.

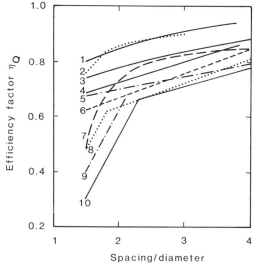

Figure 5.4   Pile group efficiencies in clay. $1. 3^2 \times$ (ST); 2, $2 \times 12d$ (SF); 3, $3 \times 30d$ (ST); 4, $3^2 \times 24d$ (W); 5, $3^2 \times 48d$ (W); 6, $5^2 \times 24d$ (W); 7, $3^2 \times 24d$ (SF); 8, $7^2 \times 24d$ (W); 9, $9^2 \times 24d$ (W); 10, $9^2 \times 48d$ (W). W = Whitaker (1957); ST = Saffery–Tate (1961), SF = Sowers–Fausold (1961) (De Mello, 1969)

In summary, design calculations for a pile group should include an assessment of the capacity both of the individual piles, and of the enclosing block. The former should generally be based on the capacity of equivalent isolated piles, only modified by the effect of installation of the surrounding piles where reliable data are available to quantify the effect. Adequate, and independent factors of safety should be ensured against either mode of failure, but the two modes should not be confused by the use of empirical group efficiency factors.

5.1.2 Lateral capacity

The lateral capacity of a pile group will rarely be critical in the design of the group. However, some discussion is appropriate concerning the limiting normal pressure which piles within a group may impose on the soil. The form of block failure relevant for laterally-loaded pile groups will generally be that shown in Figure 5.2(b), where the applied load is parallel to the individual blocks. This mode of failure will govern when the shearing resistance of the soil between the piles is less than the limiting resistance of an isolated pile. Referring to Figure 5.5, the limiting force per unit length on the back pile will be the lesser of P_u as calculated in section 4.3.1 for single piles, and $2s\tau_s$, where s is the pile spacing and τ_s is the friction on the sides of the block of soil between the two piles.

Non-cohesive soil. For non-cohesive soil, the limiting force per unit length for a single pile (at depth) may be taken as $P_u = K_p^2 \sigma_v' d$. The limiting shear stress on the sides of the block of soil shown in Figure 5.5 may be written

$$\tau_s = K\sigma_v' \tan \phi' \qquad (5.4)$$

where the earth pressure coefficient K will lie between K_0 and K_p. Thus block failure will occur at pile spacings where

$$s/d < \frac{K_p^2}{2K \tan \phi'} \qquad (5.5)$$

For most practical pile spacings, this condition will be satisfied, and the limiting force per unit length, on piles which lie in the wake of another pile,

Figure 5.5 Plan view of block failure under lateral load

may be estimated from

$$P_u = 2K\sigma'_v s \tan \phi' \qquad (5.6)$$

The value to be assigned to the coefficient K remains open to question. The stress field will vary over the space between the piles, but an average value of around unity may be taken as a first approximation.

Cohesive soil. A similar approach may be adopted for cohesive soil, taking the limiting shear stress on the sides of the soil block as the undrained shear strength, c_u. Assuming a limiting force per unit length for an individual pile of $9c_u d$, block failure will occur at pile spacings less than 4.5 diameters. The limiting force per unit length for piles in the wake of another pile is then given by the lesser of $9c_u d$ and $2c_u s$.

Failure mode of pile group. Failure of a pile group under lateral load will involve rotation of the group as well as lateral translation. Referring to Figure 5.6, piles behind the axis of rotation will fail by uplift, while those in front will fail in compression. The overall stability of the group will be determined by the axial capacity of the piles in addition to their lateral capacity. The concept of group efficiency is thus even less applicable than for axial loading, and most pile groups will have efficiencies considerably greater than unity under lateral loading, due to the geometry of the group. The group capacity, and the mode of failure, may be determined from statical considerations, but taking due account of the lateral resistance of the soil.

In practice, the lateral capacity of a pile group will seldom be a critical feature of a design. In situations where there is a large component of lateral load, it will be necessary to ensure that the maximum bending moments induced in the piles will not lead to failure of the piles. The maximum bending

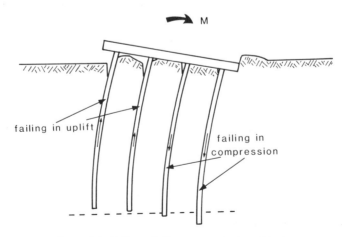

M

failing in uplift

failing in compression

Figure 5.6 Failure of pile group under lateral load

moments will generally be estimated on the basis of one of the methods described in section 5.2.3, without considering the lateral capacity of the pile group directly.

5.2 Deformation of pile groups

Over the last decade, advances in analytical techniques and careful experimental research have combined to provide a much improved understanding of the manner in which pile groups deform under load. The traditional approach of replacing the pile group by an equivalent raft foundation in order to estimate settlements, has been replaced by techniques where the group stiffness may be calculated in terms of the combined stiffness of the individual piles, making due allowance for interaction between piles in the group.

Analysis of the response of pile groups became feasible with the growth of numerical methods of analysis, in particular the integral equation, or boundary element method (Banerjee and Butterfield, 1983). Application of the numerical techniques to pile group analysis has been described in detail by Poulos and Davis (1980), who provide useful design charts showing how the response of a pile group is affected by parameters such as the number and size of piles, the relative stiffness of the piles and the soil and so forth. Butterfield and Douglas (1981) have produced a similar collection of charts giving the stiffness of common forms of pile groups under vertical and horizontal loading.

In practical terms, one of the most useful concepts to emerge from the analytical work is the use of interaction factors. An interaction factor, α, is defined as the fractional increase in deformation (that is deflection or rotation at the pile head) of a pile due to the presence of a similarly loaded neighbouring pile. Thus, if the stiffness of a single pile under a given form of loading is k, then a load, P, will give rise to deformation δ, given by

$$\delta = P/k \qquad (5.7)$$

If two identical piles are each subjected to a load, P, then each pile will deform by an amount, δ, given by

$$\delta = (1 + \alpha)P/k \qquad (5.8)$$

The value of α will of course depend on the type of loading (axial or lateral), and on the spacing of the two piles and the pile and soil properties.

The use of interaction factors may be regarded as equivalent to superimposing the separate deformation fields that each pile would give rise to by itself, in order to arrive at the overall deformation. This approach, illustrated in Figure 5.7 for axial loading, has been justified experimentally from carefully conducted tests on instrumented piles at field scale (Cooke et al., 1980), and also from model pile tests (Ghosh, 1975; Abdrabbo, 1976).

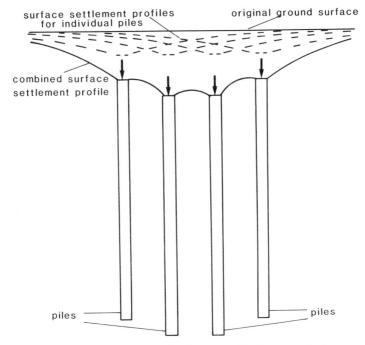

Figure 5.7 Superposition of settlement profiles for group of piles

The following sections describe how the stiffness of pile groups under axial and lateral loading may be evaluated from the stiffness of single piles and the use of appropriate interaction factors.

5.2.1 Axial loading

Randolph and Wroth (1979) describe how the solution for the axial load deformation response for single piles may be extended to deal with pile groups. Theoretical interaction factors are deduced from the form of the deformation field around a single pile, treating the pile shaft separately from the pile base. The basic approach will be summarized here, before giving results for typical pile group configurations for use in design. Under axial loading, the deformation field around the pile shaft varies approximately logarithmically with radius, r, according to

$$w = \frac{\tau_0 r_0}{G} \ln (r_m/r) \qquad (5.9)$$

(see section 4.2.1). Thus the shaft interaction factor, α_s, for a pile a distance s from the pile in question will be

$$\alpha_s = \ln (r_m/s)/\zeta = 1 - \ln (s/r_0)/\zeta \qquad (5.10)$$

where $\zeta = \ln(r_m/r_0)$ and $\alpha_s \geqslant 0$. At the level of the pile base, the deformation field due to the base load may be approximated by (Randolph and Wroth, 1979)

$$w = w_b \frac{2r_b}{\pi r} \tag{5.11}$$

giving rise to a base interaction factor, α_b, for a pile at spacing s, of

$$\alpha_b = \frac{2}{\pi} \frac{r_0}{s} \tag{5.12}$$

For groups of piles symmetrically placed around a pitch circle, where each pile responds identically, the load settlement response of the ith pile may now be obtained directly from equation (4.43), but with the factor ζ replaced by ζ^* given by

$$\zeta^* = \zeta \sum_{j=1}^{n} (\alpha_s)ij \tag{5.13}$$

where $(\alpha_s)_{ii}$ is taken as unity, and the factor η replaced by

$$\eta^* = \eta \bigg/ \sum_{j=1}^{n} (\alpha_b)ij. \tag{5.14}$$

This approach has been discussed by Randolph (1979) in relation to the response of offshore pile groups to axial loading.

One of the features of the use of different interaction factors for the pile shafts and the pile bases is that the load sharing between shaft and base will be different for piles in a group compared with single piles. Interaction between pile bases is much less than between the shafts, and this results in more load being transmitted to the base of piles in a group than to single piles—a result confirmed by model pile tests (Ghosh, 1975).

For pile groups where the piles are not all subjected to the same load, for example under a rigid pile cap, the above approach must be modified in the manner described by Randolph and Wroth (1979). The matrix manipulation involved will generally require the use of a computer. For design purposes, charts may be prepared from which the load settlement ratio of commonly occurring pile groups may be estimated.

For a given load per pile, piles in a group will settle more than a corresponding isolated pile. The increase in settlement may be quantified by a settlement ratio, R_s, giving the ratio of the flexibility of a pile in the group to that of an isolated pile (Poulos and Davis, 1980). Settlement ratios for piles in square groups containing up to 121 piles are shown in Figure 5.8 for typical pile and soil properties (taking $\lambda = 1000$, $l/d = 25$, $v = 0.3$). Curves of settlement ratio, R_s, against the square root of the number of piles in the group, n, are shown for key piles in uniformly loaded pile groups, and also

(a) $\rho = 1$

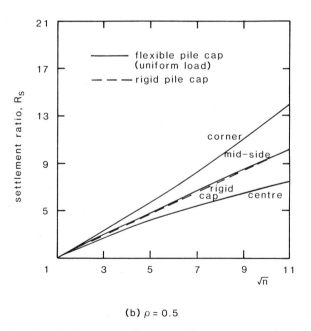

(b) $\rho = 0.5$

Figure 5.8 Settlement ratios for square pile groups under axial load for the case of pile groups loaded through a rigid pile cap. Results are shown for homogeneous soil ($\rho = 1$) and for soil with stiffness proportional to depth ($\rho = 0.5$). It may be seen that the degree of interaction (and hence the computed settlement ratio) is much less for the latter case. It is also interesting to find that the average settlement ratio for a pile group with a rigid pile cap, is virtually identical to that for the mid-side pile in a uniformly loaded group.

A convenient way of regarding the effects of interaction within a pile group has been suggested by Butterfield and Douglas (1981). The stiffness, K, of the pile group (load divided by settlement) may be expressed as a fraction η_w of the sum of the individual pile stiffnesses, k. Thus for a group of n piles,

$$K = \eta_w nk \qquad (5.15)$$

The factor η_w is the inverse of the settlement ratio, R_s, and may be thought of as an efficiency (the subscript w denoting settlement, to distinguish it from efficiency in terms of capacity). For no interaction between piles, η_w would equal unity. Butterfield and Douglas showed that plotting the efficiency η_w against the number of piles in a group gave essentially straight lines on logarithmic axes. The precise layout of the piles appeared to have little influence on the computed efficiency, rectangular groups of piles having the same efficiency as square groups at the same pile spacing.

Figures 5.9 and 5.10 show typical results for square pile groups of n piles, taking basic parameters of $s/d = 3$, $l/d = 25$, $\lambda = 1000$, $\rho = 0.75$, and $v = 0.3$. In Figure 5.9, the effect of varying the pile spacing is shown, while in Figure 5.10, the pile stiffness ratio λ is varied. Results have been computed

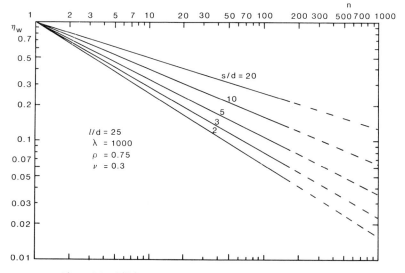

Figure 5.9   Efficiency of square pile groups at different spacing

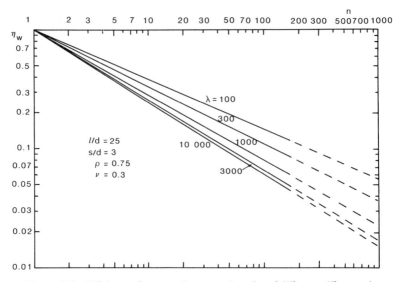

Figure 5.10   Efficiency of square pile groups for piles of different stiffness ratio

using the program PIGLET (see following section) for groups of up to 289 piles (17 × 17). The curves have then been extrapolated as dashed lines up to $n = 1000$. It may be seen that for large pile groups, the efficiency of each pile falls to 0.05 and lower, indicating that each pile is only 5% as stiff as it would be as an isolated pile. From Figure 5.9, it may be seen that a few widely-spaced piles may be nearly as stiff as many more closely-spaced piles. Thus a group of 400 piles at $s/d = 3$ has an efficiency of 0.0375, or an overall stiffness equivalent to $400 \times 0.0375 = 15$ isolated piles. A group of 36 piles at $s/d = 10$ (covering the same ground area) has an efficiency of 0.25, or an overall stiffness equivalent to 9 isolated piles. Reducing the number of piles by an order of magnitude gives a group stiffness reduced by only 40%.

Since the curves of efficiency against number of piles are approximately straight on logarithmic axes, the efficiency may be written as

$$\eta_w = n^{-e} \tag{5.16}$$

where the exponent e will lie between 0.4 and 0.6 for most pile groups. The actual value of e will depend on:

pile slenderness ratio, l/d,
pile stiffness ratio, $\lambda = E_p/G_l$,
pile spacing ratio, s/d,
homogeneity of soil, characterised by ρ,
Poisson's ratio, v.

For a given combination of the above factors, the value of e may be estimated

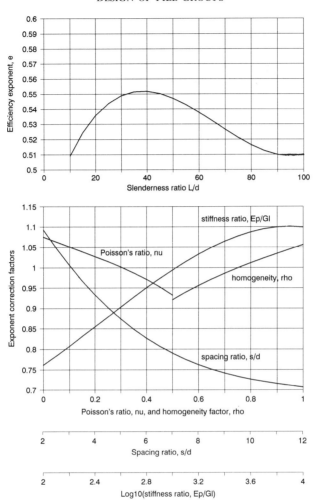

Figure 5.11 Charts for calculation of exponent *e* for efficiency of pile groups

using the curves shown in Figure 5.11. The upper part of the figure allows a base value of *e* to be chosen, depending on pile slenderness ratio (assuming $\lambda = 1000$, $s/d = 3$, $\rho = 0.75$, $v = 0.3$). The four curves in the lower part of the figure then modify this basic value according to the actual value of pile stiffness ratio, pile spacing, ρ and v.

The curves have mainly been derived by keeping four out of the five variables listed above constant (at suitable average values) and then varying the fifth variable. The average values adopted were $l/d = 25$, $\lambda = 1000$, $s/d = 3$, $\rho = 0.75$ and $v = 0.3$. Correction factors for the exponent *e* obtained from the curves will be only approximate for other values of the main parameters.

However, estimates of the final value of e for a given pile group should generally be within 5%, allowing the group efficiency to be estimated to within 10 to 20%, depending on the size of the group.

As an illustration of these curves in design, the stiffness of a 10×10 pile group will be calculated. The pile and soil properties will be taken as follows:

pile length: $l = 25\,\mathrm{m}$
pile diameter: $d = 0.6\,\mathrm{m}$
pile modulus: $E_p = 25 \times 10^3\,\mathrm{MN/m^2}$
pile spacing: $s = 2\,\mathrm{m}$
soil shear modulus: $G = 10 + 1.2z\,\mathrm{MN/m^2}$
soil Poisson's ratio: $v = 0.2$.

From equation (4.43), the stiffness of a single pile may be calculated as $P_t/w_t = 400\,\mathrm{kN/mm}$. From the first curve in Figure 5.11, the base value of e for $l/d = 41.7$ (assuming stiffness ratio of $\lambda = 1000$, at spacing of $s/d = 3$, $\rho = 0.75$ and $v = 0.3$) is 0.55. The four correction factors, which allow for the actual pile stiffness ratio, spacing, and values of ρ and v, are 0.95, 0.97, 0.96, and 1.03 respectively, giving a final value of $e = 0.50$. From equation (5.16), the group efficiency is thus $100^{-0.50} = 0.10$, giving an overall stiffness of $100 \times 0.10 \times 400 = 4.0\,\mathrm{MN/mm}$.

Load distribution. A consequence of the interaction between piles is that, for uniformly loaded pile groups, the central piles will settle more than the piles at the edge of the group, while, for piles loaded though a stiff pile cap, the edge piles will take a higher proportion of the applied load than the central piles. This feature has been demonstrated experimentally by Cooke *et al.* (1980). Typical load distributions calculated for small pile groups (Poulos and Davis, 1980) show that corner piles may take up to 3 to 4 times as much load as piles near the centre. A case history reported by Cooke *et al.* (1981) showed that piles at the corner of a large pile group containing 351 piles took 2 to 3 times as much load as the central piles.

There has been some confusion among engineers as to how to react to these load distributions. There are a number of alternatives.

(i) Lengthen the outer piles to ensure their stability under the higher loads.
(ii) Abandon the central piles since they contribute little to the stiffness of the pile group.
(iii) Lengthen the central piles in order to attract more load to them, thus evening out the load distribution among the piles.

The first alternative does not stand up to scrutiny since longer outer piles will attract even more load to the edge of the foundation, giving rise to higher bending moments in the pile cap—or alternatively greater differential settlements. For the same reason, the second alternative is not attractive, unless the weight of the structure is concentrated at the edges of the foundation. In principle, the last alternative appears a possible option.

Assuming that the loading from the superstructure is reasonably uniformly distributed, the aim of the foundation must be to transmit the load as uniformly as possible to the soil. Ideally, bending moments in the pile cap and differential settlements should both be minimized. In general, this will require more support from piles towards the centre of the foundation than towards the outside (to counteract the natural tendency of a uniformly loaded region to dish in the centre). The approach becomes particularly attractive where the upper soil layers are reasonably competent, and the pile cap may transmit some load directly to the ground. Padfield and Sharrock (1983) discuss such a design, where differential settlements of a raft foundation could be virtually eliminated by the use of a few piles beneath the centre of the raft. This is discussed in more detail in section 5.3.

In many cases, the stiffness of the structure above the pile cap will be such as to obviate the need for concern over high loads in the peripheral piles. The simple approach of uniformly distributed piles will then be adequate. The design should ensure that the pile material itself is not overstressed, and that due account is taken of possible high bending moments induced at the edge and corners of the pile cap.

Immediate and long-term settlements. It may be seen from the lower part of Figure 5.11, that the correction factor for the exponent, e, decreases with increasing Poisson's ratio. Thus the efficiency of piles in a group decreases as Poisson's ratio becomes smaller. This has consequences for the ratio of immediate to long-term settlement of pile groups. The solution for a single pile is relatively insensitive to the value of Poisson's ratio (for a given value of the shear modulus, G). Thus the immediate settlement of a single pile typically accounts for 80 to 90% of the long-term settlement (Poulos and Davis, 1980). However, for a group of piles, the immediate settlement is a smaller fraction of the long-term settlement.

For the example given above, the stiffness of a single pile for Poisson's ratio of 0.5 (corresponding to the undrained, or immediate settlement) is 433 kN/mm. Thus the immediate settlement would be 92% of the long-term value calculated with $v = 0.2$. For the pile group, the correction factor for Poisson's ratio of 0.5 would be 0.93, instead of 1.03 for $v = 0.2$, leading to a group efficiency of 0.126. The undrained stiffness for the group is thus 5.5 MN/mm and the immediate settlement would be only 73% of the long-term settlement.

Underlying soft layers. The effect of different layers of soil over the depth of penetration of the piles in a group may generally be dealt with adequately by adopting suitable values of the average shear modulus for the soil, and a value for the homogeneity factor, ρ, which reflects the general trend of stiffness variation with depth. However, particular attention needs to be paid to the case where a soft layer of soil occurs at some depth beneath the pile group, as shown in Figure 5.12. In assessing how much additional settlement may occur due to the presence of the soft layer, the average change in vertical stress caused by the pile group must be estimated.

Implicit in the solution for the load settlement response of a single pile is the idea of the transfer of the applied load, by means of induced shear stresses in the soil, over a region of radius r_m. The average vertical stress applied to soil at the level of the base of a group of piles may be estimated by taking the overall applied load and distributing it over the area of the group augmented by this amount, as shown in Figure 5.12. Below the level of the pile bases, the spread of the area over which the load is assumed to be shared may be taken as the usual rate of 1:2 (Tomlinson, 1970).

Long piles. It was noted in section 4.2.4 that long slender piles may mobilize their full skin friction over the upper part of each pile at working loads. In estimating the settlement of groups of such piles, allowance for slip between pile and soil must be made. Settlement ratios, R_s (the inverse of the efficiency, η_w), should only be applied to the 'elastic' component of the single pile settlement. An example of this has been given by Thorburn *et al.* (1983) in relation to long piles (0.25 m square by 30 m long) used to support a molasses storage tank. A pile load test showed that a single pile would settle by 8 mm, at the working load. Of this, some 4 mm was estimated as relative slip between the upper part of the pile and the soil. Analysis of the pile group indicated a group settlement ratio of $R_s = 6.4$. Thus the working settlement of the foundation was estimated as 6.4 times the 'elastic' component of the single-pile settlement (4 mm) plus the 'plastic' settlement (4 mm), giving 30 mm, which agreed well with actual measured settlements. A straightforward factoring

Figure 5.12　Use of equivalent raft for calculating effect of soft layer underlying pile group

of the total single pile settlement of 8 mm would have given estimated settlements of over 50 mm, 60% greater than measured.

5.2.2 Lateral loading

The response of pile groups to lateral loading involves not only the lateral load deformation characteristics of the individual piles, but also the axial characteristics. As the pile group tries to rotate, piles at the edges of the group will be loaded in tension and compression, providing considerable rotational stiffness to the group. Computer-based methods such as those discussed in the following section are generally required for a complete analysis of the group under such loading, although an approach suitable for hand calculations has been described, in relation to offshore pile groups, by Randolph and Poulos (1982).

As discussed in Chapter 4, a distinction should be drawn between external lateral loads applied to a pile group, and loads induced in the piles by soil movement below ground level. The situation of a piled bridge abutment (see Figure 4.47(b)) is a common one, and due account must be taken of the effect of the embankment surcharge loading when estimating the deformation of the abutment and the load distribution in the piles.

For external load applied to the pile head, Poulos (1971) introduced four different kinds of interaction factor for piles under lateral load, depending on the loading at the pile head and the type of deformation. These were

α_{uH}: interaction for deflection of free-headed piles under force loading;

α_{uM}: interaction for deflection of free-headed piles under moment loading or for rotation of free-headed piles under force loading;

$\alpha_{\theta M}$: interaction for rotation of free-headed piles under moment loading; and

α_{uf}: interaction for deflection of fixed-headed piles (restrained against rotation).

The last of these interaction factors is generally of most relevance in practice, since the rotational stiffness of pile groups is usually high.

Randolph (1981) has shown that interaction factors for fixed-headed piles may be estimated from the expression

$$\alpha_{uf} = 0.6\rho_c(E_p/G_c)^{1/7}\frac{r_0}{s}(1 + \cos^2\beta) \tag{5.17}$$

where ρ_c and G_c are as defined in section 4.4.2, and β is the angle of departure that the piles make with the direction of loading (see Figure 5.13). At close pile spacings, this expression tends to overestimate the amount of interaction, and it is suggested that, where the calculated value of α_{uf} exceeds 0.33, the

Figure 5.13   Definition of departure angle

(a) Homogeneous soil: $\rho_C = 1$

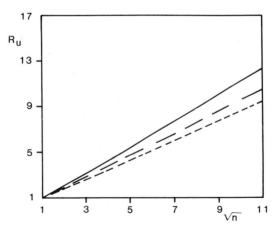

(b) Soil with stiffness proportional to depth:
$\rho_C = 0.5$

Figure 5.14   Deflection ratios for square fixed-head pile groups under lateral load value should be replaced by

$$\alpha_{uf} = 1 - \frac{2}{(27\alpha_{uf})^{1/2}} \qquad (5.18)$$

The form of equation (5.17) shows that interaction under lateral loading decreases much more rapidly with the spacing, s, than for axial loading. In addition, interaction of piles normal to the applied lateral load is only half that for piles in line with the load.

As discussed above, the lateral response of a pile group will depend on the axial stiffness of the piles in addition to their lateral stiffness, and general analysis of the group is necessary. Only if rotation of the pile cap (and thus axial movement of the piles) is prevented, do the piles deflect purely horizontally. For this case, it is possible to think in terms of a 'deflection ratio,' R_u, defined in a similar way to the settlement ratio, R_s, for vertical loading. Thus R_u is the ratio of the average flexibility of piles in a group, to that of a single pile, for lateral deflection under conditions of zero rotation at ground level. Figure 5.14 shows this ratio for square pile groups containing up to 121 piles at $s/d = 3$, for a range of pile stiffness ratios (represented by different values of slenderness ratio $s_c = l_c/r_0$—see Chapter 4). Typical values of deflection ratio are significantly lower than for vertical loading and decrease with decreasing stiffness (or slenderness) ratio. Again, values for $\rho_c = 0.5$ are well below those for homogeneous soil ($\rho_c = 1$), reflecting the lower interaction between piles for soil where the stiffness is proportional to depth.

Limitations of the elastic approach to pile interaction under lateral loading have been revealed from model pile tests carried out on a centrifuge by Barton (1982). These tests showed that leading piles of groups embedded in sand carried a higher proportion of the overall applied load than the trailing piles. This arises out of the effects discussed in section 5.1.2, whereby the limiting pressure on a pile lying in the wake of another pile is reduced. For pairs of piles at close spacing ($s/d = 2$), Barton found that the applied load was shared 60% to the front pile and 40% to the back pile. In contrast, Matlock *et al.* (1980) report results of tests on pile groups in clay, where a uniform sharing of the applied load was found. Clearly, the adequacy of the elastic approach will depend on a number of factors such as the load level, soil type and pile spacing. However, at present there is little alternative for routine design work.

5.2.3 *Computer programs for pile group analysis*

The complexity of the problem of a pile-reinforced soil mass has necessitated the use of computer-based methods of analysis. In many cases, the results of the computations are amenable to presentation in graphical, or equation form, and examples of these have been given in the preceding sections and in Chapter 4. However, estimation of the deformations and bending moments induced in a group of piles under general loading conditions will usually require computer analysis. Even the relatively simple problem of the axial response of a pile group may still require analysis by computer, if the geometry of the group falls outside the range of charts such as those given in this chapter or by other sources (for example, Poulos and Davis, 1980).

It is important to allow for the axial and lateral support of the soil in the analysis of the pile group, even where the piles are primarily end-bearing. The simplest approach is to use a standard (structural) frame analysis package, where the piles are replaced by equivalent cantilevers (Figure 5.15). Each cantilever is assumed to have the same bending rigidity as the pile it replaces, with the length chosen according to the criterion discussed in section 4.4.3 (approximately half the critical pile length under lateral loading). The cross-sectional stiffness of the cantilever is then chosen so as to give an appropriate axial stiffness (for example, by comparison with the axial stiffness calculated from equation (4.43)).

While the equivalent cantilever approach (which is essentially the same as the 'stiffness' or 'elastic centre' approach referred to in the literature) allows structural frame analysis packages to be utilized, its main limitation is that no allowance is made for interaction between piles; also, preparation of the data in order to apply the method can be rather laborious. As discussed in section 4.4.3, the equivalent cantilever idealization will generally overestimate the maximum bending moments in the piles.

An alternative and preferable approach is to use purpose-written computer programs, where the pile and soil properties may be input directly. While there are a host of individual computer programs written for pile group analysis, mention will be made here of three that are currently widely available, and for which some comparison of results has been undertaken. The three differ in the degree of rigour, and numerical complexity which is adopted and only a very brief description will be given here.

Figure 5.15 Frame analysis of pile group with piles treated as pin-ended struts

(1) *PGROUP*. Originally written by Banerjee and Driscoll (1976), this program is managed and updated by Highway Engineering Computer Branch, Department of Transport, UK. It is based on a full boundary element analysis of the complete pile group. The soil is modelled as an elastic material, and later versions of the program allow for linear variations of the soil modulus with depth. The effect of a pile cap in contact with the ground may be analysed and the piles may be raked in the plane of loading. This program is, in principle, the most rigorous of the three programs discussed, but needs considerable computing resources. The relatively coarse discretizing of the piles can lead to questionable accuracy in the results, particularly for laterally loaded piles.

(2) *DEFPIG*. Written by Poulos (1980), the program is based on integral equation techniques for the analysis of single piles and the computation of interaction factors between pairs of piles. For pile groups, these interaction factors are used together with the single pile stiffnesses to yield an overall stiffness for the group. Variations in soil stiffness with depth, and allowance for yielding of the soil, are catered for in an approximate fashion. Pile cap elements may be included for vertical loading, and piles may be raked in the plane of loading. Initial calculation of the individual pile stiffnesses and interaction factors involves a large amount of computing. Thereafter, groups of similar piles may be analysed relatively inexpensively.

(3) *PIGLET*. Written by Randolph (1980), the program is based on the approximate solutions for single piles, and expressions for interaction between piles, described in this and the previous chapter. The soil is modelled as an elastic material, with a linear variation of stiffness with depth. Piles may be raked in any direction, and the effect of any type of loading, including torsion of the group, analysed. Since the program uses previously derived solutions, computing time is insignificant, although analysis of large pile groups requires significant amounts of storage.

A detailed comparison of the results computed by the latter two programs has been presented by Poulos and Randolph (1983) (some results obtained using PGROUP are also included). As an illustration of the degree of agreement between the two programs, and also to show the type of parametric study which such programs may be used for, the effect of pile rake on a simple 3×2 pile group is presented here. The pile group details are given in Figure 5.16, the angle of rake of the piles being denoted by ψ. A flexibility matrix for the group may be written

$$
\begin{bmatrix} w \\ u \\ \theta \end{bmatrix} = \begin{bmatrix} F_{wP} & F_{wH} & F_{wM} \\ F_{uP} & F_{uH} & F_{uM} \\ F_{\theta P} & F_{\theta H} & F_{\theta M} \end{bmatrix} \begin{bmatrix} P/Gr_0 \\ H/Gr_0 \\ M/Gr_0^2 \end{bmatrix} \qquad (5.19)
$$

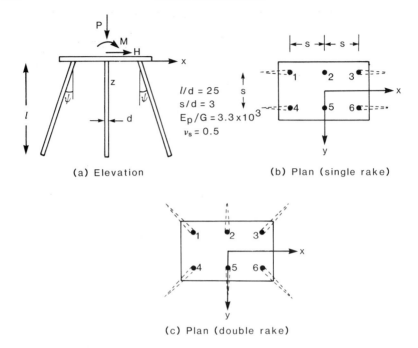

(a) Elevation

(b) Plan (single rake)

(c) Plan (double rake)

Figure 5.16   Pile group analysed in comparative study

Since the group is symmetric, the terms in F_{wH}, F_{wM}, F_{uP} and $F_{\theta P}$ will be zero. Also, from the reciprocal theorem, F_{uM} must equal $F_{\theta H}$, giving only four independent flexibility coefficients.

The variation of these coefficients with the angle of rake, ψ, is shown in Figure 5.17. Both programs give very similar results, the largest discrepancy being 18%. The most striking effect of varying the angle of rake of the corner piles is the change in sign of the group rotation under horizontal loading; groups of vertical piles rotate in the direction of loading, but the inclusion of piles raked at even modest angles can reduce this rotation and even reverse the sense of it. The effect of double rake of the piles is relatively small in the plane of loading.

5.2.4 Elastic properties of soil

Elastic idealization of the non-linear stress-strain response of soil is a practical necessity for estimating the deformation of foundations under working load conditions. Provided appropriate secant values are chosen, and additional calculations are undertaken to ensure that failure, or relative slip between pile and soil, has not occurred, this approach should be satisfactory. The usefulness of the approach, however, rests on the degree of accuracy with which values of soil modulus may be chosen.

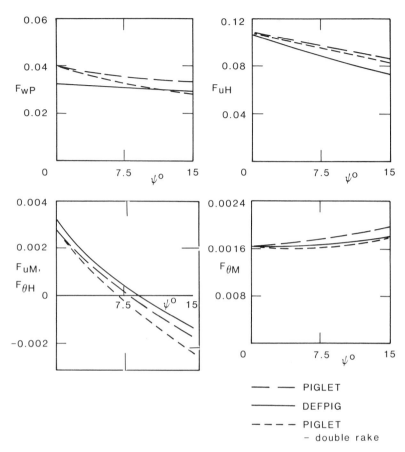

Figure 5.17 Variation of group deflection influence factors with angle of rake ψ

Modern site investigation techniques such as the pressuremeter and the screw plate test enable much better estimates to be made of the in-situ stiffness of the soil than those generally obtained from laboratory triaxial tests. Where such test results are not available, correlations must be relied upon, where the soil modulus is related to other properties or test results.

Wroth *et al.* (1979) have reviewed correlations of the shear modulus, G, with other soil properties, and have discussed the effect on the shear modulus of factors such as strain level, overconsolidation ratio, mean effective stress level and so forth. They conclude that, for most practical foundation analysis, the model of soil stiffness varying proportionally with depth is the most appropriate. In sands, this reflects the increase with depth of the mean effective stress level, while in clay and soft rocks, weathering near ground level will result in low values of soil stiffness.

For clays, typical values of G/p' (where p' is the mean effective stress level) range from 100 for normally consolidated clay, up to 200 at high values of overconsolidation ratio (OCR), the increase being essentially linear with the logarithm of OCR. These ratios would imply values of G/c_u which range from over 300 at OCR = 1, down to 50 at high OCR. In sands, the ratio G/p' appears to be much more sensitive to strain level. For strain levels relevant to pile groups at working load levels (say, 0.1%), the correlations

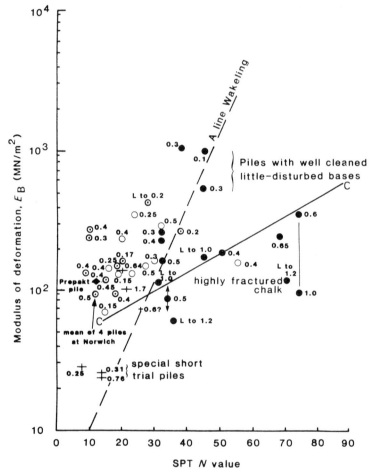

Figure 5.18 Correlation of Young's modulus for chalk against SPT value. ● Bored piles Category 4, ⊙ Driven piles category 3(a), ○ Driven piles Category 3(b) + Driven piles Category 2(b) (West Shell Piles, Portsmouth), C–C Slope derived from calculations to obtain a reasonable relationship between modulus, N value and depth.
Numerical values are stress on base (MN/m²) at the end of linear (L) portion of these curves. (Hobbs and Healy, 1979)

suggested by Wroth *et al.* give

$$G/p_a = 300 \frac{p'}{p_a} \frac{0.75(2.17 - e)^2}{(1 + e)} \qquad (5.20)$$

where p_a denotes atmospheric pressure and e is the voids ratio of the sand. Typically, this expression will lead to ratios, G/p', which lie in the range 300 to 600.

Correlations are also suggested between shear modulus, G, and results from standard penetration tests, N. Reducing the quoted correlation (due to Kovacs, 1975), for dynamic moduli, by a factor of 4 to allow for the higher strain level of 0.1%, gives

$$G/p_a = 40N^{0.77} \qquad (5.21)$$

This correlation will generally give somewhat higher values of shear modulus than the approximate correlation suggested by Randolph (1981) of

$$G/p_a = 10N \qquad (5.22)$$

In soft rocks, the appropriate stiffness will be greatly affected by the degree of weathering, and by the method of pile installation. Hobbs and Healy (1977) have correlated values of soil modulus, back analysed from pile load tests in chalk, with the SPT number, N. Their correlation is shown in Figure 5.18. There is considerable scatter in the results, much of it reflecting varying amounts of disturbance due to pile installation. Even with good quality chalk, pile driving can loosen the block structure and reduce the stiffness by an order of magnitude. Randolph and Wroth (1978) quote values of shear modulus, deduced from load tests on driven cast-in-situ piles end-bearing on good quality chalk, which are as low as 36 MN/m² for grade I–II chalk.

In choosing an appropriate soil modulus for pile design, consideration should also be given to the very high strain levels which exist around laterally loaded piles. In most cases, it may be prudent to adopt shear modulus values for lateral loading of piles, which are a factor of 2 lower than for axial loading. Alternatively, the soil shear modulus may be taken as varying from zero at the ground surface (where the strains are highest) up to the full value taken for axial loading, at a depth equal to the critical pile length.

5.3 Piled rafts

As discussed at the start of this chapter, where competent soil exists close to the ground surface, a combined pile and raft foundation may provide the most economic form of foundation. Early analytical work by Butterfield and Banerjee (1971) suggested that, although the proportion of load taken by the piles would drop, inclusion of a ground-contacting pile cap would have little effect on the stiffness of a pile group. This finding has been substantiated by experimental work (Abdrabbo, 1976; Cooke et al., 1979). The converse, however, is that the inclusion of relatively few piles beneath a raft foundation can substantially increase the stiffness of the foundation.

Numerical techniques for the detailed analysis of piled raft foundations, including assessment of bending moments and differential settlements, have been described by Hain and Lee (1978) and also by Padfield and Sharrock (1983) (although treatment of the piles in the latter approach is only approximate). Approximate treatments of mixed foundations are also available for rigid rafts using the programs PGROUP and DEFPIG, discussed in section 5.2.3. New numerical techniques, which include allowance for bending of the raft, have been presented by Griffiths et al. (1991).

A simple approach to combining the separate stiffnesses of the raft and the pile group has been suggested by Randolph (1983). The approach is based on the use of average interaction factors, α_{cp}, between the pile and pile cap. Writing the pile cap stiffness as k_c and the pile group stiffness as k_p, the overall foundation stiffness, k_f, is obtained from

$$k_f = \frac{k_p + k_c(1 - 2\alpha_{cp})}{1 - \alpha_{cp}^2 k_c/k_p} \qquad (5.23)$$

while the proportion of load carried by the pile cap (P_c) and the pile group (P_p) is given by

$$\frac{P_c}{P_c + P_p} = \frac{k_c(1 - \alpha_{cp})}{k_p + k_c(1 - 2\alpha_{cp})} \qquad (5.24)$$

The stiffnesses of the pile cap and pile group may be evaluated conventionally, while the interaction factor, α_{cp}, may be obtained from the approximate expression (Randolph, 1983)

$$\alpha_{cp} = \frac{\ln(r_m/r_c)}{\ln(r_m/r_0)} = 1 - \frac{\ln(r_c/r_0)}{\zeta} \qquad (5.25)$$

where r_m is the radius of influence of the piles (see section 4.2.1) and r_c is the effective radius of the element of pile cap associated with each pile. This may be calculated so that, for a group of n piles, $n\pi r_c^2$ equals the actual area of the pile cap.

Figure 5.19 shows how the stiffness of a single pile under vertical load is increased by the presence of a pile cap in contact with the ground. The approximate method outlined here is compared with results from Poulos and Davis (1980) and shows good agreement. The results show that a small pile cap ($r_c/r_0 \leqslant 4$) has very little effect on the stiffness of the foundation, although a significant proportion of the overall load may be carried by the pile cap (see Figure 5.20). Comparison of the approximate analysis with a more rigorous numerical analysis for the case of a 5×5 pile group shows

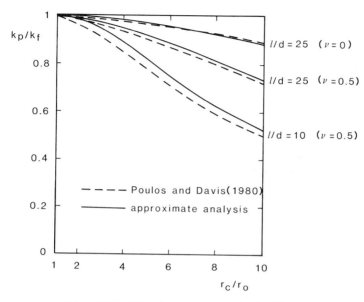

Figure 5.19   Effect of pile cap on single pile stiffness

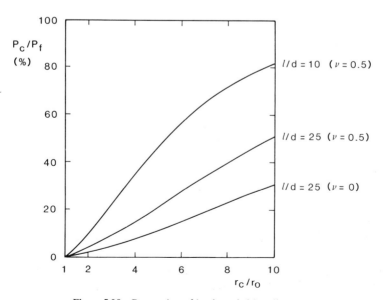

Figure 5.20   Proportion of load carried by pile cap

good agreement in terms of overall stiffness (within 6%), but a tendency for the approximate analysis to underpredict the amount of load carried by the raft by up to 30%, especially where the pile–soil stiffness ratio is low (Griffiths *et al.*, 1991).

As an example application of the approach outlined for a complete raft, the foundations for a 16-storey block of flats built on London Clay will be considered. The building, at Stonebridge Park in the London borough of Brent, was founded on a pile group consisting of 351 bored piles, each 0.45 m in diameter and 13 m long, capped by a pile cap in direct contact with ground (see Figure 5.21). The pile cap and piles were extensively instrumented to measure contact stresses between the raft and the ground, and the loads

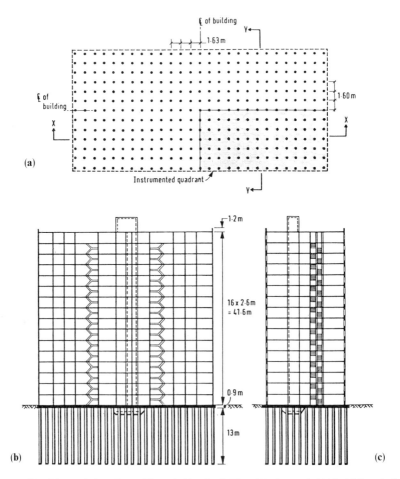

Figure 5.21   Plan and elevations of Stonebridge Park Flats (Cooke *et al.*, 1981): (a) foundation plan (351 piles 450 mm diameter and 13 m long); (b) longitudinal section x–x; (c) cross-section y–y

going into typical piles in the group. Results of the field study, and implications for the design of such foundations, have been presented by Cooke et al. (1981). A numerical analysis has also been undertaken by Padfield and Sharrock (1983).

The shear strength profile for the site may be approximated by the straight-line relationship

$$c_u = 100 + 7.2z \, \text{kN/m}^2 \qquad (5.26)$$

where z is the depth in m below foundation level (2.5 m below ground level). Adopting a ratio for G/c_u of 200 (Simpson et al., 1979) for the London Clay, the shear modulus profile is given by

$$G = 20 + 1.44z \, \text{MN/m}^2 \qquad (5.27)$$

The overall dimensions of the pile cap were 20.1 m by 43.3 m. Thus the average stiffness of the cap, acting alone, may be estimated (Poulos and Davis, 1974) as

$$k_c = \frac{2G}{I(1-v)} \sqrt{20.1 \times 43.3} \qquad (5.28)$$

where the influence factor I is 0.85 for a raft of this aspect ratio. The value of G may be taken as the value at a depth of $B(1 - 0.5B/L)$, where B and L are the breadth and length of the pile cap. This gives $G = 42 \, \text{MN/m}^2$, and a cap stiffness (taking $v = 0.1$) of $k_c = 3250 \, \text{kN/mm}$.

The stiffness of the pile group by itself may be estimated, using programs such as PIGLET, as $k_p = 5\,300 \, \text{kN/mm}$ (Randolph, 1983). The piles were laid out on a rectangular grid, with spacing of 1.60 m by 1.63 m. Thus each pile had an area of 2.61 m² of cap associated with it, from which the effective radius of each cap element may be calculated as 0.91 m. The radius of influence of each pile is $r_m = 22.2 \, \text{m}$ (from equation (4.39), taking $v = 0.1$). Equation (5.25) then gives $\alpha_{cp} = 0.70$. The overall stiffness of the foundation may then be calculated as $k_f = 5\,720 \, \text{kN/mm}$ with 24% of the load being carried by the pile cap (equations (5.23) and (5.24)).

Cooke et al. (1981) report an overall load on the foundation of 156 MN, from which the estimated average settlement may be calculated as $156/5.72 = 27 \, \text{mm}$. The field measurements indicate a long-term settlement of about 25 mm, with the proportion of load carried by the pile cap dropping from an initial value of about 50%, down to a long-term value of 23%. These values compare well with the estimated values obtained above.

Randolph (1983) comments that the piles in this foundation design were only contributing about 5% of their possible stiffness as individual piles. Halving the number of piles under the foundation would have reduced the pile group stiffness by only 13%, giving an estimated average settlement of 29 mm for the building, with some 31% of the load being taken by the pile cap.

Settlement reducing piles. Where a primarily raft foundation is planned, a few piles may be incorporated beneath the central part of the raft, in order to reduce the differential settlements to an acceptable level (Burland *et al.*, 1977). Such piles provide some measure of reinforcement to the soil and help prevent dishing of the raft in the centre (see Figure 5.22). From a design point of view, they may be regarded as absorbing some part of the overall load applied to the raft. Under working conditions, the full shaft capacity of the piles may be mobilized over most of the length, excluding a region just beneath the raft where the shear transfer will be reduced (Combarieu and Morbois, 1982).

Padfield and Sharrock (1983) discuss an alternative design of the foundation for Stonebridge Park (see Figure 5.21) which uses as few as 40 piles, situated near the centre of the raft, at a spacing of 3.2 m (7.1 pile diameters). Such a scheme would appear to eliminate virtually all differential settlements, even for a flexible raft. An estimate of the average settlement of the foundation might proceed as follows.

The shaft capacity of each pile was estimated by Padfield and Sharrock as 1200 kN. Assuming each of the 40 piles mobilizes most of its shaft capacity, say 1000 kN to allow for a zone of reduced shear transfer just beneath the raft, the net load on the pile cap would be $156 - 40 = 116$ MN, giving an average settlement of $116/3.25 = 36$ mm. An additional component of settlement due to the load carried by the piles may be estimated in the manner described at the end of section 5.2.1. Assuming slip occurs between pile and soil at a movement of 1% of the pile radius, i.e. 2.3 mm, and calculating the settlement ratio, R_s, for the group as 7.0 (Randolph, 1983), the overall elastic settlement in the soil close to each pile will be $7.0 \times 2.3 = 16$ mm. The average settlement between the piles may be obtained

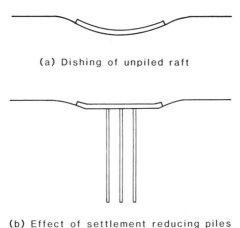

(a) Dishing of unpiled raft

(b) Effect of settlement reducing piles

Figure 5.22   Use of settlement reducing piles to minimize differential settlement by multiplying this figure by the interaction factor α_{cp}, which, for this new group, is 0.54 (r_c is now $2 \times 0.91 = 1.82$ m). This gives an average additional settlement of 9 mm, and an overall settlement of $36 + 9 = 45$ mm. This compares with a settlement of 49 mm computed by Padfield and Sharrock.

It should be emphasized that the above method of calculation is only approximate, but it represents a physically meaningful approach to the design of raft foundations which use a few piles as settlement reducers, in the manner suggested by Burland *et al.* (1977). Settlement of the pile cap is estimated first under a reduced loading and is then augmented by an average 'background' settlement due to load taken by the piles. The above example serves to demonstrate that significant savings may be made in foundation design by eliminating piles which are not needed from a stability point of view. The design question becomes not 'how many piles are needed to carry the weight of the building', but 'how many piles are needed to reduce the (differential) settlements to an acceptable level'?

There are of course many considerations that come into the design of a piled foundation. Combined pile and raft foundations are only viable where the surface soil layers are reasonably competent, and even then the number and spacing of piles may well be determined by the spacing of the columns carrying the structural load. The use of one pile beneath each column is a convenient way of minimizing bending moments in the pile cap. However, the piles may still be designed to take only a proportion of the column loads, the remaining load being shed into a pile cap cast directly on the ground.

5.4 Discussion

Recent years have seen major advances in understanding the manner in which pile foundations interact with the surrounding soil. This and the preceding chapter have presented some of the modern analytical methods available, concentrating on those that may be used directly (that is without excessive computational effort) in design calculations.

The choice of piles of sufficient axial capacity is still generally the key issue in pile design. However, interpretation of the word 'sufficient' must be made in the light of the foundation requirements. In particular, settlement considerations now play a more dominant role in pile design. As will be discussed further below, the choice of what 'factor of safety' is appropriate for pile foundations will depend to a large extent on how much settlement may be tolerated.

Practical considerations, such as those discussed in Chapters 2 and 3, will play an important part in pile design, particularly as regards choice of pile type, installation method and even the overall layout of the pile group. The role of analysis is then to optimize the foundation and to assess, through parametric studies, the effect of uncertainties in assumed soil and other design parameters. It may be helpful to review here some of the main advances in analytical methods that have been made over the last decade or so.

(1) In calculating the axial capacity of a pile simple 'total stress' methods have been augmented by a better understanding of the effective stress changes that occur during and after pile installation. No complex pile design would now be complete without an assessment of the effective stress state implied by a total stress calculation of capacity.
(2) The concept of 'group efficiency', based on a comparison of the capacity of the individual piles in a group with that of the surrounding block, is generally considered inappropriate. Individual piles may be considered more or less efficient when part of a group, due to differences in the stress state around the pile. However, this is a separate consideration. The two modes, of individual pile and block failure, should be appraised independently.
(3) The concept of 'efficiency' of a group is more appropriately used in respect of the stiffness of the foundation. Solutions for the axial load settlement response for single piles, together with a method for estimating the average decrease in stiffness due to the presence of surrounding piles, enable the stiffness of a group of piles to be readily estimated. Such an approach obviates the need for the traditional 'equivalent raft' approach, which takes no account of the number or spacing of the piles in the group.
(4) The lateral response of piles may be estimated simply using a consistent approach, treating the soil as a continuum. This approach avoids uncertainty in the choice of coefficient of sub-grade reaction, and enables group effects to be assessed within the same framework.
(5) Computer methods of pile group analysis under general loading conditions are now widely available. The methods take proper account of the lateral support offered by soil around the piles and enable more realistic estimates to be made of the load distribution in the piles under working conditions. In particular, unnecessary use of raking piles to take any lateral component of load may be avoided.

5.4.1 Factors of safety

The question of what factor of safety should be applied in pile design deserves particular consideration. Traditionally, it has been customary to design piles with ultimate axial capacities between 2 and 3 times the required working load. For example, for bored piles, Burland and Cooke (1974) recommend the use of a factor of safety of 2.0 on the combined shaft and base capacity, or, for underreamed piles, the use of partial factors of unity on the shaft capacity and 3.0 on the base capacity.*

*A factor of safety of 2.0 is often deemed sufficient where test piles have been loaded to failure. However, a higher factor of safety, 2.5, is recommended where only proof loads are applied to working piles.

In many areas of design, the size of the factor of safety reflects the confidence with which the ultimate capacity may be estimated. However, in many circumstances the allowable settlement of a pile foundation will be the overriding criterion in design, and the factor of safety against collapse will not be relevant. In general, it will be found that a tight settlement criterion will lead to factors of safety which are in excess of 3.

In situations where the settlement is not a key issue, for example where piles are used to support storage tanks or other structures which may undergo substantial settlement without damage, then there are strong arguments for adopting relatively low factors of safety. Maintained load tests on piles generally show that the creep or consolidation settlement of the pile starts to increase significantly once the load reaches about 70% of the ultimate capacity. This finding may be used to advantage to control more precisely the load distribution within a group of piles. Thus, by designing piles with a factor of safety of 1.5, higher loads in piles at the edge or corners of a group may be avoided, since the stiffness of such piles will start to decrease at higher load levels. This approach relies on relatively uniform ground conditions.

In Sweden, there has been a recent move towards what is termed 'creep piling' (Hansbo and Jenderby, 1983). In essence, piles are designed at the load at which they will start to 'creep' significantly, which corresponds to a factor of safety of about 1.5. It has been found that the load distribution among the piles in a foundation may be controlled more precisely using this approach, giving resulting savings in the structural design of the pile cap.

In summary, then, it may be argued that the factor of safety should be determined by the required settlement characteristics of the structure. Low factors of safety may be used where large settlements may be tolerated, even to the extreme of a 'factor of safety' of close to unity, where piles are used as settlement reducers beneath a primarily raft foundation (see section 5.3). For large groups of piles, the settlement criterion will almost always dominate the design. Thus, provided there is a minimum factor of safety, which may be as low as 1.5, the design should be approached in terms of satisfying the settlement criterion. For another possible approach, see section 10.3.

5.4.2 Use of raking piles

Raking piles are often used to provide the necessary support to structures which are subjected to horizontal or inclined loads. The extent to which they are used is partly due to early analytical methods for estimating the load distribution in pile groups (see section 5.2.3), which ignored any lateral support offered by the pile. Modern analytical methods enable better estimates of the lateral performance of piles to be made, and more sophisticated analyses are now available for groups of piles. Such analyses indicate that, in certain circumstances, particularly where the vertical component of load is relatively small, raking piles may be used to advantage to reduce bending moments in the piles and to minimize horizontal deflections.

There are, however, certain circumstances where raking piles should be avoided.

(1) Raking piles should not be used in soil which is consolidating, since vertical movement of the soil may lead to overstressing the pile in bending.
(2) Raking piles should be avoided in any structure which may undergo significant vertical settlement. Again, where raking piles are forced to move vertically, the lateral component of movement may lead to overstressing of the pile.

Where raking piles *are* used, it should be borne in mind that the standards of alignment tolerance are generally lower than for vertical piles. Thus the centreline may deviate from the design by up to 1:25, compared with 1:75 for a vertical pile, and the position in plan of the pile will also be subject to a greater variation than for a vertical pile. The technical and practical difficulties in the use of raking piles are discussed further in Chapter 10 (section 10.3.2).

6 Retaining walls

6.1 General

Although for many years sheet piles have provided the commonest way of supporting deep excavations, there are many instances in which for reasons of vibration, or noise, or the sheer difficulty of driving, or because the piles can not be withdrawn for re-use, the use of sheet piling is less attractive than it might otherwise be.

The practical alternative solutions are to use contiguous reinforced concrete bored pile, secant pile, or diaphragm walls; the last-mentioned might in practice be regarded as a particular form of non-circular secant pile wall. The use of bored piles for a retaining wall often goes hand in hand with the use of bored piles for the support of the main structure on a site.

Contiguous piles are constructed in a line with a clear spacing between the piles of 75 to 100 mm as a rule and in consequence cannot easily be used for water-retaining structures. Their main use is in clay soils where water inflows are not a problem, though they have also often been used to retain dry granular materials or fills. Clearly, where water is not a problem the spacing of the piles can be adjusted so long as the gap between piles is such as to prevent soil collapse between them.

Secant piles are constructed so that there is an intersection of one pile with another, the usual practice being to construct alternate piles along the line of the wall leaving a clear space of a little under the diameter of the required intermediate piles. The exact spacing is determined by the construction tolerances which can be achieved. These initially placed piles do not necessarily have to be constructed to the same depth as the intermediate piles which follow, depending on the way in which the wall has been designed and reinforced. The concrete may also be chosen to have a slower rate of strength development in order to ease the problem of cutting one pile into another.

The intermediate piles are usually formed through a heavy temporary casing which can be rotated by an oscillator device, the cutting edge of the casing being toothed in a way which enables the casing to cut into the concrete of the initial piles on either side.

Obviously these piles can be used to form a continuous watertight wall or nearly watertight wall, but this depends on the constant control of tolerances for plan position and boring direction. A failure to ensure intersection will quickly make the wall non-watertight.

The introduction of continuous-flight augers in the last 10 years has provided yet another variant in the area of bored pile walls. Using this type of equipment it is possible to construct 'interlocking' bored pile walls in much the same way as for secant piles. The difference is that rotary bored 'soft' concrete piles, frequently made from concrete with the addition of a few per cent of hydrated bentonite mixed into it, are first installed. The strength of the soft material is generally in the region of a few newtons per square millimetre at 7 to 14 days old, at which time the structural piles are installed to intersect with them and form a continuous wall which is capable of retaining water and soil pressures. The permanence of such soft concretes is to some extent unproven, but this is an area of active research to manufacture concretes which are both soft and durable. Because of the ability to use continuous-flight augers, the process is rapid and economical, but the maximum available pile diameters are more limited than for traditional secant bored piles.

Guide walls are sometimes constructed to ensure the correct plan position of the tops of contiguous and secant bored pile walls, and in the construction of diaphragm walls they are essential.

In this latter form of diaphragm construction a grab is used to cut a series of slots between a pair of parallel guide walls set at just over the wall thickness apart. The guide wall depth is usually not less than one metre and during excavation of each slot the stability of the ground is ensured by means of a

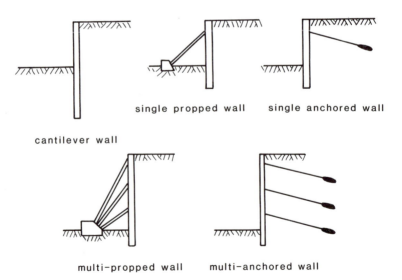

Figure 6.1 Types of retaining walls

bentonite suspension. When the trench reaches the required depth a 'stop end' is placed at one end of it, occasionally at both ends depending on the sequence of working, the function of this long element being that it can be removed after concreting to leave a close contact face for the concrete of the next adjacent panel (Sliwinski and Fleming, 1975).

When a panel has been completed and the concrete has set, the construction of the next panel in the line may be commenced in exactly the same way.

The maximum depth to which contiguous bored piles may be constructed is in practice about 18–20 m though in certain circumstances and particularly with wide pile spacings it can be more. Secant piles can be bored to depths of the order of 30 m but the difficulties of construction greatly increase as pile depth goes below about 20 m and this is especially so in soils where the piles cannot be made 'dry' during construction. Diaphragm walls can be constructed using heavy rope-operated grabs to depths of the order of 50 m or more, though the difficulties of extracting very deep stop ends where these are necessary are considerable. This problem can to a large extent be overcome by the installation of permanent precast concrete stop ends.

Clearly the selection of a walling method depends very much on whether a wall is required to retain water or not, and it is expensive to choose a watertight type of wall when it is not required.

The reinforcement of all these types of retaining wall is conventionally by round steel reinforcing bars, but there are occasions, especially in secant piling when the use of structural steel I sections is preferred.

The way in which walls are used has a substantial influence on their dimensions and cost. In practice the majority of bored piles are constructed as cantilever walls, but penetration length, deflection and bending moments may be reduced by propping or tying the wall back by means of ground anchors. However, props tend to obstruct the working area of a site, and when other operations such as excavation, piling and slab construction have to be carried out with the prop in place, it is clear that many contractors would prefer to spend more on the wall and less on the following works.

This problem can be alleviated by the use of ground anchors, in which case a clear unobstructed working space can be given on the side of the wall where excavation has to be carried out, but because of adjacent structures or services or because other complications arise it is frequently not acceptable to use ground anchors.

As the required retained height of a wall increases various solutions present themselves for strutting it, for example by partially constructing each floor in a basement as the excavation is carried out, and using that section of floor as a stiff waling beam around the site. Alternatively, flying shores across a site are also used quite frequently.

The complications of strutting or tying back walls are such that the costs of deep retaining walls increase rapidly with the depth of wall required, and of course if the wall is of a diaphragm or secant type and is being used additionally as a water cut-off, the simple act of tying walls back or propping them may save nothing on the length of pile which has to be constructed in the ground.

Where the ground-embedded length of wall is in dense or medium-dense sands and gravels, in stiff clay, or in soft rocks, walls of up to about 10 m in retained height have been constructed as cantilevers, but this requires concrete walls of perhaps 1.2 m or more in thickness. Commonly 500 mm or 600 mm diameter contiguous cantilevered pile walls can be used for retained heights of approximately 5 m.

Tied or propped walls may be essential where hard rock occurs at such a depth below the dredge level that the penetration of a cantilevered pile wall is either impracticable or excessively expensive.

A special case arises with deep circular excavations, often used for pump houses for example, and sometimes for car parks. In this case a diaphragm wall can be used to surround the excavation and this can be constructed and excavated to substantial depths without the need for internal props or ties, simply relying on the self-propping action of the continuous ring so formed for support.

In terms of the performance of bored pile or diaphragm retaining walls it is evident that the requirements are related to some extent to the purpose for which the walls are used. It is not possible to construct any retaining wall without some deflection occurring in the whole soil system as a result. The consequences of failure of a wall depend on its height, the dangers to life that would result, and the damage to adjacent property. Clearly the attention and care in design and construction has to be greater in some cases than in others, and a sound and adequate design depends on a fundamental understanding of the soil mechanics of the problem and on experience of performance in similar circumstances.

The whole question of the design of such retaining walls is a matter which has received little attention until recent years, and the complexities are such that at the present time a variety of different design methods are in use, each with its shortcomings. For major excavations, detailed analyses of the resultant soil movements have been made using finite element computer programs, but these only yield information of a quality corresponding to the input information. However, there seems little to be gained in the majority of cases by using over-complex solutions, and the procedures outlined in this chapter are not intended to apply to those special cases where a number of solutions need to be compared and evaluated in order to minimize damage in sensitive surroundings.

6.2 Soils and the data needed for wall design

The point has been made that a first priority in every piling problem is a proper understanding of the soil conditions, and in this respect the design of retaining walls makes its own special demands.

It is first of all necessary to fully appreciate the conditions of the site, its general topography and layout. If the site is one which is basically unstable in the first place, then the installation of a retaining wall may be a waste of time and effort, or this factor may influence the design so that a much stronger or different type of wall is used. Sloping hillsides, for example, at slopes of perhaps 10 degrees or more in stiff clay, may be subject to surface soil drift in the top 2 m or so, and a result can be that passive rather than active soil pressures apply near the wall head. The site may be close to a river and the water levels in the ground may be dependent on the fluctuations in its level. Even the cutting down of trees in overconsolidated clays immediately beside a wall on the retained side may have significant effects on wall deflection and soil pressures in that vicinity.

The geological conditions need to be appreciated—the wall may be required to retain overburden over a rock stratum, in which case the direction of dip is important. If in such a case the rock is so hard that bored piles cannot penetrate it by a sufficient amount at reasonable cost, then a propped or tied wall must be used. If for example the piles or diaphragm walls are to be constructed using bentonite suspensions, the fluid could be lost, to great embarrassment, in a cavernous rock formation. These are examples of situations which sometimes arise, where circumstances may control or dictate the type of solution which is possible.

The history of the site is also of interest—clearly it is very relevant to know if a site was excavated or filled in the past, or whether its ground has been severely contaminated by chemicals; perhaps it is underlain by old sewers or tunnels, or indeed it may have had old basements with heavy floors which, although no longer seen, would lead to extra expense in piling for new works.

In practice the fill which occurs on many sites is not examined in any detail in site investigations, and yet for a retaining wall it may be one of the most important factors in deciding the basic dimensions of a wall and its cost. Soil testing is often not easy in such conditions, but good descriptions can enable reasonable assumptions to be made about the strength parameters which have to be used.

Most site investigations are concerned primarily with the bearing capacity of the ground which is to support a new structure, and are therefore concerned with the immediate strength of the soils rather than the long-term strengths, which because of the compressive stresses are bound to be increased. The differences between short and long-term strengths are really a function mainly of clay behaviour, and it is here that the most important problems lie.

The strengths of clay soils depend mainly on their geological history; the amount of compression they have undergone in the past and any subsequent release or partial release of that compression. Normally consolidated clays are basically as laid down, whereas overconsolidated clays have been subject to a cycle or cycles of overburden change in the past.

In the type of problem for which bored pile walls can provide an appropriate solution, the piles are usually installed in the ground and the earth is then taken away on one side. It is unusual, though not impossible, for the soil retained by such a wall to be placed after the construction of the wall, and setting this case aside, the effect of excavation in a clay is to release some of the applied vertical stress. In consequence, softening takes place on the passive side, with its most immediate effects being near the excavated level. Likewise, the deflection of the wall, however small, has a softening effect on the active side of the wall.

It is very unusual in the design of walls to choose to rely on passive pressure from soft normally consolidated clays in order to maintain the stability of a wall. In the first place the soil strengths are low and would require deep penetration, but in addition the deflection of a wall in the long term in such conditions is likely to be larger than can be tolerated.

A site investigation carried out before any excavation has been done and giving quick undrained strengths for a firm to stiff clay is not very relevant to the long-term stability of a wall after excavation. It is for this reason that most engineers now agree that the design of retaining walls in overconsolidated clay soils should be carried out using effective stress parameters derived either from undrained triaxial tests with porewater pressure measurements or from drained triaxial tests.

In carrying out such tests the strain rate should be slow enough to allow virtually all excess porewater pressures to equalize or dissipate in the sample as the test proceeds. The strength parameters as defined by ϕ', the frictional component, and c', the cohesive component, will be obtained, and since c' is a powerful parameter in the calculation of earth pressures with significant influence on the results, exaggerated values for this would lead to unsafe retaining wall designs. Such exaggerated values are obtained if the tests are made at too fast a rate of strain.

The choice of an appropriate stress level for testing in these clays is of some importance. In the top of the passive zone in front of a wall after excavation the conditions are such that the samples should be tested at low lateral stress levels, but in practice it is not very satisfactory to test at confining pressures below about 50kN/m^2 using standard laboratory equipment. The range of confining pressures in which interest lies is generally between say 50 and 150kN/m^2, and a great deal of valuable reference data can be obtained from specialist papers on soils such as Chalk, Barton Clay, Gault Clay, Glacial Tills, Mercia Mudstone, London Clay, Oxford Clay and the Upper Lias Clay (see Table 6.1).

A good deal of debate centres around the effective stress parameter c', as stated above, since it appears to be a parameter of transitory importance. Its value depends on the degree of overconsolidation, the stress level of the test, the degree of chemical weathering to which the soil has been subjected, and the amount of sample swelling that occurred before the test. A small part of the inferred value of c' is due to real interparticle bonds and these are destroyed by the dilation of the sample at strains beyond the peak value in the triaxial test.

Table 6.1 Reference sources for specialist papers in various geological formations

Chalk	Hobbs and Healey (1979)
Gault Clay	Samuels (1975)
Glacial Tills	Gens and Hight (1979)
	Skempton and Brown (1961)
	Vaughan et al. (1975)
	Weltman and Healy (1978)
Mercia Mudstone	Chandler (1969)
London Clay	Bishop (1971)
London Clay	Bishop et al. (1965)
	Skempton (1970)
	Skempton (1974)
Lower Oxford Clay	Parry (1972)
Lias Clay	Chandler (1972)

This parameter is therefore one to be used with a degree of caution, but to eliminate it entirely from wall designs where the strains are small does not seem to be justified. In selecting parameters for wall design, c' should be reduced below any measured laboratory values in those strata where the largest strains are to be expected, and this means appropriate reduction in the upper part of the passive zone and probably in the active zone down to the dredge level. This practice has been successfully followed in many designs in combination with appropriate selection of factors against wall failure by rotation.

It is suggested that on the passive side of a wall the value of c' should be taken as zero at the dredge level where, in the case of both cantilever and propped walls, the strain will be at its maximum value due to both swelling and wall movement, and that it may gradually be increased to the peak laboratory value at the toe of the wall or over a depth of $1-2$ m below the dredge level whichever is the lesser depth.

On the active side of the wall because of the wall deflection and consequent softening, it is suggested that above the dredge level c' should be reduced to about half its laboratory value.

It is sometimes argued that for temporary works in stiff clay, the undrained triaxial test results may be used in design, perhaps modified or reduced by some arbitrary factors to take into account the partial reduction of strength with time. The rate of change of intergranular stress in the soil is dependent on the soil fabric; fine sandy or silty partings, fissures, laminations, and joints. Failed cuttings and trenches in stiff clay are often seen to have a blocky appearance. The material beneath the surface of a block may have a strength near that which would be measured in an undrained triaxial test, so that comparatively small volumes of absorbed water only may be involved in softening along joints, and yet a face has failed which would not have done so at normal undrained strengths. For this reason the time taken for the transition between short- and long-term strength is very unpredictable, and in designing walls for a temporary state it seems wiser to use realistic long-term effective stress parameters together with a low factor on stability, rather than undrained strengths and a much higher but uncertain factor.

As in the case of soft clays, it is usually undesirable to rely on loose sands for wall support in the passive zone because of the large strains which may occur. In order to arrive at an appropriate angle of internal friction for a sand the current practice is usually to carry out Standard Penetration Tests, bearing in mind the corrections which may be necessary in the N value for tests at shallow depth and for fine soils in submerged conditions.

Various proposals have been made for shallow depth correction but it is suggested that the corrections given on Figure 2.8 are appropriate.

The correction of N for submerged conditions applies only to very fine sand or silt below the water table, and is due to the excess porewater pressures set up during driving of the sampler. In this case

$$N_{corrected} = 15 + \tfrac{1}{2}(N_{measured} - 15)$$

The approximate relationship given by Peck *et al.* (1974) may then be used to determine the angle of internal friction ϕ, but as this does not fully take into account various features of cohesionless soils it is probably better to use the system proposed by Stroud (1989) (Chapter 2, section 2.5.4).

Having established the basic soils parameters ϕ' and c' the other parameters which have to be considered apart from the physical geometry of the system are

   (i) Active side wall friction angle δ_a
  (ii) Passive side wall friction angle δ_p
 (iii) Wall adhesion c_w

6.2.1 Wall friction and adhesion

Piles are frequently required to carry load by shaft friction and the existence of such friction is unquestionable in all manner of soils. In the case of a load-bearing pile it is the lateral stress which is locked in between the soil and the pile which gives rise to a force acting through friction on the pile surface in the axial direction. Bored-pile and diaphragm retaining walls have surfaces of similar roughness and are acted upon by lateral earth forces in the active and passive senses. Hence, provided there is differential vertical motion between the wall and the soil, frictional forces must be mobilized which affect the stability of the wall, and these forces should be taken into account in wall design.

Corresponding to the shear strength parameter ϕ', the forces acting on a wall are related to the parameters δ_a for the active side, δ_p for the passive side, and corresponding c', c_w as wall adhesion. To neglect these factors is certain to lead to uneconomic designs.

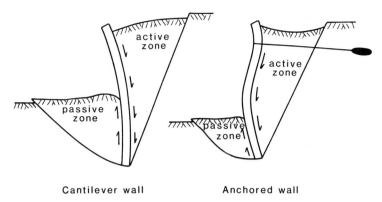

Cantilever wall   Anchored wall

Figure 6.2   Directions of soil movement at failure

In order to justify the reduction of active and the increase of passive earth pressure due to these effects it is first necessary to ensure that the movements of the soil around the wall in any potential failure mechanism are directionally such as to give benefit.

The soil on the active or retained side of the wall has to move down relative to the wall as shown in Figure 6.2 in order to mobilize friction in the beneficial direction and, on the passive side, the displaced soil has to move upward. These conditions arise naturally with most walls, but if the wall itself is going to suffer settlement under vertically applied load, then δ_a will diminish and similarly if a heavy strip footing is placed in front of the wall on top of the passive zone it may prevent upward movement of this wedge and reduce δ_p. In this latter case however, the surcharge of the footing would probably compensate for this effect on earth pressure, increasing the resistance to failure.

The question of evaluation the parameters δ_a, δ_p and c_w raises the problem that these forces have not been measured in practice but are derived rather by inference. Experience with load-bearing bored piles in granular soils suggests, from a good many instances of load testing, that the ratio δ/ϕ' is unlikely to be less in practice less than 0.7. Current practice in virtually all soils is to take $\delta_a = \frac{2}{3}\phi'$ and this leads to satisfactory analysis.

The most vital part of the soil mass from the point of view of both wall stability and deflection is the passive zone in front of the wall, and because errors on this side are of much greater consequence than on the active side, it is current practice here to take $\delta_p = \frac{1}{2}\phi'$.

There is no ready means of inferring values of c_w in relation to c' in clays, but by analogy with adhesion factors α, which are used in pile design in stiff clay, where α has values of the order of 0.5, so also the assumption is made that $c_w = \frac{1}{2}c'$. Because c' must be influenced by remoulding of the clay in the area immediately adjacent to the wall, it is felt that c_w should be limited in stiff clay soils to a maximum of about $15\,\text{kN/m}^2$.

6.2.2 *Sensitivity analysis*

Having determined the basic values of all the soil parameters to be used in a design, it is often useful to consider potential errors and mis-selections in these values in the specific design in question, in order to ensure that in a worst credible situation the factor against rotational instability of a wall exceeds unity. If however, this practice is taken to extremes and combinations of several concurrent worst credible situations are applied simultaneously, then the result will be uneconomic walls.

6.3 The basic design principles for bored pile or diaphragm walls

Before the installation of any retaining wall commences, there exists a state of lateral stress in the ground at that location. In normally consolidated soils, expressing the ratio of lateral effective earth pressure to vertically applied effective earth pressure at any point as K_0, then $K_0 = 1 - \sin \phi'$ in cohesionless soils and is perhaps a little lower in clay soils.

In overconsolidated clays which have been subject to high overburden pressures in the past, then the value of K_0 is related to the normally consolidated K_0, and the overconsolidation ratio (OCR), as indicated in Figure 6.3. The OCR is expressed as the maximum vertical stress which the soil has experienced in the past divided by the present vertical effective stress.

In practice the value of K_0 in stiff clays may be as high as 3, but lower in the near surface layers because these may have failed in a passive sense in the course of overburden removal, (i.e., K_0 cannot exceed K_p where K_p is the coefficient of passive earth pressure).

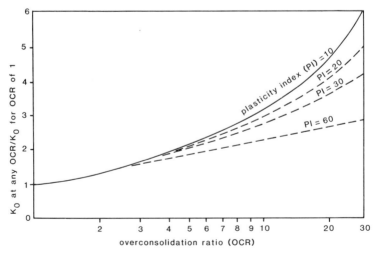

Figure 6.3 Values of K_0 related to overconsolidation ratio for soils of varying plasticity. After Brooker and Ireland (1965)

If, in the course of carrying out excavation in front of a retaining wall, steps are deliberately and carefully taken to prevent movement of the wall, then the forces which would apply to that wall would correspond to the earth pressures at rest. Such a case might arise for example in a deep circular excavation supported by 'rigid' diaphragm walling panels, but in practice no system is so rigid as to fully maintain this pressure state and estimates have to be made of wall stresses based on some intermediate assumption—ensuring that if the worst happens a collapse will not occur. Even in the circular diaphragm wall case, the very act of digging the trench will have had significant reducing effects on the locked-in soil stresses.

In the majority of cases it is either impractical or uneconomical to resist movements in the ground, and walls can happily be permitted to undergo small movements so long as these produce no unfortunate consequences in other structures adjacent to or connected with the wall. The economic design is one which permits movement but limits it to the values necessary to develop active pressure conditions on the retained earth side.

Considering an element of soil in the ground subject to an effective principal stress σ_v', this element may be failed either by reducing the horizontal stress acting on it σ_h' to a value where the Mohr Circle intersects the failure envelope as shown in Figure 6.4, or by increasing the effective stress until failure similarly occurs.

If this concept is applied to the earth in front of and behind a retaining wall then the soil fails in front of the wall in the *passive* mode and behind the wall in

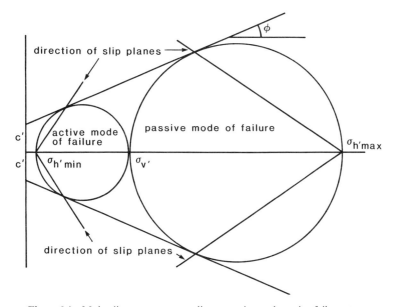

Figure 6.4   Mohr diagram corresponding to active and passive failure states

Figure 6.5 Smooth retaining wall mobilizing full active and passive pressures

the *active* mode as shown in Figure 6.5 and where the cohesion intercept $c' = 0$, then

$$\frac{\text{active effective pressure}}{\text{vertical effective pressure}} = K_a = \frac{1 - \sin \phi'}{1 + \sin \phi'} = \tan^2\left(45 + \frac{\phi'}{2}\right).$$

Likewise,

$$\frac{\text{passive effective pressure}}{\text{vertical effective pressure}} = K_p = \frac{1 + \sin \phi'}{1 - \sin \phi'} = \tan^2\left(45 - \frac{\phi'}{2}\right).$$

It will be observed that the influence of $c' > 0$ is to increase the size of the Mohr Circles, to decrease active pressure, and to increase passive pressure. The decrease of active pressure due to change in c' from zero is

$$2c' \sqrt{K_a} = c' K_{ac}$$

The increase of passive pressure is correspondingly

$$2c' \sqrt{K_p} = c' K_{pc},$$

and the general expressions for earth pressure, as developed by Rankine, may be expressed as

$$\text{active earth pressure } P'_a = K_a(\gamma z - u) - K_{ac} \cdot c'$$

$$\text{passive earth pressure } P'_p = K_p(\gamma z - u) + K_{pc} \cdot c'$$

where γ is soil bulk density, z is depth below surface in a uniform soil and u is the porewater pressure. While the Rankine method displays the principles involved, the method cannot readily be extended into particular cases where wall friction exists on the faces contacting the plastic zones. Coulomb adopted a different and more versatile approach to the same problem; analysing the stability of a non-plastic active soil wedge at the point of failure, as shown in Figure 6.6 for a smooth wall and for the level retained ground case, he obtained a solution identical to that of Rankine. Happily the form of the active and passive pressure equations may be retained for cases where inclined surfaces and wall friction are both taken into account.

Figure 6.6 Forces acting on Coulomb active wedge at failure

The equations developed from the Coulomb approach and due to Mueller-Breslau show that the active earth pressure coefficient for horizontal pressures, K_a, is

$$\frac{\cos \delta_a \cdot \sin^2 (\alpha - \phi')}{\sin^2 \alpha \cdot \sin (\alpha + \delta_a) \left\{ 1 + \sqrt{\dfrac{\sin (\delta_a + \phi') \cdot \sin (\phi' - \beta)}{\sin (\alpha + \delta_a) \sin (\alpha - \beta)}} \right\}^2}$$

and the passive earth coefficient for horizontal pressure and level ground on the passive side, K_p, is

$$\frac{\cos \delta_p \cdot \sin^2 (\alpha + \phi')')}{\sin^2 \alpha \cdot \sin (\alpha + \delta_p) \left\{ 1 - \sqrt{\dfrac{\sin (\delta_p + \phi') \cdot \sin \phi'}{\sin (\alpha + \delta_p) \cdot \sin \alpha}} \right\}^2}$$

where ϕ' = angle of the internal friction
  α = inclination of the wall to horizontal (frequently $90°$)
  δ_a = angle of wall friction on the active side of the wall (soil downdrag)
  δ_p = angle of wall friction on the passive side of the wall (soil rising).
  β = angle of retained soil surface to horizontal.

While the values of K_a calculated by the Coulomb method are widely accepted and, for $\alpha = 90°$ and $\beta = 0°$, are very similar to values for K_a from other sources, the values of k_p have been shown to be significantly too high except in cases where $\delta_p = 0$, and work by Sokolovski has led to values of K_p which are as shown in Table 6.2.

The angle of the surface of soil in the passive zone, β, is expressed in Table 6.2 in relation to the horizontal, being positive for soil and rising with increasing distance from the wall face. The coefficients K_p represent the coefficient of horizontal pressure on a vertical wall.

The reason for the need to calculate K_p by means other than the Coulomb wedge method is that as δ_p increases above $0°$, the lower boundary of the passive wedge becomes curved and goes deeper into the soil—rather like a bearing capacity type of failure mechanism. This leads to a reduced value for

Table 6.2 Passive earth pressure coefficients. After Sokolovski

ϕ'		10°			20°			30°			40°		
β	δ	0°	5°	10°	0°	10°	20°	0°	15°	30°	0°	20°	40°
− 30°	K_p	1.04	1.10	1.14	1.26	1.47	1.62	1.49	2.01	2.42	1.86	2.98	4.15
− 20°	K_p	1.18	1.28	1.33	1.51	1.80	2.00	1.90	2.69	3.29	2.50	4.42	6.30
− 10°	K_p	1.31	1.42	1.50	1.77	2.16	2.41	2.39	3.50	4.35	3.37	6.36	9.42
0°	K_p	1.42	1.55	1.63	2.04	2.51	2.86	3.00	4.46	5.67	4.60	9.10	13.9
+ 10°	K_p	1.49	1.64	1.73	2.30	2.88	3.32	3.65	5.62	7.29	6.16	13.06	20.4
+ 20°	K_p	1.53	1.69	1.80	2.53	3.26	3.79	4.42	7.13	9.27	8.34	18.3	29.9
+ 30°	K_p	1.52	1.70	1.82	2.76	3.61	4.24	5.28	8.76	11.69	11.3	26.7	43.4

K_p. The Mueller Breslau value for K_p without any modification is substantially too optimistic where the value of ϕ' is above about 20 to 25°.

The relationship between K_a and K_{ac} and similarly between K_p and K_{pc} may be as expressed as in the Rankine method of analysis, but when wall adhesion is also present a further modification has to be made.

Thus the values for the coefficients which refer to soil cohesion become

(i) $\quad K_{ac} = 2\sqrt{K_a\left(1 + \dfrac{c_w}{c'}\right)}$.

(ii) $\quad K_{pc} = 2\sqrt{K_p\left(1 + \dfrac{c_w}{c'}\right)}$

With the respective values for K_a, K_{ac}, K_p and K_{pc} known it is then possible to use the vertical effective stress at any point adjacent to the wall, to calculate the horizontal limit of active or passive earth pressure on that element.

However, in addition to considering earth and water pressures, pressures from other imposed effects such as surcharges must be added to the problem before wall stability can be examined. The methods of dealing with some of these features are outlined as follows.

6.4 The treatment of groundwater conditions

Although superficially water might appear to be easier to deal with than earth pressure, there are a number of problems associated with it which are difficult to quantify.

On many sites the best way of determining true groundwater levels is to use piezometers, rather than to rely on short-term borehole observations. This is especially so in the case of clay soils.

The stability of a retaining wall depends not only on the earth pressures applied to it, but also on the groundwater regime that exists around it, and the installation of a wall can itself influence and bring about changes in this. In the case where a wall is installed in a soil with a groundwater gradient, the wall, if it is impermeable, may act as a dam to produce a higher water level than existed prior to its installation and this is a difficult condition to assess. In such cases it is usual to install drains behind the wall in order to collect and divert water which would otherwise overtop it, and this can define the design condition on the 'active' pressure side. On the contrary, in the case of a permeable contiguous bored pile wall, vertical drains are sometimes installed in the gaps between piles, leading to a toe drain, and the effective water level on the 'active' and 'passive' sides will to all intents and purposes equalize.

To fully analyse the water pressures on a wall and their effects on the potential failure mechanism, it is proper to draw a flow net and, using the porewater pressures derived from this, to analyse a series of active wedges, relating the shear stresses along the planes to the net effective normal pressures. This rigorous procedure is not normally followed in practice because it would lead to complex and tedious analysis, particularly in cases where the soil is multi-layered, with strata of differing permeability. Two conditions are therefore worth considering for computational convenience in the case of impermeable walls.

(i) It is assumed that the final water levels on the active and passive sides of the wall can be defined, and the water pressure diagram is drawn as in Figure 6.7 This condition would apply if at the toe of the wall as constructed there existed an impermeable stratum, but elsewhere above this the ground is considered to be permeable. This reduces the groundwater to a simple model which is widely used.

(ii) It is assumed that the wall is surrounded by a soil of uniform permeability, in which case with the water levels on the active and passive sides defined, a flow net can be drawn to indicate the long term seepage around the wall (Figure 6.8).

Rather than become involved with analysis of this kind it is convenient to make an approximation and to assume that the water head $(H_1 - H_2)$ is

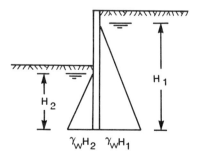

Figure 6.7 Water pressure diagram on active and passive sides of the wall

230   PILING ENGINEERING

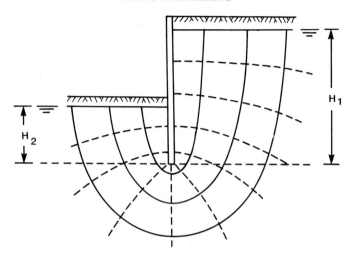

Figure 6.8 Seepage around an impermeable wall in soil of uniform permeability

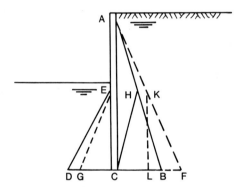

Figure 6.9 Derivation of water pressure diagram

dissipated uniformly along the surface length of the wall $H_1 + H_2$, so that a water pressure diagram as shown in Figure 6.9 is derived.

It will be observed that this now eliminates the water pressure jump that occurs at the toe of the wall C, and the pressure at the wall toe on the active side CB equals that on the passive side CD. The lines AF and EG represent the condition of the approach described above. The net water pressure diagram now is represented by AHC whereas the system outlined in (i) above would produce a net diagram $AKLC$. The total water pressure applied to the active side of the wall is therefore significantly reduced.

However, it will also be observed that this is equivalent to reducing the bulk density of water γ_w on the active side of the wall and increasing it on the passive side, and this in turn increases the soil bulk density on the active side of the wall and reduces it on the passive side. Thus because of upward seepage pressure

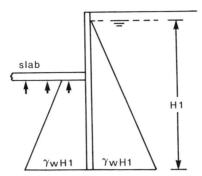

Figure 6.10 Impermeable slab subject to uplift water pressure

the passive soil pressure is reduced and because of downward seepage the active soil pressure is increased. The question which arises therefore is one of how much difference these assumptions make in any particular design case.

It has also to be remembered, in the case of deep basements for example, that where an impermeable slab is cast in the bottom, this slab may be subject to uplift pressures in the long term unless provision is made for drainage. Thus seepage may cease and full water pressure on both sides of the wall may return to its original active side level as indicated in Figure 6.10.

Examination of cases involving cantilever walls in clay soils reveals that where the difference in water level is say 3 or 4 metres between active and passive sides, the effective difference in the finally determined wall dimensions and bending moments between methods (i) and (ii) is negligible. Where the difference is say 6 metres, the case (ii) analysis leads to slightly deeper walls and slightly greater bending moments.

The difference in approach probably is of most concern in relation to walls which are propped or tied back by ground anchors, which are generally of greater retained height than normal cantilever walls, and which are often subject to greater water differential pressures than cantilever walls. In this case toe penetration below the dredge or excavation level may be relatively small depending on the selected position of the tie or anchor, and although the uplift seepage effect on the passive side may be small, it is prudent to recognize the sensitivity of penetration to this effect and to ensure that toe penetration is adequate to deal with any uplift seepage pressure that may exist.

In general terms therefore method (i) above is widely used and for most walls proves to be a satisfactory way of dealing with water pressures, though the assumptions made are far from rigorous.

In the case of contiguous bored pile walls, where the wall does not provide an impermeable barrier, and where high water pressures cannot normally exist on the active side of the wall, it is usual only to take into account a small nominal difference of water head between active and passive sides.

6.4.1 *Tension cracks*

On many sites, there is overburden of fill or granular material on top of clay on the active side of a retaining wall, or alternatively the area behind the wall has been built on or paved over. These features should prevent the formation of shrinkage or tension cracks in the clay behind the wall in periods of dry weather. However, in some instances the surface of clay is exposed, and where one uses a value of cohesion c' in the design it is implied that a negative active earth pressure can exist in the vicinity of the wall head. This feature in these circumstances could lead to surface cracking in the clay and the cracks could fill with water, which would in turn cause local pressure corresponding to the head of water retained in the crack. The possibility of such cracks would seem mainly to be confined to cantilever walls rather than to propped or anchored walls where the position of the anchoring force is located usually at some distance below the wall head, in which case the deflected form of the wall is such as to produce a closure of any cracks so formed.

For reasons associated with the requirements of the old British code of practice for retaining walls (CP2), it is common to stipulate in designs a minimum equivalent fluid pressure on the active side of retaining walls of $4.7\,kN/m^2$ per metre depth measured below the wall head, and while this feature is not of much other effect in wall design when using effective stress parameters, it does serve the useful purpose of providing a reasonable positive-pressure condition where tension cracks might otherwise occur in the immediate vicinity of the wall head.

6.5 Earth pressures due to ground surcharges

6.5.1 *Uniform surcharges*

A uniform surcharge acting on the active side of a wall produces a uniform vertical stress throughout the retained soil, and this is assumed to give rise to a lateral earth pressure equal to the surcharge multiplied by the active earth pressure coefficient or coefficients on the whole of the active face of the wall.

Occasions arise in which, for example, a road extends across the width of the active soil wedge, and in this case it is normally assumed that standard road loadings apply and these are taken as a uniform surcharge. In clay soils where the design parameters are chosen on a long-term basis, this approach may be rather too conservative and some modified lesser surcharge may on occasions be justified.

6.5.2 *Point loads and line loads*

The presence of a point, line or strip load resting on or in the soil being retained by a wall gives rise to additional lateral pressures on the wall. The usual

Figure 6.11 Lateral pressure distribution produced by point load Q_p. Rigid unyielding wall leads to double the value of σ_h

methods for assessing the lateral stresses due to these conditions are those of Boussinesq and Terzaghi. The figures given by Terzaghi (1954) are in fact a modification, based on some experimental work, of the Boussinesq method.

The Boussinesq theory is formulated to give the stresses in a semi-infinite elastic medium due to the application of a load at the surface, and if the Boussinesq formulae are used with reference to loads applied near a rigid boundary the effect is to constrain the deformation of the elastic medium and to double the horizontal stresses acting on the boundary. Thus for a point load, and taking a Poisson's ratio of 0.5, the Boussinesq relationship is as shown in Figure 6.11.

Figure 6.12 Lateral pressure due to point load (Boussinesq method modified by experiment). After Terzaghi (1954)

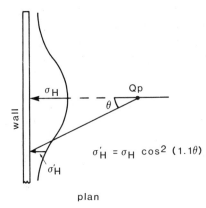

plan

Figure 6.13 Lateral pressure due to point load (Boussinesq equation modified by experiment). After Terzaghi (1954)

Figure 6.14 Lateral pressure due to line load (Boussinesq method modified by experiment). After Terzaghi (1954)

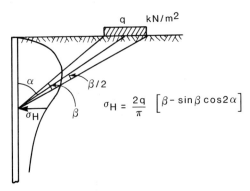

Figure 6.15 Lateral pressure due to strip load (Boussinesq equation modified by experiment). After Teng (1962)

However, it is clear that the deformability of the wall and how it relates to elastic theory is dependent on the wall type. Thus a cantilever wall may deform much more than a multi-strutted wall, and the section of an anchored wall above the top anchor may act very much as if it were rigid. It is therefore important, if using the Boussinesq method, to consider carefully how it should be applied.

The lateral soil pressures acting on a wall due to a point load and as calculated by Terzaghi are shown in Figure 6.12 and Figure 6.13; similarly the lateral pressures due to line loads are shown in Figure 6.14. Lateral pressures due to a strip load, based on the Boussinesq method as modified by experiment and given by Teng are shown in Figure 6.15.

6.6 The use of berms

The passive pressure acting on the embedded part of a wall is a function of the vertical soil pressure acting in the potential failure zone near the wall, and this can be increased significantly by leaving a berm against the foot of the wall at dredge level. Since in retaining-wall design the temporary condition during construction is often worse than the permanent condition, and since berms can often be accommodated at reasonable cost in that condition and can be removed either in short sections as permanent work proceeds or at a stage when propping from the permanent works is possible, they offer a valuable means of minimizing costs in certain instances.

There are various ways in which the effects of berms can be assessed, and while little detailed research has been carried out into their behaviour, experience confirms that the approximate methods used in design are adequate. It would in theory be possible to analyse the passive failure mechanism for a wall including a berm, by an iterative method involving a series of trials with an appropriate plastic failure mechanism, but this would lead to a protracted method and would be inconvenient in everyday design. For this reason, several approximate methods are used in practice as indicated in the following paragraphs.

Figure 6.16 Relation of effective ground level to berm dimensions

Figure 6.17   Treatment of berm as surcharge

(i)   An empirical method is to treat the berm as causing an increase in the
      effective ground level on the passive side of the wall. In this method the
      height of the berm is treated as not more than $1/3 \times$ the berm width as
      shown in Figure 6.16, and the effective ground level is then taken as half
      the height of the berm at the point where it contacts the wall. Any stable
      soil existing above the $1:3$ slope may then be treated as a surcharge acting
      on the effective ground surface over the approximate width of the
      potential passive failure mechanism. While this method is open to
      criticism on the grounds that it is highly empirical, it appears to yield
      relatively satisfactory and conservative results by comparison with other
      methods, both in terms of wall depth and of bending moment. Unless
      such a berm is wide, it is unwise to terminate any tied wall above or very
      near to the final excavation level on the passive side.

(ii) An alternative method of treating a berm is to convert it to an effective
      surcharge acting on the potential passive failure zone. In this method the
      effective self-weight of the berm is calculated and is then distributed over
      the approximate width of the passive failure mechanism at the general
      final site excavation level (see Figure 6.17), thus increasing the passive
      pressure available accordingly. In addition to the enhanced passive
      pressure, some shear forces must exist in the berm above the dredge level
      and these will contribute to the stability and to moment reduction in the
      wall. If the shear forces acting within the berm are not taken into account,
      and it is perhaps difficult to do so reliably because of softening of the
      material involved in the berm, then this second method must be regarded
      as conservative. Berms are usually and sensibly blinded with concrete in
      order to diminish the risk of softening in clay soils and to inhibit erosion
      and damage in granular soils.

6.7 Wall stability analysis

It is important to recognize that a wall can only be as stable as is the soil mass
in which it exists. If there is a possibility that, as a consequence of the
excavation, the whole body of soil in which a wall is installed can become
unstable, then the solution to this problem becomes an overriding requirement

and must be analysed using normal slope stability methods. This potential failure mechanism is most likely to occur in cases of shoreline or waterfront structures and with walls which are placed in sloping hillsides. For waterfront structures the possibilities of scour at the dredge level must also be borne in mind, since the construction of a wall may in itself give rise to water currents which will in time erode the zone of passive resistance.

Having established the general stability of a site, one then must return to consideration of the wall and to the simple need to satisfy the equilibrium of the vertical, horizontal and rotational forces.

Two basic concepts have been used in calculating the stability of walls, namely the so called 'free earth' and 'fixed earth' methods. In the 'free earth' method the stability of a wall is considered without making the assumption that the deflected form of the wall is controlled by fixity in the ground. In the 'fixed earth' method it is assumed that the rotation of the toe of a wall is constrained by fixity in the soil mass and a consequence of this can be a reduction of moment in propped walls, but an increase in the embedment length. However, for such a wall there is no relevant mechanism which may be used to analyse the condition of rotation about the prop, and the method relies on an empirical approach.

In a cantilever wall, equilibrium is not possible without involving a reaction force at the wall toe and in practice the wall is lengthened after the basic analysis, in order to provide this. Thus the analysis in this case involves 'fixed earth', but here a mechanism of failure can be postulated to accord with the rotation of the wall at some point below the excavated level and hence the analysis can be ascribed a physical meaning.

The idealized and simplified soil pressure diagrams associated with the analysis of cantilever and propped walls are as shown in Figures 6.18 and 6.19. In the initial stage of calculation this is simplified further (Figure 6.20).

The procedure adopted in the calculation of retaining wall stability is in general to progressively increase the embedment of the wall, analysing the disturbing and restoring moments, until a state of stability is reached, at which point the equilibrium of horizontal forces determines the value of the reaction

Figure 6.18 Cantilever wall: idealized pressure diagram for rotation about point *O*

Figure 6.19   Simplified pressure diagram at rotational failure

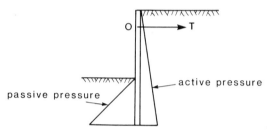

Figure 6.20   Anchored or propped wall: idealized pressure diagram for rotation about point O

R (in the case of a cantilever wall) or the reaction T (in the case of an anchored or propped wall). This procedure establishes the penetration of the wall below excavation which represents an incipient failure condition. However this is of passing interest in terms of the required wall penetration for stability and is used for reasons to be described later, only in the establishment of bending moments in the wall.

The procedure is then continued until some desired ratio is achieved between the restoring moment acting on the wall and the disturbing moment. This ratio is frequently described as a Factor of Safety, but for the reason that at this stage in design the wall length has not finally been determined in the case of a cantilever, and because there are other ways in which this factor can be expressed in other methods of analysis, the term 'Factor of Safety' is not entirely satisfactory and hence it should be regarded rather as a method related 'rotational stability factor'.

In practice this factor F_p can conveniently be introduced by simply dividing the passive earth pressure at all levels by F_p, but of course the water pressures acting on the passive side of the wall are not so treated.

Once the notional rotation point has been found in the case of a cantilever under this condition, the force R is determined, and reversing the roles of active and passive pressure, so that passive pressure appears on the active side below this point, an additional length of wall is added to provide the balancing force R.

In the case of the anchored wall, the rotation point in the design is

Table 6.3

Effective stress parameter	Range of F_p values	
ϕ'	Temporary works	Permanent works
40	2.0	2.25
35	1.75	2
30	1.5	1.8
25	1.35	1.65
20	1.2	1.5

constrained to be at the anchor level, and hence the length of the wall can be determined simply using a similar factor F_p applied to the passive earth pressure. Once this has been done, the force in the prop or anchor is determined as before from the horizontal equilibrium conditions.

When using an effective stress method of design such as is outlined above, the values F_p which are applicable to both cantilever and anchored walls are necessarily related to the soil properties and appropriate values for F_p are in the ranges shown in Table 6.3.

6.7.1 *Walls with two or more props or ties*

While the sections above deal with the cases of cantilever walls, or walls which are in part supported by a single prop or tie, it is frequently found that to reduce bending moment in walls it is necessary to insert more than one prop or tie as the excavation proceeds. This introduces a degree of redundancy into the problem and means that a simple statical analysis is no longer adequate.

Obviously, if a wall were allowed to rotate or translate to such a degree that full active and passive pressures were mobilized, the problem would resolve itself to that of a beam on elastic foundations in conventional structural terms and the stiffness of the props would govern the bending moments in the wall and the forces in the props.

However, some forms of construction such as that termed 'top-down', where floors are inserted to prop the wall as downward excavation proceeds, preclude this degree of movement, and indeed such movements would rarely be tolerable. Retained soil pressures, when sufficient movement to lead to active pressures is restrained, will lie somewhere between earth pressure at rest and active pressure, depending on the stiffness of the wall and the stiffness of the propping system.

It is for this reason that a number of computer programs have been developed, aiming to characterize the wall and props by their elastic properties and the soil as an elasto-plastic medium, with the limits of elasticity prescribed by the full active and passive states.

In the last few years a number of Finite Element programs have been developed permitting such calculations to be carried out and several of these can be used in conjuction with modern desk-top computers. It should be pointed out, however, that there are significant difficulties in defining accurate numbers for all the stiffnesses, especially that of the soil. Some props themselves rely on soil reactions, and concrete materials also have time-dependent stiffnesses. The best way to arrive at soil stiffnesses is to use back analysis of cases which closely resemble the problem in hand and in which the bending and prop forces have been monitored. A range of programs are available.

CRISP-90. This is a Finite Element program developed at Cambridge University with the support of the British Transport and Road Research Laboratory. CRISP stands for 'Critical State Program' and is approved by the Department of Transport for the analysis of geotechnical problems. It is written in American National Standards Institute Standard Fortran and requires a desk-top computer with an INTEL 80386 processor (2 to 4 m Bytes RAM), a maths co-processor, VGA screen and hard disk. It uses an incremental tangent stiffness approach and has a variety of other features.

FREW. This is a program in the OASYS suite, developed by Ove Arup and Partners, London. The wall is represented by a line of nodal points and 3 stiffness matrices, relating nodal forces to displacements. The soil may be described by any one of three methods:

(1) A subgrade reaction method.
(2) Stiffness matrices calculated by a Finite Element program.
(3) The Mindlin method, representing the soil as an elastic solid—this method is restricted to the case of constant stiffness with depth.

The program presents results for earth pressure, shear forces, wall-bending moments, strut forces and displacements. Full details are given by Pappin *et al.* (1986).

FLAC. This is a program for the solution of general geomechanical problems based on a Finite Difference method. The letters of the title stand for Fast Lagrangian Analysis of Continua and the program has been developed by ITASCA Consulting Group, Inc., PO Box 14806, Minneapolis, Minnesota 55414. Amongst other uses, it is capable of solving a range of earth retention problems, and any type of non-linear soil stress/strain relationship can be followed. It is suitable for the solution of retaining wall, tunnel-lining and rockbolting problems. It is claimed that the Finite Difference method is as flexible as the Finite Element method and requires less computational effort. The published works of Wilkins (1963), Jaeger and Starfield (1974), Cundall (1976), and Marti and Cundall (1982), are relevant to this program.

WALLAP. This program is available from Geosolve, London and offers a limit equilibrium analysis for calculating 'factors of safety', and a Finite Element analysis for calculating bending moments, displacements, and strut forces. The stability analysis is carried out only for cantilever or singly-propped walls. The finite element analysis is elasto-plastic, with limits defined by active and passive states. Minimum effective earth pressure can be prescribed as an equivalent fluid density in accordance with certain codes of practice and fluid pressure within tension cracks can be specified. This program can be used on most modern desk-top computers.

It should be emphasized that the results obtained from such programs are very dependent on the relative stiffness of the chosen components of the model and that simple arbitrary rules for determining soil stiffnesses can lead to large variations. Depth, soil grading, and overconsolidation ratio can have considerable influence on the stress/strain properties of soils and simple rules for arriving at these parameters should be shunned in favour of proven experience. Programs, apart from needing to be mathematically correct, also need to be calibrated against known case evidence.

Considerable experience has been gained through studies carried out at Imperial College, London where a close association has been maintained with a series of major excavations. Their experience can be utilized through the Imperial College Finite Element Program (ICFEP).

Each stage of any multi-propped excavation must be studied, mimicking the stages of excavation and propping as closely as possible. In this way envelopes of bending moment, prop force and required penetration can be derived for the purpose of calculating, for example, reinforcement in reinforced concrete walls. Care should be taken in the matter of wall penetration below the excavation level, since upward seepage can lead to significant reduction in passive pressure over the permanently embedded length of a wall, even in stiff clays. In several major cities the effective groundwater table is rising due to the decrease in industries which previously extracted water from boreholes. This warrants consideration and a report has been produced on the subject by the Construction Research and Information Association (CIRIA).

6.8 Structural wall design

6.8.1 *General considerations*

When a retaining wall has been analysed for stability, it remains to design the structural form of the wall, and at this stage appropriate bending moments and shear forces must be used. This aspect of design has an important influence on the economy of the design as a whole. Temporary and permanent states must be considered. For example, in the case of many cantilever walls,

Figure 6.21   Force lines indicating redistribution of pressures on wall with time

a ground-bearing slab is cast to abut the wall and has the effect of propping the wall in the post-excavation stage. In many circumstances this effectively stops wall deflection and further moment development. In most instances, therefore, the temporary state proves to be the critical one. It is unusual to design such walls for pressures on the retained side other than the water pressure and the (effective stress) active pressure. Even in overconsolidated clays it would appear that the horizontal stresses in the soil mass, which are relieved at the excavation stage, do not become restored. The force lines in the soil mass almost certainly reorientate around and beneath the excavation (Fig. 6.21).

Experience with many walls in such circumstances in stiff clay appears to indicate that wall bending moments in the slab vicinity do not in the long-term significantly exceed those calculated at the excavation stage when only active and passive soil states, are normally considered, although the position of maximum moment may change a little.

6.8.2 Determination of bending moments and shear forces in a cantilever wall

The conventional method of determining maximum bending moment in a cantilever wall is to consider the active and passive pressure diagrams as used in the stability calculations. The full ultimate passive pressure is used on the excavated wall side, rather than any reduced value. This is because maximum passive pressure must be mobilized near the excavation level through wall rotation, and because excavation will, by reduction of the vertical soil stresses at that level, have produced an ultimate or near-ultimate passive state. Hence in practice maximum bending moment appears in most circumstances not far below the excavation level.

This means that from the head of the wall down to the point of maximum bending moment the design moments correspond most nearly to ultimate conditions.

bending moment diagram
obtained using full
passive pressure

bending moment and shear
force from line drawn
as tangent to moment
diagram passing
through wall toe

Figure 6.22 Determining moments below the maximum point

However below the point of maximum bending moment, because of the extension of the wall to provide a rotational stability factor, the full ultimate pressures on the passive side will no longer apply and in consequence if the ultimate passive state were assumed here, the moments would be curtailed too rapidly and excessive shear forces would be evident.

The method of determining moments below the maximum moment point is therefore empirical and as shown in Figure 6.22. It should however be pointed out that where a significant change in the stiffness of the soil occurs below the excavated surface level, for example if a wall toe is embedded significantly into rock, then this simple method may have to be modified in the light of judgement and of the fixity which the increased stiffness may provide near the wall toe.

6.8.3 *Determination of bending moments and shear forces in tied or propped walls*

For walls with a single prop or tie, the earth pressures and prop or tie force can be used simply to determine the bending moment and shear forces in the wall. It is commonly assumed for the purpose of bending moment determination that full passive pressure acts on the wall below the excavation level, rather than some reduced pressure which might be derived because of the extension of the wall to provide a 'factor of safety' or rotational stability factor. This is because of the usually significant reduction of effective overburden pressure at and below the final excavation level, which initiates the passive condition.

For multi-propped or tied walls, where the moments have been derived from one of the available Finite Element or similar computer programs, the significant stiffness difference between anchor ties and, for example, floor props will already have been taken into account and the bending moments will be available. While steel reinforcement may be varied and curtailed to some extent in reinforced concrete walls, the temptation to believe in bending moments being exactly certain in value at given levels should be resisted, since very often the variation of, say, the stiffness of one prop relative to another can alter the diagrams in a significant way. It is often the case, however, that the spacing between floors which prop a wall in the final condition is less than the spacing between the props or ties in the temporary stage, and consequently bending moments and shear forces can be at their most severe in the temporary state.

It will be noted that, for relatively flexible walls, pressures are increased at the propping points and diminished midway between propping points; the soil effectively arching between the props. This feature will be apparent in the analysis if carried out by one of the appropriate computer programs.

Most reinforced concrete diaphragm walls are not designed in the structural sense in accordance with conventional codes applicable to above ground conditions. Historically this is because concrete for use in piles and walls has to be very free-flowing because, if stiff, it would not flow properly around steel reinforcement. Shear does not appear to be a problem in such cases. It is notable that to the author's knowledge, no case of shear failure in such walls has ever been reported. This may be because shear in a wall would presuppose shear in the soil behind it, and it would appear that in this respect, the soil and the wall may therefore act in a composite manner.

6.9 Retaining wall deflection and associated soil movements

Although the act of excavation to install a bored pile or diaphragm wall panel must in itself lead to some small ground deformation, it is the excavation which subsequently takes place adjacent to the wall, and the following changes in the general soil profile which account for most of the movement which occurs.

When a deep excavation is carried out, this affects the stresses in the whole immediate area, both within and without the plan area of the works involved. Such stress changes lead to general movements in the soil mass, and for large excavations, particularly in clay soils, this is a time-dependent and significant feature.

Secondly, relief of lateral pressure due to excavation adjacent to a retaining wall leads to lateral wall movements, and the magnitude and timing of these are very dependent on soil properties, the way in which excavations are staged, the stiffness and position and timing of installation of props or ties, the prestress if any which is provided in the lateral support system, the presence of berms and how and when they are removed, and the stiffness and general dimensions of the wall.

It is clear therefore that there is no single simple answer to the questions which may be asked regarding predicted wall movements but at the same time there are some methods of analysis which range from an advanced finite element type to simple reliance on experience. However good and thorough the analysis, however, any failure to comply with the design concepts in final construction can completely invalidate any conclusions which might have been drawn from calculation.

The simplest model for assessing the magnitude of general ground movements due to excavation is probably based on the whole mass acting elastically. Movements may be estimated by treating the excavation as negative load as for an embankment with vertical sides (Poulos and Davis, 1974). The analysis will then be based on Young's modulus E and Poisson's ratio v_u, the undrained parameters if short term conditions are being considered, or E' and v' where long-term analysis is sought.

It is useful to remember that where G is the shear modulus,

$$\frac{E'}{2(1+v')} = G' = G_u = \frac{E_u}{2(1+v_u)}$$

For undrained conditions $v_u = 0.5$, whereas v' is in the range 0.1 to 0.33 and may be taken as about 0.2 for stiff clays.

In so far as the local movements in the vicinity of the wall are concerned the general pattern of behaviour is as indicated in Figure 6.23.

The interaction between the soil and a retaining wall is complex. Rigorous analysis is not feasible at the present time. The conventional calculations for stability are based on lumped factors of safety and provide no explicit information regarding deflection. It has been shown by Bica and Clayton (1989) that there are some 25 variants on the normal stability calculation method, so it is hardly surprising that deflection is difficult to determine even on an empirical basis.

A recent method for assessing the deflection of retaining walls of the stiff cast-in-place type in overconsolidated clays has, however, been proposed by

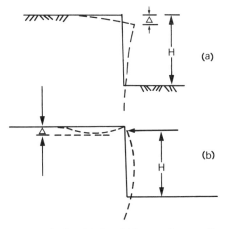

Figure 6.23   Local movements in the vicinity of (a) a cantilever wall, and (b) a propped wall

Table 6.4 Examples of recorded wall deflection

Location	Ground conditions	Max. ret. ht.	Corresponding wall depth	Wall type	Anchor/ props	Wall lateral deflection	Distance measured from wall head	Source of information
COHESIONLESS SOILS								
Buffalo, N. York	Loose sand to 5.5 m. Dense sand and gravel below	11 m	15 m	Soldier piles (tied)	2 levels	2.5 mm	0 m	Peck (1969)
London, Ontario	10 m of med. dense sand over very dense sand	15 m	17.7 m	Sheet piling	5 levels	97 mm / 230 mm	0 m / 15 m	Bauer (1975)
Zurich	Approx 3 m fill over silts and fine sands with some gravel. Some 5 to 15% clay content	15.5 m	24 m	0.8 m thick prestressed diaphragm wall	1 level ties plus berm	−6 mm / +22 mm	wall head / excavation level	Gysi et al. (1977)
COHESIVE AND GRANULAR SOILS								
Oakland, California	Loose sand to 4.3 m, dense sand and very stiff sandy clays below	19 m	29 m	Soldier piles (propped)	4 levels pre-stressed struts	−6 mm / +7 mm	0 m / 13 m	Peck (1969)
Seattle	Very stiff clays and dense sands	23.8 m	24.5 m	Soldier piles (tied)	8 levels	115 mm	0 m	Peck (1969)
Chicago Water Tower Place	Fill and fine sand to 7.3 m, then silty clay-soft at top but with increasing strength with depth	13.4 m	18.9 m	0.75 m thick slurry trench wall	1 level anchors and berm	100 to 150 mm / 50 mm	0 m / 13.4 m	Gnaedinger et al. (1975)
London, UK	Fill and Thames gravel to 8.5 m, then London Clay	8 m approx from ground level	10 m approx from ground level	0.5 m thick diaphragm wall-wall head 4 m below ground level	1 level anchors	3 mm	at wall head	Hodgson (1975)
London Chelsea Town Hall	Fill and brick-earth to 4½ m, sand and gravel to 7m, brick earth to 8m, then London Clay $c_u \simeq 100$ KN/m²	13 m below ground level	18.9 m below ground level	0.5 m thick diaphragm wall. Wall head 4.5 m below ground	2 levels struts	27 mm	at wall head	Corbett et al. (1975)
London Guildhall	Approx 9.5 m Thames Gravel over London Clay	10.3 m below ground level	12.15 m below ground level	0.5 m thick diaphragm wall. Wall head 2.4 m below ground	2 levels ties	9 mm / 4 mm	wall head / excavation level	Littlejohn and Macfarlane (1975)

Site	Soil	Excavation depth	Depth below ground	Wall	Support	Movement	Location	Reference
London Vauxhall	Thames gravel to 11.4 m, then London Clay	14.5 m below ground level	16.7 m below ground level	0.6 m thick diaphragm wall. Wall head 1.7 m below ground	3 levels ties	22 mm / 7 mm	wall head / excavation level	Littlejohn and Macfarlane (1975)
London YMCA	Fill and gravel to 7 m overlying 17 m London Clay	16 m	18 to 19 m	0.6 m thick diaphragm wall	1 level ties plus berm	5 mm / 19 mm	wall head / excavation level	St John (1975)
London Charing Cross Road	6 m fill and gravel over London Clay	11 m	18 m	0.8 m thick diaphragm wall	3 levels struts	20 mm / 30 mm	wall head / excavation level	Wood and Perrin (1984)
London New Palace Yard	10 m sand and gravel over 34 m London Clay, over Woolwich and Reading Beds	18 m	30 m	diaphragm wall-wall head 2 m below ground	5 levels struts	2 to 17 mm / 30 mm	wall head / 15 m	Burland et al. (1979)
Reading A 329 (M)	3 m sand and gravel over 17 m London Clay, then Woolwich and Reading Beds	8 m	15 m approx.	1 m thick diaphragm wall	None	16 mm / 8 mm	0 m / 8 m	Burland et al. (1979)
Humber Bridge	6 m soft silty clay, 4 m sand, silt and clay, 12 m firm to stiff silty clay, 5 m sand and gravel, 7 m stiff to hard fissured silty clay	29.5 m below ground	31.5 m approx. below ground	0.8 m thick diaphragm wall. Wall head 4.5 m below ground	5 levels struts	20 mm / 22 mm	at wall head / at final excavation level	Busbridge (1975)

SOFT TO MEDIUM CLAYS/SILTS

Site	Soil	Excavation depth	Depth below ground	Wall	Support	Movement	Location	Reference
Oslo	Soft to medium clay (wall toe at rock)	10 m approx.	15 m approx.	sheet pile wall	5 levels ties	25 mm	10 m	Peck (1969)
New York	Fill and organic silt to 12 m, then sand to 15.2 m hardpan to 19 m and rock below	18.5 m approx.	21 m approx.	0.9 m thick diaphragm wall. Wall head 1 m below ground	6 levels ties	up to 70 mm / 25 mm	wall head / mid height	Saxena (1975)

STIFF CLAYS

Site	Soil	Excavation depth	Depth below ground	Wall	Support	Movement	Location	Reference
Houston, Texas	Stiff clay c_u approx. 150 KN/m²	18.3 m	22.9 m	0.9 m r.c. piles at 1.37 m centres	2 levels struts berm	19 mm / 6 to 7 mm	wall head / mid height	Peck (1969)
London Brittanic House	London clay to approx. 24 m, then Woolwich and Reading Beds	14 m	17 m	diaphragm wall	large berm	40 mm plus	wall head	Cole and Burland (1972)
London Neasden Lane Underpass	London Clay to depth of over 30 m	8 m	13 m	0.6 m thick diaphragm wall	4 levels ties	50 mm / 40 mm	wall head / excavation level	Sills et al. (1977)

Bolton *et al.* (1989). The basis of the method is (a) to calculate the soil strength which is mobilized to hold a proposed wall in equilibrium; (b) to use stress/strain data to deduce an average soil shear strain; and (c) to use this average shear strain to attribute strains to the various active and passive zones and hence calculate the rigid body wall movements. The method is relatively simple to operate but requires laboratory data describing the mobilized shear strength as a function of strain. Ideally, samples should be tested to simulate the stress paths experienced by soil elements in the field, and obviously the soil behind and in front of the wall would need to follow their separate stress/strain states in order to represent correctly the behaviour of each major soil zone. In practice this kind of information is unlikely to be available from conventional triaxial testing, but can be generated from special triaxial tests as described by Bolton and Powrie (1988).

Provided the deflected form of a wall can reasonably be established, either by calculation or experience of similar cases, then reasonable deductions may be made concerning the deformations which occur in the retained soil, using the method proposed by Bransby and Milligan (1975) and Milligan (1983). This method predicts that significant movements will occur within a zone bounded by a line at 45° from the base of the wall to the ground surface, and that any point within that zone will have equal displacement components in the vertical and horizontal directions. Hence the soil surface deformation is a direct reflection of the deflected shape of the wall.

If the soil is not going to behave in an undrained manner, then the zones of deformation are reduced in extent, with the ratio of horizontal to vertical movement near the surface in the retained soil increasing to as much as 3 in the case of a cantilever wall and falling to as little as 1/3 behind a propped wall.

Examples of recorded wall deflection are shown in Table 6.4, together with appropriate references for further information, but it will be observed that most of the published data refer to major excavations and large retained heights. Other data on wall movements may be obtained from the British Standards Institution publication *Recommendations for ground anchorages — a draft for development* (1982).

Practical experience shows that in the fields of bored pile and diaphragm wall construction, a large volume of work relates to cantilever walls with retained heights varying from perhaps 3 m to 8 m, and these walls have not attracted so much attention in terms of instrumentation and performance measurement. Experience in Britain suggests that if the passive zone is situated in moderately dense to dense sands or in stiff clays or soft rocks then lateral deflection of a wall head is normally in the range 15 mm to 30 mm provided the parameters used in design are realistic. This experience seems to apply through a wide range of retained soil heights, because as the height increases, so also does the depth and stiffness of the required wall. However some warning should be given that if the soil in the passive pressure zone consists of loose sands, soft or firm clay, or even backfilled material, then such deflection values could be exceeded.

The use of capping beams on cantilever walls, particularly where the plan shape of the wall involves changes of direction, is effective in restraining lateral and vertical deflections and unifying wall behaviour, provided such beams are adequately reinforced for the shears and tensions which arise.

It should also be noted that the use of berms and their removal just before casting a slab up to the wall can significantly reduce deflections of the wall in both the short and the long term, and that increase of the embedment of walls below dredge level in the case of tied walls helps to reduce the deflections near the bottom of the wall.

In so far as ground settlement is concerned, on the retained side of walls some approximate guidance on expected values may be obtained from the relationships suggested by Burland *et al.* (1979) as follows in relation to Figure 6.23.

Soil type	Δ/H for preliminary design
Soft clays	2%
Loose sand or gravel (beware groundwater or vibration)	0.5%
Stiff clay	less than 0.15%

Similar figures have been given by Peck (1969).

6.10 Measurement of the deflection of retaining walls

In recent years the availability of good-quality instrumentation has allowed many researchers to undertake detailed measurements of wall movements during following stages of construction. This has provided much worthwhile information relating to the back analysis of wall behaviour, and has improved the general knowledge of how walls behave in practice.

In using modern inclinometers it is usual to instal the necessary ducts with the reinforcement cages during construction of the retaining wall, and it is essential to avoid damage to these ducts thereafter. It is not always easy to ensure that duct heads will remain unmutilated in following construction work, so that the purpose of the duct must be explained clearly to those who may be concerned with its subsequent preservation. Inclinometer ducts may also be placed in boreholes in the ground to measure movements of soil away from the wall position.

It is then necessary to establish the true plan position of each duct with accuracy, bearing in mind that total deflections to be recorded may be no more than say 20 mm. Thus one must seek to achieve positional measurement accuracies within 1 or 2 mm, and this may necessitate a sturdy reference point close to the duct position which will remain accessible throughout the works, the use of a good-quality theodolite and the use of invar tapes in accordance with precise surveying practice.

The inclinometer consists of a pendulum arrangement within a watertight torpedo-shaped housing. The pendulum is strain gauged so that the output from the torpedo presents a record of the angle at which the pendulum is held. The instrument is so designed that the component of inclination is recorded in the plane defined by four wheels on the torpedo, and the wheels are guided in vertical slots formed in the specially extruded duct tube. The wheels on the probe are lightly spring-loaded.

The gauge length of instruments may vary but is usually between 0.5 m and 1.0 m and the probe is generally lowered one gauge length at a time so as to obtain a continuous and accurate measure of any duct deviations. Resolution is often of the order of 0.1 mm on the gauge length and it has been found that, with careful work, accuracies of about 25 mm can be achieved over lengths of 300 m run. Errors can arise due to spiralling of the duct, lack of repeatability of inclinometer positions, temperature changes and operator skill.

In the field, a check can be made on the accuracy by taking two sets of readings in the same vertical plane, but separated by 180°, and the sum of these readings should be constant if an error does not exist.

When used in walls it is normally assumed that the movement of the wall is in a plane at right angles to the wall, so that the duct must be correctly orientated.

Vertical movement of walls and of the soil around them is also of interest during the processes of excavation and thereafter. In order to measure vertical movements at or near the surface, no more is required than well established reference points in the soil, preferably founded below the immediate surface, and good-quality surveying methods. If however it is required to measure vertical ground movements at various depths below the surface, then magnet extensometers may be most suitable. These instruments consist of a number of small ring magnets, secured at required positions in the ground by a combination of springs and grout, with an internal PVC tube passing through them. The position of each magnet is recorded by the operation of a reed switch that is lowered down through the tube with a steel tape. The system and movements within it can be related either to the surface survey or a deep magnet reference point as required.

It is fairly obvious that to fully instrument and record data for a large retaining wall is expensive, but quite apart from the improvement of knowledge which it may bring, it can be used to establish whether significant movements may have occurred in adjacent properties and this can sometimes help resolve contentious claims.

7 Problems in pile construction

With any civil engineering work, care is necessary at all stages of design and construction. In particular, workmanship and supervision is of prime importance. Construction of piled foundations is a specialist activity, calling for considerable expertise and reliable workmanship, the more so as the completed element can rarely be inspected for defects. Remedial work to piling found to be wanting at a later date can be extremely time-consuming and expensive, if not well-nigh impossible.

Two main methods of installation of piles are in common use: driving using a pile hammer or boring. Each system has particular advantages and disadvantages, and selection of the correct pile type calls for considerable experience.

The various methods of constructing and installing piles are discussed in Chapter 3: each system gives rise to particular constructional problems.

7.1 Driven piles

Preformed driven piles take the form of prefabricated elements driven into the ground. Materials commonly used for the manufacture of piles include steel, concrete and timber. This type of pile has the advantage that the pile element can be thoroughly inspected before driving and that quality control is easy to organize and enforce.

7.1.1 Design and manufacture

The piles should be designed to withstand handling and driving stresses. Stresses arising from these temporary conditions are usually more severe than the stresses in service.

During the manufacture of concrete piles adequate curing of the piles is necessary, and a continuous water spray is recommended. The formation of shrinkage cracks can lead to corrosion of the reinforcement. Similarly, premature lifting of the pile can cause cracking. Such defects are usually easy to detect. However it is rather subjective to decide when a crack is too wide. Piles with major cracks due to excessive bending moments should be condemned.

Experience and personal judgement are necessary when assessing the severity of cracking; the aggressiveness of the environment should also be considered. It is not necessary to reject all cracked piles, and a series of transverse hair cracks of a width 0.15 to 0.3 mm (DIN 4026, BS 8110) may be considered acceptable.

With segmental pile systems special precautions are necessary to ensure that the ends of the segments are square. Out-of-squareness will cause difficulty in making the joints and will cause eccentric transfer of load across the joints. A lack of fit of the joints will rapidly become unacceptable, owing to the slender geometry of pile segments. It is suggested the limit on out-of-squareness should be less than 1 in 300, in good practice. The first section driven determines the direction of pile motion and following sections must follow the same path. Unless the surrounding soil is very weak the straightening of the pile caused by soil forces will cause cracks in the section adjacent to the joints.

The design and manufacture of steel H-piles is quite straightforward. If heavy driving conditions are expected, the pile may be fitted with a shoe or a rock point. Preparation of tubular or hexagonal piles may involve a considerable amount of welding. As far as possible, the weld should be carried out in the fabrication shop under the supervision of suitably qualified personnel. Reference may usefully be made to British Steel Corporation's publication *Weld Quality Acceptance Standards for Site Butt Welds in Steel Bearing Piles*. Pile driving will subject the welds to high repetitive stresses and any defects can easily lead to the propagation of brittle fractures. It is important that the pile sections are correctly aligned, as lack of straightness will cause problems on driving and result in high bending stresses being locked into the pile.

Buckling of slender piles is not usually a critical design consideration, as even soft alluvium provides sufficient lateral restraint to prevent buckling of the pile. However, the selection of unduly light sections can cause problems in handling and pitching of the piles.

Similarly there are few problems with the manufacture of timber piles, provided the correct species of wood is selected for the job in hand. The piles should be fitted with a steel or cast-steel shoe to protect the toe of the pile. A steel band is normally fitted to the head of the pile to prevent 'brooming' during driving.

7.1.2 *Installation of driven piles*

7.1.2.1 *Pile driving.* The most obvious cause of damage during driving of concrete piles is distortion or spalling of the head of the pile (Figure 7.1). Such damage is frequently the result of insistence on driving to a predetermined length. The driving of piles is usually controlled by a specified final set, and driving to a length may entail the adoption of expedients such as jetting or pre-boring.

Figure 7.1   Damage to head of precast concrete pile. Photograph courtesy of W.A. Dawson Ltd., Luton

Depending on the soil profile, heavy driving may be necessary to penetrate thin dense layers. The piles selected should be capable of accepting this heavy driving, indeed it may be considered necessary to drive test piles to check the performance of the pile in this respect. Steel piles will accept heavy driving as they rarely break. However, severe distortion of the pile head may occur. Damage to the pile toe may also occur, but this is not obvious unless piles are extracted (Figure 7.2).

Damage as a consequence of pile driving may be minimized by suitable selection of piling plant. Selection of the correct hammer weight is particularly important, and recommendations are summarized in Chapter 3. In general a heavy hammer with a low drop results in lower impact stresses than a light hammer with a higher fall, giving the same rated energy per blow. Typical driving helmets and dollies for precast concrete and steel piling are shown in Chapter 3.

The purpose of the dolly is to reduce the peak impact stresses arising from the hammer blow, and selection of the correct packing materials requires considerable experience. It should be noted that the resilience of the packing changes with constant use, and most commonly-used materials harden appreciably. Selection of packing for driving steel piles is not so important. However, the helmet should be designed to suit the particular section driven, to reduce distortion of the pile head. Driving precast shell piles requires

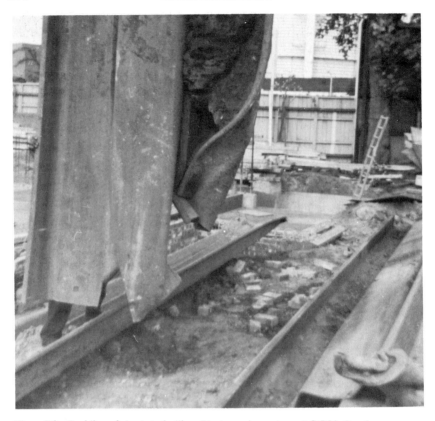

Figure 7.2 Buckling of sheet steel piling. Photograph courtesy of CIRIA, London

special equipment as the energy supplied by the hammer has to be distributed between the mandrel and the shells. Normally such piles would be installed by a specialist subcontractor experienced in the use of this type of piling.

It is not usually practicable to inspect the toe of a driven pile, except in the case of tubular piles. If there is any doubt as to the ability of the pile toe to penetrate the various strata, a rock point should be fitted to assist penetration. Problems can be expected when piles are driven through strata containing boulders or other obstructions. In such ground concrete piles may be broken and steel H-section piles deformed or split (Figure 7.3).

It is important that the designed geometry of the pile group is maintained. Positive control is possible only at the ground surface, and piling frames should be set on stable foundations. If guide trestles are used they should be of substantial construction, and positive guides for the piles should be fitted. It is not possible to correct deviations once pile driving has started. Piles, particularly flexible piles, may wander off line considerably below ground level. An extreme case of an H-pile turning through 180° has been reported. However, a greater danger is that piles will collide in a closely spaced group.

Figure 7.3 Buckling of steel H-pile driven into boulder clay. Photograph courtesy of Sir Robert McAlpine & Sons Ltd., London

Misaligned piles will not generally reduce the bearing capacity of the group significantly. However, O'Neill *et al.* (1977) reported that the stiffness of experimental pile groups was reduced by 30% over that of the designed geometry. Very little can be done to prevent piles wandering, except to choose a stiff pile section and possibly experiment with different types of shoe. If obstructions in fill are causing the problem, pre-boring may be resorted to.

Table 7.1 Pile driving records for a typical steel H-pile: Denton Island Bridge, Newhaven, Sussex. (Courtesy East Sussex County Council)

Date	Depth (m)	Driving resistance (blows/0.25 m)	Comments
29.3.83	12.5 to 27.5	3 to 5	Hammer: 4 tonne drop weight, fall about 1 m
11.4.83	27.5 to 32.5	8 to 10	Soil profile: 0–30 m silty clay
11.4.83	32.5 to 36.0	12 to 15	alluvium
11.4.83	36.0 to 41.25	9 to 12	30 m —upper chalk SPTs in top 6 m
14.4.83	41.25 to 47.0	15 to 18	of chalk gave $N = 35$ to 50
	47.25	68	blows/300 mm, i.e.
	47.5	59	Grade I material
	47.75	63	Below 36 m no data
	48.00	47	
15.4.83	48.25	81	Design length of pile
	48.5	78	pile 36 m
	48.75	68	
	49.00	43	Pile completed

7.1.2.2 *Control of final set.*   In some soil types difficulty may be experienced in achieving the desired set, and both increases and decreases of the set some time after piling have been noted. The pile driving process frequently results in an increase in pore-water pressures within a few diameters of the pile. Dissipation of these pressures leads to an increase in adhesion, and hence set, with time. If the set-up is rapid, cessation of driving to splice the pile, etc., may cause difficulty in re-starting the pile. In extreme cases refusal can result (Figure 7.4).

Piling into chalk or other soft carbonate rocks can present special problems, particularly with low-displacement piles, such as H-piles. Breakdown and liquefaction of a thin layer of rock around the pile may lead to very low driving resistances. Table 7.1 shows the driving records for steel H-piles in Upper Chalk at Newhaven. Driving resistances were very low compared to the values expected from the Standard Penetration Test results. Pile tests showed that the

Figure 7.4   Driving records for two piles in the German North Sea. (McClelland *et al.*, 1969)

piles had an inadequate bearing capacity at the designed length, when tested 14 days after driving. The piles were lengthened by about 10 m, and re-testing proved them to be satisfactory.

In dense silts and some weathered rocks, the reverse effect may be observed. It is postulated that negative pore-water pressure may be induced by pile driving, leading to a temporary increase in strength and driving resistance. Subsequent dissipation of the negative pore pressure is then reflected in a loss of set. Such an effect was recorded on Teesside, where tubular steel piles were driven into highly fractured hard Mercia Mudstone. The problem was overcome by re-driving the piles after a few days. In some cases several re-drives were necessary and the loss of set was eventually reduced. Unfortunately no pile load tests were undertaken in this case and the effect of the reduction in set on the bearing capacity could not be assessed.

Occasionally piles may buckle or telescope under very heavy driving (Figure 7.3). Such conditions are very difficult to detect as the driving resistance may not change appreciably. On any site a constant check should be made on the driving resistances encountered and compared with previously driven successful piles. Any pile thought not to be satisfactory should be investigated by integrity or load tests, or extracted and examined.

7.1.3 Associated ground movements

When large groups of displacement piles are to be driven, heave of the ground may be a significant factor. Careful levelling of the pile heads and the ground surface should be instigated if significant heave is expected. Cases have been reported of heave reducing the bearing capacity of piles, or in extreme cases inducing tensile failure in the shaft. Provided the pile shaft is not damaged, bearing capacity can usually be recovered by re-driving the piles to an acceptable set. If tensile failure of the pile shaft occurs, repairs can be expensive and may involve drilling down the axis of the pile to grout the fractures. The problem of heave can be alleviated by driving the centre piles of a group first and working out to the perimeter piles, selecting a low displacement type of pile, or pre-boring holes to remove a volume of soil equal to 50–100% of the volume of the piles. It should be noted that pre-boring can lead to a loss of skin friction, although this may not be significant in large groups of closely spaced piles.

The effects of heave may extend several metres beyond the perimeter of a pile group and damage to adjacent structures has been recorded.

When piling close to sensitive structures it is usual to adopt low-displacement piles to minimize heave.

In sensitive clays heave is localized and the remoulded soil around the pile is extruded at the ground surface. Reconsolidation of the soil in this situation may result in downdrag on the pile. Similarly, in loose sands and gravels little heave is recorded as the pile driving operation usually causes compaction of the deposit. Insensitive clays and silts are the soil types most prone to heave, and the volume of the soil displaced is approximately equal to the volume of the piles. Based on a study of case histories, Hagerty and Peck (1971) suggested that the surface heave for such soils would be equivalent to one-half the displaced volume of soil, the remainder of the displaced volume causing ground heave in the soil surrounding the pile group.

In particular circumstances lateral soil displacement may occur, leading to the damage of adjacent structures or the displacement of previously driven piles. The walls to the Clyde Dry Dock at Greenock were overloaded and cracked when cast-in-place piles were driven adjacent to the wall (Geddes *et al.*, 1966). In this instance high pore pressures were also thought to have been induced in the silty alluvium. Significant lateral soil displacement may also occur when piles are driven through river banks or other sloping sites.

In addition to soil displacement, heavy ground vibrations are transferred to the soil close to the pile. In silty alluvium, substantial pore pressures may be developed as a result of pile driving. On one site, piezometers adjacent to the piling were seen to overflow for a short period. Dissipation of the excess pore pressures may lead to the development of downdrag forces. In a recent case, pore pressures set up by pile driving re-activated an old landslip, causing 600 mm displacement of the pile heads.

7.1.4 *Noise and ground vibration*

7.1.4.1 *Noise from piling operations.* Pile driving is an inherently noisy operation and severe environmental restrictions may be included in the conditions of contract. Noise levels of over 85 decibels within 10 m of the piling plant are quite common. Typical data on noise levels produced by piling operations have been published by CIRIA (Report No. 64). These data are reproduced in Table 7.2.

Investigations into the sources of the noise have shown that a large proportion arises from secondary effects, rebound of the hammer, rope slap, engine noise, etc. Improved design of the components of the piling rig can considerably reduce the high-frequency components of noise. To dampen the noise sufficiently to be acceptable in urban situations, it may be necessary to enclose the hammer or the leaders in an acoustic chamber. Several piling companies have developed such chambers to suit their particular piling machines. The use of such devices usually results in some reduction in the efficiency of the pile hammer and creates difficulties in handling and pitching piles.

Driving steel sheet piling is exceedingly noisy, as the driving cap usually involves steel to steel contact. In areas where severe restrictions are placed on noise levels, pile driving vibrators or the Pilemaster hydraulic pile driver may be adopted. Such machines involve the use of an auxiliary power unit which may, in itself, emit a high level of sound.

Table 7.2 Noise data for different pile types (after CIRIA Report 64)

Pile type	Dia. (or width) (m)	Method	Size	Dolly type and condition	$L_w(A)$ (dB(A))	Soil	Leq (dB(A)) at 10 m (1 cycle)
Sheet steel	0.48	Diesel hammer	310 MN m energy	None	133	Clay	104
H-section steel	0.36 square	Diesel hammer	620 MN m energy	None	125	Fill/clay/ sandstone	96
Precast shell	0.535 dia.	Drop hammer	6 tonne, 0.5 m drop	Wood (good)	124	Fill	83
Precast concrete	0.275 square	Drop hammer	4 tonne, 0.5 m drop	Wood (average)	116	Chalk/clay	87
Cast-in-place driven casing	0.4 dia.	Drop hammer	4 tonne, 1.0 m drop	Wood (good)	119	Fill/clay/ sand-marl	89
Impact bored cast-in-place	0.5 dia.	Tripod winch	18 kW	None	104	Rough fill/ clay/limestone	76
Bored cast-in-place	0.75 dia.	Crane-mounted auger	Crane engine 99 kW Donkey engine 125 kW	None	117	Clay/silt	79

7.1.4.2 *Ground vibration caused by piling.* Related to noise is the potentially more serious problem of ground vibrations, and the consequent hazard posed to adjoining structures. Many investigations have been made into the problem, firstly to predict the likely peak ground accelerations, velocity, or displacements, and secondly to establish safe limits of ground vibrations for various types of structure. To date, none of these investigations has proved very conclusive with respect to piling operations. Prediction of peak-to-peak acceleration or velocity in real situations is most complex. Firstly, the energy transfer to the soil is poorly understood and attenuation of high-frequency components is rapid. Secondly, the response of various forms of construction to the ground vibrations is difficult to predict, and some structural details, e.g. floor spans which resonate, may lead to a considerable magnification of the effect.

Several indices have been proposed as a measure of the intensity of ground vibrations (Geiger, 1959). The most widely accepted of these criteria are based on the peak particle velocity or the energy intensity of the vibrations induced in the soil adjacent to the foundations of a building. Empirical guidelines have been drawn up using these criteria to define various levels of damage. One of the most popular of these is included in the German Codes of Practice DIN 4150 (1970) (Table 7.3).

The recommendations in Table 7.3 have not been drawn up specifically for ground vibrations induced by piling, and it is considered that they are overly stringent. For structures which are not of great prestige or historical value it is considered that the limits in this table could well be doubled. When considering reasonable limits on ground vibrations, the background level of vibration should be assessed. In built-up areas heavy traffic can cause surprising intensities of vibration, and peak-to-peak velocities exceeding 3 mm/s have been recorded 10 m from a road.

The use of empirical limits on velocity or acceleration necessitates the use of field instrumentation to observe the induced vibrations. Suitable equipment is expensive to purchase, and several levels of recording are possible. The simplest scheme is manual recording of peak-to-peak phenomena, and the most complex is a full record of the ground vibrations to enable a frequency analysis to be carried out. Siting of the accelerometers is important, as

Table 7.3 Maximum allowable velocity, from DIN 4150

Class	Description	*Max. velocity (mm/s)
1	Ruins and buildings of great historic value	2
2	Buildings with existing defects (visible cracks in brickwork etc)	5
3	Undamaged buildings in technically good condition except for minor cracks	10
4	Strong buildings, industrial buildings of R.C. concrete or steel	10–40

*Particle velocity in soil adjacent to foundations of structure.

amplitudes differing by a factor of 10 can easily occur due to structural magnification. It is also highly desirable to use triaxial accelerometers, as it is not usually clear in which direction the most damaging vibrations will propagate. The human body is extremely sensitive to vibrations and very low intensity shock waves may be detected. For this reason, people frequently conclude that ground vibrations are more severe than measurements show them to be. To limit unreasonable claims for damage to adjoining structures, it is highly desirable to carry out a detailed structural survey of the buildings. Good-quality photographs of sensitive areas or existing defects form useful records to counter or confirm particular assertions.

A common situation is for piles to be driven through a soft soil into a bearing stratum, with adjacent buildings founded on the soft material. Practical experience in such situations suggests that the soft layer effectively damps the ground vibrations. As a crude rule little or no structural damage is likely to be experienced if the pile and rig is at least 5 m from an adjacent building. The occupants of such a building may, however, suffer severe nuisance. Various expedients may be adopted to reduce the intensity of ground vibrations caused by pile driving. These include pre-boring, the use of low-displacement piles, or in particularly sensitive areas, the adoption of alternative pile types (see also section 10.4.5).

7.2 Bored piles

7.2.1 *Excavation of the pile bore*

In favourable soil conditions, i.e. stiff clays, bored piles can be constructed without casing, except for a short guide length. Under less satisfactory conditions temporary or permanent casings may be used to support the pile, or bentonite drilling fluid may be resorted to. Excavation of the soil is by augers, buckets, or a variety of grabs (Chapter 3). Continuous-flight auger equipment, which can place piles in many soil conditions without the use of any casing, has recently become popular.

7.2.1.1 *Overbreak*. The formation of cavities or overbreak outside the nominal diameter of the pile is a significant problem in cohesionless soils (Figure 7.5), particularly below the water table. Under such conditions it is essential that a temporary steel casing is used and that the cutting edge of the casing is driven below the base of the advancing bore. If the casing is not at the base of the hole, slumping or washing of soil into the bore can occur, with subsequent problems on concreting the pile. The cavities may or may not be filled with water. Even with well-controlled drilling it is possible for the inflow of water through the base of the advancing bore to seriously loosen the surrounding soil, with a consequent loss of skin friction.

In very clean sands or rounded gravels, a soil arch may not form around the

Figure 7.5 Bulbous projection on pile shaft caused by overbreak. Photograph courtesy of CIRIA, London

borehole, even in dry conditions. Under these circumstances collapse of soil into the bore will form large irregular cavities. However, the problem is easily overcome by the use of a temporary casing.

The problem may be alleviated by drilling under water and maintaining an excess water head in the casing or by resorting to the use of a bentonite drilling mud. Alternatively the boreholes may be permanently cased.

Severe boring problems may occur, with resultant overbreak in pile shafts, in boulder clay. In such heterogeneous materials it is very difficult to assess the maximum size of any rock fragments from site investigation data. It is frequently advantageous to adopt large-diameter piles, drilled with heavy equipment, to enable the largest boulder to be removed from the bore. The alternatives of chiselling to break up boulders or to redrill the pile at another position, can be expensive and serious delays to the piling programme may result.

7.2.1.2 Base of boreholes. To develop the available end-bearing capacity it is necessary to ensure the base of the bore is clean and undisturbed. In clay soil the boreholes can be cleaned satisfactorily with special tools, but, in stiff fissured clays with sand partings and perhaps with water seepage, blocks of material may fall into the borehole. Underreamed piles are particularly liable to this problem (Figure 7.6).

The base of boreholes should be inspected before concreting and any debris removed. High-intensity spot lamps now available may be used for this purpose. The importance of cleaning the base of boreholes cannot be overstressed: cases of pile failure have been attributed to neglecting this step.

In granular soils, loosening of the base is almost impossible to avoid, even when drilling under water or bentonite. For piles founded in such deposits it may be wise to reduce the contribution of end bearing when computing the allowable capacity of the pile. Special measures may be adopted to consolidate the loosened soil. These usually involve placing a stone fill pack at the base of the borehole, followed by high pressure grouting after the pile is concreted. The main effect of the grouted packing is to preload the pile and reverse the direction of the shaft friction. The overall load/settlement performance of the pile is enhanced, but the technique has little effect on the ultimate bearing capacity of the pile. Sand or silt may accumulate at the base of a pile before concreting, leading to the effect shown in Figure 7.7. The fines may arise from local seepages into the borehole, or from sediment falling from the drilling water or mud.

Figure 7.6 Blocks of clay fallen from underream of pile. Photograph courtesy of CIRIA, London

Figure 7.7 Toe of small-diameter pile filled with silt prior to concreting. Photograph courtesy of CIRIA, London

For piles deriving bearing capacity from a rock socket, cleaning the base in a water-filled hole is obviously crucial to the satisfactory behaviour of the pile. In such circumstances final cleaning may be effected by the use of an air-lift or by adopting the reversed-circulation drilling technique.

Debris such as small tools, boots, bags and timber can easily fall down a borehole and care should be taken to avoid this. A temporary steel casing projecting above the ground should be used at the top of all holes. Empty bores should be covered at all times with a substantial cover to prevent personnel accidentally falling into the shaft.

7.2.1.3 *Effects of water in boreholes.* The ingress of small quantities of water into pile bores is not a serious problem. However, a build-up of water at the base of the hole can cause problems on concreting. A surprisingly small amount of water can result in segregation of the concrete at the toe of a pile. Figure 7.8 illustrates the defects found in practice arising from this cause.

If water is seen to accumulate at the base of the borehole, the depth should be measured. For accumulations more than a few centimetres deep the water should be removed before concreting, or alternatively a tremie pipe may be used to place the concrete. Expedients sometimes adopted, such as placing a dry batch of concrete, are not usually successful and should be discouraged. Water will cause softening of the surface in clay soils, and for this reason piles

Figure 7.8 Toe of pile after concreting with water in bore. Photograph courtesy of CIRIA, London

should be concreted as soon after boring as possible. If concreting is delayed, slumping of softened soil from the shaft may occur in extreme cases, and low values of adhesion will certainly result.

7.2.2 Concreting the pile

Concreting cast-in-place piles is a special operation calling for considerable skill and the correct design of the concrete mix. More problems are caused by using a mix which is too stiff than by using one with too high a slump.

7.2.2.1 Quality of the concrete.
The desirable properties of the concrete for forming cast-in-place piles are discussed in Chapter 3. It is important to appreciate that the method of placing concrete in bored piles places severe restraints on the type of concrete that may be used. In general it is preferable to ensure the pile shaft is free from defects by using a highly workable mix, than to specify one with a low water/cement ratio in the expectation of producing a concrete of high durability. Using concrete with too low a workability is a major cause of defects in bored pile construction (Figure 7.9).

It is not practicable to vibrate concrete in a pile shaft, hence reliance is placed on the energy of the falling stream of concrete to do this. The free-fall method of placing concrete will result in the segregation of the concrete if the mix is unsuitable.

In general terms concrete suitable for constructing cast-in-place piles should be free flowing yet cohesive; the mixes are designed to be 'self-compacting'. The slump test is quite adequate to assess the former property; however, cohesiveness or resistance to segregation is difficult to define. If available, rounded aggregates should be chosen and a reasonably high sand content adopted. Plasticizers may be used to improve the workability of the mix with advantage. Experiments with super-plasticizers have shown that these additives are not worthwhile owing to their high cost and the limited time for which the effect occurs. Similarly, air-entraining agents are not effective, except for concrete to be placed by tremie pipe, as the entrained air is lost on impact of the falling concrete.

The durability of the concrete is difficult to reconcile with the high workability required. Very few cases of deterioration of cast-in-place piles have been reported in the literature, and no cases of failure are known. The site investigation should reveal the presence of aggressive ground conditions. In aggressive conditions it is recommended that the cover to the steel is increased, and the outer 50 mm or so of the shaft may be considered as a protective coating. Sulphate attack, the most common problem, may be overcome by using sulphate-resisting or suitably blended cements. Extremely aggressive conditions may be encountered in sabka deposits of the Middle East, with highly saline groundwaters. In these situations an expedient, albeit expensive, is to case the pile with a permanent plastic sleeve, mainly to prevent chloride

Figure 7.9 The effect of low slump concrete. Photography courtesy of CIRIA, London

attack and steel corrosion in that part of a pile which is near to or above the ground water level. The combined presence of water, chlorides and oxygen can lead to rapid steel corrosion in such an environment.

7.2.2.2 *Placing concrete.* Concrete is normally placed in the bore by allowing it to fall freely from the ground surface. It is important to make arrangements to prevent the concrete impinging on either the reinforcement or on the sides of the hole. Segregation and inclusion of soil debris can be the result. The concreting process should be continuous and be completed without a break.

If prolonged delays between batches occur, water and laitance bleeding from the previously placed concrete may cause a weakness in the pile shaft. This emphasizes the need for good mix design.

Raking bored piles present special problems and it is suggested such piles should be concreted using a tremie pipe or chute.

For boreholes filled with water or drilling mud, the tremie method of concreting is a necessity. Underwater bottom-discharge concreting skips are not normally accepted because of the difficulty of ensuring complete discharge at each withdrawal of the skip.

7.2.2.3 *Extracting temporary casing.* After concreting, extraction of the temporary casing can create problems particularly if delays occur and partial separation of the pile shaft may result (Figure 7.10). Friction between the concrete and the casing is obviously aggravated by the use of dirty or dented casings. With unsuitable mix designs, it is not unknown for the casing to be withdrawn still filled with concrete. Difficulties arise in pulling the casing if a low-slump concrete is used, or if the aggregate is particularly angular. Insistence on vibrating the concrete can lead to arching within the casing. However the use of external casing vibrators has been found to be beneficial.

Figure 7.10 Separation of pile shaft caused by extraction of casing. Photograph courtesy of CIRIA, London

Figure 7.11   Schematic presentation of formation of water-filled cavities. Courtesy Cementation Piling and Foundations Ltd., Rickmansworth

Figure 7.12   Defect in pile shaft caused by water. Photograph courtesy of CIRIA, London

Bleeding of mixing water from the joints of screwed casings may cause local consolidation of the concrete, leading to disturbance of the reinforcement cage and difficulty in extracting the casing.

Cavities outside the casing can lead to serious defects in the pile, particularly if they are waterfilled. Figure 7.11 shows the mechanism of the formation of the defect and Figures 7.12 and 7.13 show examples of defective piles. If airfilled voids exist outside the casing, defects are not structurally as serious (Figure 7.14). Normally the concrete in the shaft would slump to fill the cavity resulting in bulbous projections. There is a danger that debris can be dragged into the pile shaft, particularly if large cavities exist. Dry voids may form where thin gravel layers exist or where piles are formed in fill. The formation of such voids can be reduced by the use of a vibrator to insert the pile casing. Obviously, if there is any doubt that satisfactory pile shafts will be constructed, thin permanent casings should be used to support the soil.

Where slumping of concrete into voids has occurred, topping up the pile shaft with fresh material frequently leads to problems (Figure 7.15). Debris

Figure 7.13 The effect of a large water-filled cavity. Photograph courtesy of CIRIA, London

Figure 7.14 Defect caused by concrete slumping into a dry cavity. Photograph courtesy of CIRIA, London

may be dislodged by removal of the casing or by the placement of additional concrete and incorporated in the pile shaft.

7.2.2.4 *Problems in soft ground.*

A variety of defects may arise when forming cast-in-place piles in very soft alluvium exhibiting undrained shear strengths less than 15 or 20 kN/m^2. These defects relate to the lateral pressure exerted by the fresh concrete, which in turn is dependent on the workability of the mix. The lateral pressure of the concrete can easily exceed the passive resistance of soft soils, and bulges on the pile shaft will almost certainly occur. Such defects may be detected by a close check on the volume of concrete used or by sonic integrity testing.

Near the head of the pile the lateral pressure of the concrete may be low, and further reductions in pressure can be caused by friction as the casing is removed. In such situations it is possible for soft soil to squeeze the pile section, leading to local waisting of the concrete (Figure 7.16). Again, the likelihood of this type of defect may be overcome by casting concrete to ground level or by the use of permanent light steel casings.

Figure 7.15　The consequence of 'topping-up' after extraction of casing. Photograph courtesy of CIRIA, London

Figure 7.16　Waisting of a cast-in-place pile. Photograph courtesy of CIRIA, London

These risks can be analysed if reliable soil information is available and steps can be taken to avoid them.

7.2.2.5 Effects of groundwater. The main effect of groundwater seepage, particularly in soils containing thin bands of silt, is to cause local slumping of the bore. The resulting debris may not be completely removed before concreting, or the material may become included in the body of the concrete.

Leaching of cement from fresh concrete is most unusual as high water velocities are required. Occasional cases of damage have been reported in made ground (Figure 7.17). A more common effect is illustrated in Figure 7.18, where the pile cut-off level is below the groundwater level. For this situation to develop, the water pressure would have to exceed twice the head of the concrete at the base of the casing. The problem may easily be overcome by overfilling the pile or by filling the casing above the pile with a little sand and topping it up with water.

If groundwater causes instability of the pile bore, drilling using a permanent casing or bentonite slurry should be resorted to.

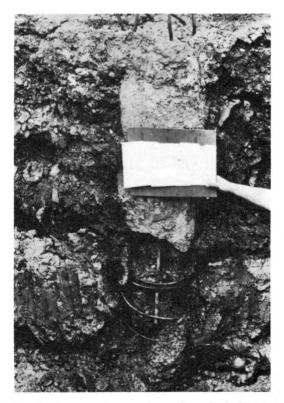

Figure 7.17 Erosion of wet concrete by groundwater flow: pile in fill. Photograph courtesy of CIRIA, London

Figure 7.18 Effect of water inflow at head of pile

Figure 7.19 Concrete retained by reinforcement cage in underreamed pile. Photograph courtesy of CIRIA, London

7.2.3 Design of pile reinforcement

A relatively common problem with the construction of cast-in-place piles is unsuitable detailing of the reinforcing cage. In some instances the spacing of the bars may be so close as to completely retain the concrete (Figure 7.19). Even if this is not so, free flow of the concrete may be sufficiently impeded to prevent complete embedment of the steel. This problem is related to the workability of the concrete and is more likely to occur with low-slump concrete. A few large-diameter bars are to be preferred, with a minimum spacing of at least 100 mm.

The lateral outward pressure on the main steel caused by the flowing concrete may be considerable, and hoop steel should be adequate to restrain the main steel. Displacement of the reinforcement due to this cause has been observed; the defect is usually evidenced by the downward movement of the projecting bars at the pile head. Attempts to compensate for a slumped cage by the insertion of individual free bars which are pushed into the wet concrete can result in the defect shown in Figure 7.20.

Figure 7.20 Defects associated with displacement of reinforcement. Photograph courtesy of CIRIA, London

Where the pile shaft contains a large amount of steel, e.g. in tension piles, widely-spaced welded steel bands may be used instead of the normal helical binding. Such welded bands will also prevent twisting of a heavy cage which can occur when setting the steel or concreting of the shaft. The use of steel bands is quite common practice in some European countries.

Experience has shown that reinforcement cages should extend at least 1 m below the level of any temporary casing. This will prevent movement of the cage caused by drag as the casing is extracted.

7.2.4 *Piles constructed with the aid of drilling mud*

Excavation of bored piles with the aid of a bentonite-based drilling mud may result in particular defects associated with the method of construction. The conditions under which such defects can occur are discussed in the following sections.

Slurries with a concentration of bentonite clay in the range 4 to 6% are normally used to support the borehole. The fresh slurry can become contaminated with soil arising from the excavation, resulting in an increase in density and viscosity of the drilling fluid. The borehole is basically stabilized by the pressure exerted by the drilling mud. If this pressure is less than the active soil pressure, failure of the sides of the bore will result. Such failures are most likely to occur near the head of the pile, especially if the water table is high. Short guide casings should be used and the level of the drilling mud should be maintained at least 1 m, and normally 1.5 m above the groundwater level to avoid problems of instability.

Excessive viscosity of the drilling fluid may occur with heavy contamination of soil. In this situation rapid movement of the drilling bucket may lead to a 'piston effect'. On withdrawal of the bucket, significant reductions in pressure can occur and local collapse of the borehole may result. Clearly control of the viscosity of the drilling fluid should be enforced to avoid this problem and digging buckets should be provided with a fluid by-pass.

Cleaning the pile base should be carried out with extra care as it is not possible to inspect the base. Cleaning may be effected by means of a special bucket or with an air lift, and loosening of the base should be avoided. Fine sand and silt suspended in the bentonite slurry will gradually settle to the base of the pile, hence final cleaning of the hole should be carried out shortly before concreting.

An interesting example of bored piles constructed under bentonite slurry in Italy was quoted by Fleming and Sliwinski (1977). In this case several methods of construction were adopted and the piles were load tested. Details of the tests and the soil strata penetrated are shown in Figure 7.21. It is clear from the test results that both bucket excavation and drilling of the bores developed a loosened layer of sand at the base. The maximum available performance was only achieved after grouting the base of pile no. 6 to consolidate the loosened

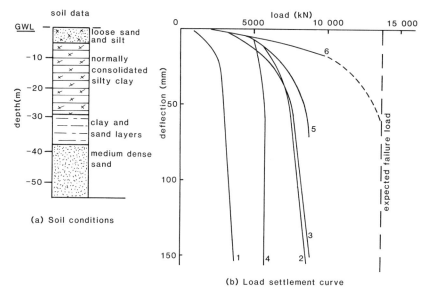

Figure 7.21   Results of tests on piles bored with drilling mud at Porto Tolle, Italy. Maintained load test:
(1) Pile dia. 800 mm, depth 42 m, bucket excavation
(2) Pile dia. 1000 mm, depth 42 m, bucket excavation
(3) Pile dia. 1000 mm, depth 42 m, bucket excavation
(4) Pile dia. 1000 mm, depth 43.5 m, bentonite circulation
(5) Pile dia. 1000 mm, depth 43 m, bentonite circulation
(6) Pile dia. 1000 mm, depth 46 m, bucket excavation with grouted toe. After Fleming and Sliwinski (1977)

base soil and partially reverse the direction of shaft friction. The groundwater level was very high at this site and problems with the stability of the excavation are the most likely cause of observed behaviour.

For placing concrete under a bentonite drilling mud, the tremie method is used and it is important that a suitably designed mix is adopted. In other respects, concreting bentonite-filled holes gives rise to problems similar to those of concreting open bores (section 7.2.2).

It has been found that the process of forming a pile under bentonite suspension does not materially reduce the shaft friction. Normally a very thin and weak filtercake layer will form on the wall of a bore in any reasonably fine permeable soil, and the rising column of concrete will largely remove this and any loose debris from the hole. However, if the bore is left filled with slurry for long periods, or if the bentonite suspension is heavily contaminated with soil particles, an appreciable thickness of filtercake may accumulate. The filtercake may be strong enough to resist abrasion during the placing of concrete, giving rise to a significant reduction in shaft friction on the completed pile. In clay soils the concrete becomes contaminated with debris and intermixed with slurry. It is therefore necessary to overfill the piles sufficiently to ensure sound concrete is present at pile cut-off level. In practice, overfilling by about 1 m has been found satisfactory (see ICI Piling Specification).

7.3 Driven cast-in-place piles

There are many proprietary systems of constructing driven cast-in-place piles. In those systems which incorporate a permanent lining of steel or concrete, defects in placing concrete are likely to be few, providing the lining system is watertight. Such piles are very similar in behaviour to driven precast piles. Piling systems which use a driven temporary casing, basically to form a hole, may also suffer from some of the defects associated with bored cast-in-place piles. One of the major advantages of driven cast-in-place piles is that the driving records provide a valuable, if crude, check on the stiffness of the bearing stratum.

As discussed in section 7.1.3, driving steel casings for cast-in-place piles may result in soil heave or lateral displacement of adjacent piles. Once the piles have been cast it is not possible to correct the effects of heave by re-driving the piles, and pre-boring or other measures may have to be adopted to alleviate the problem.

In some instances casings are driven by an internal drop hammer acting on a gravel or dry concrete plug at the base of the casing. If the dry mix is of the wrong consistency, driving may not cause the plug to lock in the casing. In this situation any end plate will be broken off, and ingress of soil and water into the casing cannot be prevented and the tube may have to be withdrawn and re-driven using a correct mix. Heavy driving with an internal hammer may result in bulging of the casing, or in extreme cases, splitting of the steel if it is of inadequate thickness.

With some proprietary piling systems, the internal hammer is used to compact low-slump concrete as the casing is withdrawn. With this system the hammer may catch on the reinforcement and damage it, or over-vigorous compaction may cause the cage to spread. It is highly desirable to use a robust steel cage and to check the level of the reinforcement of such piles before and after concreting.

Concrete placed in driven cast-in-place piles is liable to suffer many of the problems associated with bored piles and the reader is referred to sections 7.2.2 and 7.2.3. With this type of pile it is recommended that concrete should be cast to ground level in order to maintain a full pressure head and prevent contamination of the pile.

7.4 Continuous-flight auger piles

The main problem with continuous-flight auger piles is related to the matching of concrete supply with the withdrawal rate of the auger and it is desirable to over-supply concrete at all stages of pile construction to a limited degree. It is for this reason that major specialists in the UK now fit instrumentation to all the machines used for this type of work. The items normally measured are the rotations of the auger, the concrete pressure at the auger head, the concrete volume being supplied and the depth of auger penetration below ground. All of these items are measured against a real-time base and, on completion of any pile, a record of the main parameters can be printed in the machine cab. The data are stored and can be analysed in greater detail using an office-based computer.

Pressure alone is not an adequate measure, since it is frequently lost when the auger is withdrawn to a shallow depth (say 7 or 8 metres below ground), due to the considerable head of concrete effective at the auger base and the relatively loose state of the soil within the auger flight at this stage.

Most of the problems which have occurred in the past with this type of pile are believed to be due to over-rapid extraction of the auger in the latter stages of construction. When concrete is seen to issue from the ground, wrongfully giving the impression that all soil has been displaced, there is a tendency to extract the auger too quickly. If this is done, soil may be carried back into the body of the pile by the reverse motion of concrete on the auger and the upper part of the pile can become contaminated.

Other problems which can occur are related to excessive rotation of the auger when it encounters a hard stratum and particularly when the ground above consists of water-bearing sands or silts. Slow excavation relative to the number of rotations will cause the auger to start loading from the sides rather than from the base, thus loosening the surrounding ground and, in some cases, creating water-filled cavities. A great deal can be done to avoid this situation by the selection of a drilling machine with sufficient power to drill the harder strata at a steady and sufficient rate. A well-designed boring head is also necessary.

Care must be exercised when concrete placing commences so that the auger is not lifted too rapidly in the initial stages. Failure to do this may result in significant loss of end-bearing or possible segregation of the concrete in water-bearing ground.

The use of integrity testing, coupled with good on-board instrumentation and an understanding of the process, provides a high level of assurance in this type of work. The monitoring of the concrete supply to ensure that concrete volume supplied always exceeds the volume vacated in the ground by the auger, remains the most important item in the process. Where upper soils are very soft, it is possible to use short permanent casings with this method, though this is rarely necessary.

This type of pile is usually constructed using a pumpable concrete with a slump in the region of 150 mm. It is necessary to ensure that the supply pipe does not permit grout leakage, if line blockages are to be avoided. Reinforcement cages should be robust and preferably welded or otherwise firmly fixed at bar intersections, main steel bars should be heavier than in conventional bored piles, depending on the depth to which they have to be inserted, and lateral reinforcement or hoops should be reduced to a reasonable minimum except in the top 3 or 4 diameters of pile where shear due to lateral loading may be an important consideration. A generous number of spacer wheels should be provided to ensure that the cage remains concentric in the pile. For long cages, say more than 10 metres, it may be necessary to use a vibrator to assist insertion.

7.5 General comments and conclusions

Many piling systems are available and some of these systems have been developed to deal with particular ground conditions. It goes without saying that an adequate site investigation should be carried out and that an experienced engineer's advice be sought as to the most suitable piling systems to be adopted. In a few marginal ground conditions often associated with organic soils, there is always a risk of defective construction with cast-in-place piles unless permanent casings are adopted. Similarly, problems may be experienced with driven piles in boulder clays or other soils containing obstructions.

Piles are normally lightly reinforced, and damage to the head of the pile may occur due to impact with or soil movement caused by heavy plant. Such damage can lead to protracted arguments between the main contractor and the piling subcontractor. Similarly, incautious excavation around the piles may lead to a slope failure with consequent damage to the piles. Such a case was recorded during excavation for a conical sedimentation tank in soft alluvium. Piles previously cast to support the edge beam were sheared when failure of the excavation occurred.

In practice, defects in piles are not very common, and it is hoped that as integrity methods of testing are more widely used, defects may be detected at an early stage. The wider use of such methods should also encourage higher standards of construction and supervision of piling contracts. Piled foundations usually comprise a large array of piles, and defects in a few piles, although highly undesirable, will not have a significant effect on the overall behaviour of the foundation. Failure of piled foundations as a consequence of failure of the piles is extremely rare.

Cementation Piling and Foundation Ltd have recently carried out a survey of bored piles, using TNO sonic equipment. The results of the 1981 and 1982 surveys are given in Table 7.4. From this table it is evident that the major cause of defects was damage to the piles after construction as a consequence of incorrect breaking down or overloading by site traffic. Damage due to construction defects was about $\frac{1}{2}$ per cent. Obviously the experience of one contractor may not be representative of the industry as a whole. Nevertheless, such surveys contribute to confidence that quality of construction is high, and the accumulation of such data over a period should

Table 7.4   Results of bored pile surveys*

	1981	1982
Number of piles tested	5000	4550
Number of piles to show faults	73	88
Fault		
Soil contamination 0–2 m	24%	5%
2–7 m	9%	9%
Poor-quality concrete	6%	3%
Voids adjacent to pile shaft	3%	2%
Damage subsequent to construction	58%	80%
Total percentage of piles with defects	1.5%	1.9%
Percent failure due to construction defects	0.6%	0.4%

*Courtesy Cementation Piling & Foundations Ltd.

encourage a higher standard of workmanship. No such comprehensive data on the quality of driven piles is available, although defects in the pile body before installation are more readily detected.

Construction of piled foundations is a skilled operation, and calls for a high level of training and experience on the part of the contractor. Inexperience on the part of the operatives, incorrect specifications or wrong selection of pile type are the sources of many problems. Specialist contractors can avoid many of the pitfalls that lie in wait for the unwary, and in difficult ground conditions advice from such firms should be sought at the design stage.

8 Integrity testing

8.1 Introduction

Over the past two decades, there has been a steady improvement in the quality of both bored cast-in-place and displacement types of piling. This has been brought about by a gradually increasing awareness of the ways in which defects in piles may occur, better specifications, a greater degree of supervision, and an interest by the leading piling contractors in improving methods and efficiency. The introduction of various integrity tests may also have promoted self-imposed improvements in installation techniques by piling contractors, although it should not be forgotten that certain leading piling contractors have been actively involved in the introduction and use of integrity test methods.

The cost of rectifying defective piles discovered at a late stage in building construction can be enormously high, and for this reason alone, an early indication of problems from the use of a suitable integrity test is highly cost effective, and a strong incentive for the development of straightforward and reliable tests.

Whilst established piling practice has improved, there are a number of newer methods of pile installation developed in recent years, and much reliance has to be placed on the skill of the specialist contractors who employ these techniques. Integrity tests, provided they are appropriate to the pile type and soil conditions, have a role here, permitting a measure of quality control.

Before the introduction of integrity testing, the only check on pile quality was to load-test a small proportion of the piles in a group. Exceptionally, a small group of piles would be tested in total, when, for example, the load/settlement performance was critical. The cost in the latter case was high, and in the former case, the check was limited. If the proportion of defective piles in a group was high, the selection of a small proportion for load test would be more likely to detect a defective pile, than if only a very small proportion were defective, as would perhaps be the normal situation. The load test would be useful in establishing the short-term load/settlement characteristic of the piles, but it would not be a quality check. The purpose of a load test should therefore be established—it is primarily a method of establishing the settlement characteristics of a pile. It is not a good method of quality control, except as an incentive to good installation when it is known that piles are to be selected at random for load testing. Integrity testing on the other hand is primarily a quality check, and as the majority of a pile is buried and is not accessible for visible inspection, quality checks by indirect methods are appropriate.

The statistical probability of detecting a faulty pile in a group with a given number of defective units for various testing levels is shown in Table 9.1 (Chapter 9).

Notwithstanding, there has not been a wholesale acceptance of integrity testing as a routine procedure. The reason is that the tests themselves require interpretation, particularly when they involve electro-mechanical and electronic techniques generally outside the experience of engineers engaged in civil engineering construction. Other less sophisticated integrity tests, for example core drilling a pile shaft, are unlikely to detect defects such as lack of cover to reinforcement on the perimeter of a pile, and hence, despite considerable expenditure, the engineer has only achieved a partial check.

By understanding the merits and limitations of the methods of integrity test available, it may be possible to select a test which is appropriate to the pile type and ground conditions, and where a particular type of defect is considered more likely than another, to specifically relate to this particular problem, be it transverse cracking in driven concrete piles, or lack of end bearing in bored cast-in-place piles.

8.1.1 *Occurrence of faults in piles*

The occurrence of faults in piles is infrequent, but when they are discovered, perhaps in one pile, doubt is then perhaps cast on the other piles on a site. The result may be delays to the contract, dispute as to causes and remedies, and probably an increase of cost. As shown by Table 7.4, the largest single cause of damage is that inflicted on pile heads after completion of the pile.

It is wrong to imagine that modern integrity testing can discover each and every fault that may exist in a pile. For example cover to reinforcement, small zones of segregated concrete or small voids may be difficult to detect, but most of the successful systems can give strong indications of the presence of major problems. The consequences of a major defect in an important pile are such that costs of remedial measures may far exceed the cost of an integrity survey, whereas early detection may led to relatively inexpensive repairs or indeed modification of the method used so as to obviate further difficulties.

8.2 Types of integrity test

In some kinds of integrity test, pre-selection of the pile(s) to be tested is necessary. For example, access tubes for sondes have to be cast-in. The cost of this operation is generally significant, and equipping all the piles in a foundation in this way is usually prohibitively expensive. A proportion of the piles is therefore selected, and while this is usually a higher proportion than that usually selected for load testing, the selection process carries with it the same disadvantages, i.e. a risk of missing a defective pile, and the possibility of greater care being taken with the installation of the piles to be subjected to testing. Therefore, the confidence that can be placed on the results of a particular type of test does not depend on the technical merit of the system alone.

The tests which are generally recognized as 'integrity tests' are as follows:

(i) Acoustic tests
(ii) Radiometric tests
(iii) Seismic (sonic echo tests)
(iv) Stress—wave tests
(v) Dynamic response tests
(vi) Electrical tests

Other, more straightforward, examination of the integrity of a pile can be made by excavation around the pile head (or adjacent to a pile by excavation of a shaft), or exploratory drilling of the shaft, supplemented by a television inspection of the drill-hole.

8.2.1 Acoustic tests

(i) Single-hole test.   In order to carry out this test, a sonde has to be traversed the length of the pile. A tube has either to be cast in the pile for this purpose, or a hole drilled in the pile. The sonde contains a transmitter of acoustic energy at one end, and a receiver at the other, with an acoustic isolator between. The dimension of a typical sonde is 55 mm diameter by 2 m long (see Figure 8.1). The sonde is raised and lowered on an electrical cable which supplies power and also carries electrical signals. In order to obtain good acoustic coupling to the pile shaft, the hole in the pile (which should be around 70 mm diameter for the sonde as described), is fluid-filled. Water is suitable for this purpose, but may be subject to a high rate of leakage, and a drilling mud may then be more suitable.

The sonde is generally lowered to the base of a pile, and raised at a steady rate. Ultrasonic signals from the transmitter radiate in all directions, some arriving at the receiver via the pile medium. If this is sound concrete, the delay time from transmitted pulse to the received signal is small. The compressive wave is received first, followed by the shear wave. If there is a void, porous concrete or an inclusion of soil, the delay time is increased. At the receiver, a transducer converts the pulses to electrical energy which is received at the surface as an amplitude modulated (a.m.) voltage. For interpretation, the a.m. signal is used to produce a variation in intensity of an oscilloscope scan. A

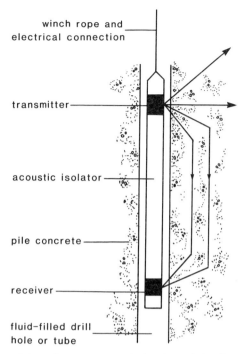

winch rope and
electrical connection

transmitter

acoustic isolator

pile concrete

receiver

fluid-filled drill
hole or tube

Figure 8.1 Schematic diagram of a single tube sonic logging apparatus

peak in amplitude of the signal corresponds to a brightening of the trace, and a trough to a darkening. A voltage which is proportional to the depth of the sonde is derived from the winch unit. The final display may appear as shown in Figure 8.2. In this example, defects are indicated at depths of 4 m and 7.5 m, as signified by the increased time of transmission. The profile from a sound pile shaft is also shown.

The cost of drilling access holes is high, and is approximately equivalent to the cost of forming the original pile, metre for metre. The prior installation of an access tube in a cast-in-place pile is cheaper, but then a pre-selection has to be made. Steel tubes should not be used, as steel is a better transmitter of acoustic energy than concrete. For the same reason the tube or drill hole should not be adjacent to a reinforcing bar. Plastic tubes can fail under the hydrostatic head of wet concrete combined with the elevation of temperature that occurs in a massive concrete section as the concrete hydrates. High quality ABS tubes may be necessary, rather than uPVC.

In the single-tube test, the radial extent of the zone of concrete examined increases as the separation between the transmitter and receiver is increased. The resolution decreases however, and the limiting size of a detectable defect increases. With a spacing between the transmitter and receiver of 30 mm, Kennet (1971) quotes a vertical resolution of 50 mm. A number of traverses can

Figure 8.2 Acoustic test—display of results. Photograph courtesy of Testconsult Ltd., Lymm

be made at different separations to increase the volume of concrete included in the inspection, enabling finer detail to be detected in the zone immediately around the access hole. There is, however, some doubt as to the actual volume of material effectively scanned in the single sonde test, and the method is not in common use.

(*ii*) *Double-hole test.* This test is essentially similar to the single-hole test, except that the transmitter and receiver are traversed in individual holes, whilst being held level with each other, or obliquely. Defects which are present between the two tubes are well defined, but there is no radial scanning as with the single tube method. If the transducers are held obliquely, additional information may be obtained, which can assist in the delineation of the extent of a zone of anomalous response.

For this test, holes are usually preformed with steel or plastic tubes, as it is unlikely that a pair of holes could be drilled sufficiently parallel to each other over the length of a pile. The test arrangement is shown in Figure 8.3. Some errors can occur at the base of the pile if the holes are not of the same depth, if the cables are not precisely paired in length, or if the tubes are of significantly larger diameter than the transmitter or receiver units. The test is, however, in fairly common use.

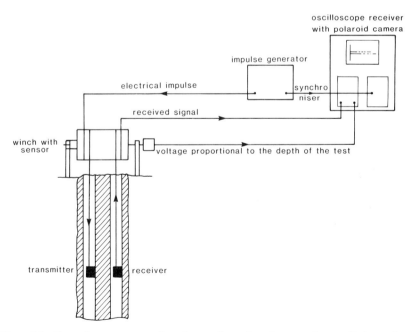

Figure 8.3 General arrangement of equipment for twin tube acoustic test. Redrawn from material provided by Testconsult Ltd., Lymm

8.2.2 *Radiometric tests*

These test methods employ nuclear radiation techniques for the detection of
variations in density of the pile material, and hence the location of apparent
defects.

In many respects, radiometric test methods are similar to those employed in
acoustic tests. Both generally require pre-selection, as access tubes require to
be cast in the piles for test. As a probe traverses the pile, radiation is received
and measured, either from a source mounted on the probe in the same tube, or
from a source and detector in separate tubes. The method can however be
employed in fresh concrete. At the present time, radiometric methods are not
commonly used in the UK.

8.2.2.1 *Backscatter test (gamma-ray).*

When a single tube is used (or a
number of individual tubes in the same pile), the probe is constructed as shown
in Figure 8.4. A probe of this type is described by Preiss and Caisserman
(1975); overall dimensions of 48 mm diameter by 600 mm long are quoted. A
54 mm-diameter access hole is required. Such a hole could be drilled in a pile
by rotary percussion; or more usually preformed with a steel tube.

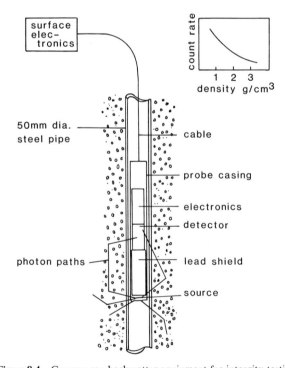

Figure 8.4 Gamma ray backscatter equipment for integrity testing

The radioactive source produces gamma rays which travel into the concrete of the pile. A proportion of the rays emitted are scattered in the pile material and reflected back towards a detector which is separated from the source by a lead shield within the probe. The proportion of backscattered particles counted by the detector varies with the density of the pile material. Low-density zones, such as may occur at an inclusion of soil, increase the count rate, as less radiation is adsorbed. A density profile with depth can be produced in this way. The probe is lowered on a cable, and count rate observed and recorded at the surface. With a probe of the dimensions described, a zone around the access tube of approximately 100 mm radial extent is examined. In order to satisfactorily examine the majority of the concrete in a large-diameter pile, a number of parallel access holes would be required.

8.2.2.2 *Neutron test method.* A neutron source is employed for checking between pairs of tubes in a pile, as a gamma-ray source capable of being used in this way could pose a radiation hazard. Neutron transmission varies with the moisture content of the material through which it passes. This is also suitable for detecting soil or water-filled/empty voids in a pile shaft. In order to carry out the test, the source and detector are raised simultaneously from the base of the pile, and a moisture content profile obtained from the traverse.

8.2.2.3 *Precautions.* The radiocative materials employed in these methods will require that the equipment is used in accordance with the Ionising Radiations (Sealed Sources) Regulations, 1969. Further assistance may be obtained from the National Radiological Protection Board, Rowstock, Oxon.

8.2.3 *Seismic test methods* (*sonic echo method*)

A hammer blow struck at the top of a pile produces a stress wave which can be detected by a surface mounted accelerometer at the pile head. At the moment the pile is struck, the adjacent accelerometer first picks up the surface Rayleigh waves, and later responds to the wave which has travelled to the base of the pile (or a discontinuity) and has been reflected back. A constant of proportionality (C) may be assumed between the velocity of the surface waves V_r and the velocity of the longitudinal waves V_l. Hence surface wave measurements at the pile head may be used to deduce the longitudinal wave velocity, assuming that the concrete in the pile is consistent. This may not be the case as the concrete at the pile head in a cast-in-place pile may be less dense. The relationship is given by $V_l = C \times V_r$, where C is approximately equal to 1.90. In some circumstances, the surface waves are not well defined and V_l may be assumed to be of the order of 4000 m/s. The distance to the base of the pile or a discontinuity (l) is given by $2l = t \times V_l$, where t denotes the delay time.

The recording system comprises a triggering device to start the time trace, with the accelerometer connected to an oscilloscope by way of an amplifying

Figure 8.5 Seismic integrity test (sonic echo method) in operation. Photograph courtesy of Cementation Piling and Foundations Ltd., Rickmansworth

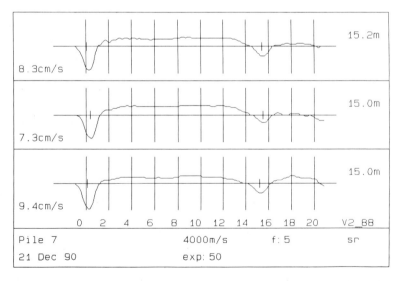

Figure 8.6 Result of integrity test on cast-in-situ pile carried out using TNO-FPDS equipment. Courtesy of Cementation Pilings and Foundations Ltd

filtering system. Signals from the accelerometer are digitally processed such that background 'noise' is filtered out and a clean trace, suitably amplified, may be obtained by striking the pile head a number of times with a small nylon-headed hammer. It is necessary only to remove loose or weak embedded material from the surface of the pile in order to obtain an acceptable signal. The analysing equipment is compact, comprising a portable computer with an integral printer (Figure 8.5). For each pile tested, a fully-annotated record is produced which shows the estimated length of sound pile. The form of the trace provides information on the geometry of the pile, such as reductions or increases in cross-section. Typical outputs are shown in Figure 8.6. The test may be carried out after the concrete in the pile has cured for at least 4 days.

The method is not generally suitable for precast jointed piles or piles with jointed permanent casings. The method was developed originally by TNO in the Netherlands and also by CEBTP in France at a later stage. Whilst the interpretation of the pile 'signature' requires special expertise there is now a considerable fund of experience to call upon both in the UK and abroad.

8.2.4 Stress wave tests

In the stress wave test, a substantial force is applied to the head of a pile (e.g. by the drop of a pile hammer) or a purpose-made sliding mass as shown in Figure 8.7. The force at the pile head is monitored by strain gauges on the pile, and the displacement is simultaneously measured using an electronic theodolite. The method has been developed by the Case Organisation in the USA and the Institute TNO. The results of the test permit calculation of the pile capacity in terms of shaft adhesion and end bearing by the stress wave principle (see section 9.3.1). A disadvantage of the test is the cost of setting it up, which is considerable unless it is carried out as part of a driven pile contract, when a suitable hammer will be available. It is claimed that the results are comparable with those available from conventional static load tests, but a site-specific correlation between the results of a stress wave test and static loading test should be established.

8.2.5 Dynamic response tests

In the vibration test, an electrodynamic vibrator is placed on the prepared pile head, and the pile is subjected to a continuous vibration over a broad frequency band. Although the method of pile excitation is different, the results of the dynamic response test are considered generally similar to those from the seismic (sonic echo) test. Both tests provide similar information regarding the basic pile geometry. The differentiation between the tests lies largely in the electronics involved in the signal processing. The response of the pile is monitored via a velocity transducer. Resonance peaks are produced, which

Figure 8.7 Arrangement of sliding mass (2000 kg) for stress wave integrity test. Photograph courtesy of Cementation Piling and Foundations Ltd., Rickmansworth

may be used to deduce the effective length of the pile and to reveal apparent defects. The method was developed by CEBTP, and is now widely used.

A block diagram of the test arrangement is shown in Figure 8.8. The vibrator supplies a constant-amplitude sinusoidal input over a frequency band of, say, 20 to 1000 Hz. The voltage from the velocity transducer is fed to the y axis of an x-y plotter which is provided with a signal on the x axis proportional to the frequency, f. By dividing the magnitude of the pile head velocity, v, by F_0, the applied force, the mechanical admittance of the pile is obtained. In the idealized response curve for a pile shown in Figure 8.9, the slope of the early part of the curve, m, is indicative of the stiffness of the pile-soil system. This is related to the behaviour of the pile under load, but as the forces involved are at least an order of magnitude less, good correlation is not to be expected. The distance between resonance peaks at higher frequencies is related to the

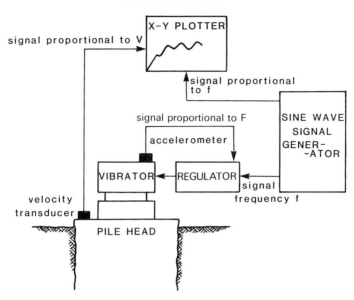

Figure 8.8   Schematic diagram of the test arrangement for the vibration method of pile integrity testing

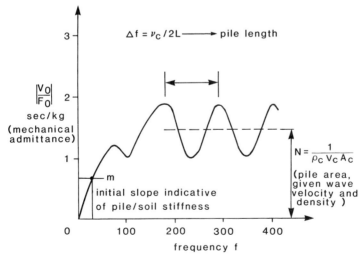

Figure 8.9   Idealized response curve from a vibration test

effective pile length, l, by the expression:

$$f_2 - f_1 = V_c/2l.$$

where V_c is the velocity of longitudinal wave propagation in concrete. Hence l, may be estimated. A value for the cross-sectional area of the pile, A_c, may be obtained from the mean mechanical admittance of the pile head, N, over the higher frequency range, from the relationship

$$N = 1/\rho_c \times V_c \times A_c$$

where ρ_c is the average density of the pile concrete.

A development of this test is to record the transient response of the pile to a single blow, and to carry out a Fast Fourier Transform analysis of the waveform. A test of this type is now offered commercially by Messrs Testconsult Ltd.

An advantage of this type of test is that it can be carried out only 4 days after casting. The pile head is prepared 24 hours in advance of a test by rough levelling using a mortar screed where necessary, and final levelling with a thermosetting resin just prior to the test. It is usual to carry out the test with the pile head freely accessible, but in some soil conditions it may be possible to carry out the test from the pile cap. The advice of the specialist testing house should be sought if the cap has been cast and the pile is to be tested in this condition. Expert interpretation of the results of the tests is required, but as with the seismic test, there is by now a great deal of experience in the use of the test, to guide the interpretive process.

There are some soil type/pile length-diameter limitations with the test. In stiff clays, a minimum length to diameter ratio of 20:1 may be found, with a maximum of the order of 30:1. A general figure of 30:1 can be used, possibly extending to 50:1 where soft materials overlie a stronger founding strata. Both cast-in-place and driven piles are suitable for testing, but jointed piles will give erratic results, depending on the final condition of the pile joints, i.e. whether they are in tension or compression. Major changes in the cross-section of a cast-in-place pile, whether by design (at an underream) or accident (because of a defect), will produce resonance peaks and hence given an indication of their distance from the pile head.

8.2.6 Electrical testing methods

Various methods of using the response to electrical inputs of a pile and/or the pile-soil system, or detection of electrical currents inherent in the pile-soil system are available. The following techniques are generally recognized:

(i)   self-potential
(ii) resistivity testing
(iii) resistance to earth
(iv) induced polarization.

Of these, the methods based on measurement of resistance to earth have been most promising. The other methods are, however, worthy of brief description. A virtue of electrical methods is the lack of any special preparation of the pile, and the tests can be carried out on fresh concrete, although the results may

Table 8.1 Summary of methods of carrying out integrity tests on piles

General method	Type	Relative cost	Advantages	Disadvantages
Acoustic (section 8.2.1)	Single hole ultrasonic logging	High cost if inspection hole has to be drilled	The system scans the pile concrete radially for a distance of approximately 100 mm around the drilled hole or installed tube. The method is relatively rapid, some tens of tests being possible daily	Pre-selection is generally required as it is usual to install purpose made tubes in the pile for the equipment. It is necessary to employ plastic tubes for this purpose, with the attendant risk of collapse, steel tubes being unsuitable for *single* hole logging. The installation of the tube adds to the reinforcement steel in restricting free concrete flow and may in itself lead to defects. Poor bonding of the tube to the concrete may yield an anomalous response. Unless a hole is drilled, it is not possible to inspect right to the base of a pile by this method. Specialist interpretation is generally necessary
	Twin tube ultrasonic logging	Medium cost, depending on expense of casting-in inspection tubes	By using a suitable number of inspection holes, the major portion of a pile core may be inspected. The inspection may be carried out to the base of the pile. Speed of testing similar to single tube method (per traverse) at about 500 to 1000 m/day	As for single tube method except that inspection to the base of the pile is routine. The number of inspection tubes required increases with pile diameter, and multiple tubes are usually necessary for large-diameter bored piles

Table 8.1 (*Contd.*)

General method	Type	Relative cost	Advantages	Disadvantages
Radiometric (section 8.2.2)	Single tube radiometric logging; backscatter technique (gamma ray method)	Medium to high cost if hole to be formed	The system may be used on freshly cast piles and is fairly sensitive to voids or inclusions. Almost the full depth of the pile may be inspected. A zone approximately 100 mm radially from the inspection hole is examined (not uniformly). For the backscatter method, the radioactive source is weak, and no special safety precautions are required. The method is rapid and permits a large number of piles to be checked in one day	It is usual to use cast-in tubes with the disadvantage of pre-selection and added congestion in the pile in the pile shaft. The number of inspection holes required increases with pile diameter. Specialist interpretation is required
	Twin-tube radiometric method; transmission technique (Neutron method)	Medium to high cost	By using a suitable number of inspection holes, the major portion of a pile core may be inspected. The inspection may be carried out to the base of the pile. The method is suitable for use on freshly cast piles. The method is rapid and a number of piles may be checked in one day	As for single tube method with the added disadvantage that a higher intensity gamma ray source is required than with the backscatter technique requiring special safety precautions
Seismic (section 8.2.3)	Sonic echo	Low cost	Rapid method which does not require pre-selection. Cracks in long slender piles detected	Only suitable for continuous piles i.e. not jointed types) Concrete cast-in-place piles should be at least 1 week old

Stress wave (section 8.2.4)	Stress wave test	High cost	No pre-selection is required. An approximate estimate of the bearing capacity of the pile may be obtained	The method may require special preparation of the pile head and may be rather costly and time consuming. The method is not widely used in the UK. Expert interpretation is required
Dynamic response (section 8.2.5)	Vibration test	Medium to low cost	Method does not require pre-selection. The pile may possibly be tested 3 to 4 days after casting. From the results it is possible to compute pile length or distance to a major discontinuity. In certain circumstances, pile/soil stiffness at low stresses may be estimated	Some preparation of the pile head is necessary 24h before the test. The pile should be trimmed level with the surrounding ground and be clear of blinding concrete or other external structures
	Transient response test	Medium to low cost	Rapid method. Other advantages as for vibration test	There are limitations on the geometry of the pile to be tested which vary with the ground conditions. Specialist interpretation is necessary
Electrical (section 8.2.6)	Self-potential resistivity testing.	Not applicable	Rapid methods using standard geophysical equipment may be used on low cut-off piles and possibly on freshly cast piles	Experimental methods, not in routine use. The methods are unsuitable for piles reinforced in the top few metres only. Specialist interpretation required

vary with the age of the concrete. In addition, electrical methods tend to be sensitive to exposure of the pile reinforcement rather than to inclusions in the pile shaft for instance. When the pile reinforcement is exposed, the pile is liable to fail in the long term, even if its immediate load-carrying capacity is adequate. However, exposed reinforcement may well be associated with a voided pile, and hence the electrical methods are possibly biased towards detecting the most significant types of defect.

8.2.6.1 *Self-potential.* A small potential is found to exist between the reinforcement in a pile and a non-polarizing electrode placed in the ground some distance away. This may be measured using a sensitive millivolt meter. The results are likely to be influenced by the length of reinforcement in the pile, and the groundwater table/zone of capillary rise. Exposed reinforcement may however produce a detectable anomaly in the measured voltage.

8.2.6.2 *Resistivity testing.* This method of test simulates the Wenner electrode configuration employed in geophysical exploration, with the pile reinforcement acting as an outer electrode. The other outer conductor comprises a non-polarizing electrode. A potential is applied between the outer electrode, and the voltage difference which is induced between two additional electrodes placed symmetrically between the outer electrodes is measured with a Meggar Earth Tester. Anomalous readings may be indicative of a defect.

8.2.6.3 *Resistance to earth.* Some success with this method of testing was first reported by Bobrowski *et al.* (1975). It has been developed further by Forde and Whittington at the University of Edinburgh. In order to measure the resistance to earth, a known current is passed between the pile reinforcement and a non-polarizing return electrode situated some multiple of the pile length away.

The potential difference is then measured between the pile and a second electrode set at some point between the pile and the return electrode. The arrangement is shown diagrammatically in Figure 8.10.

To overcome problems with stray earth currents inherent with direct current methods, Forde and Whittington have employed low-frequency alternating currents. Using analogue experiments it has been found that only the reinforced section of the pile contributes to the response, and defects below a partially reinforced pile would not be detected. The ratio of the resistivity ρ_c of the concrete forming the pile, and the resistivity ρ_s of the soil around the pile, has a marked influence on the sensitivity of the test to detecting defects. From the analogue work, Forde and Whittington have deduced that the ρ_c/ρ_s should be large or small, but not near unity.

In order to analyse the results of such tests, the ratio of electrode distance from the pile d, and pile length, l, i.e. d/l, is plotted against resistance, R. Typical results are shown in Figure 8.11.

Figure 8.10 Arrangement of electrodes and measuring equipment for the 'resistance to earth' method of pile integrity testing

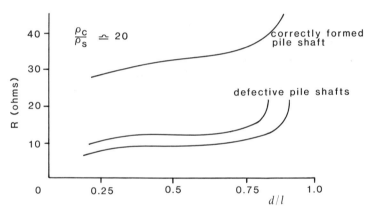

Figure 8.11 Plot of ratio between electrode distance from the pile (d), and pile length (l) v. resistance, R

Figure 8.12 Test arrangement for the 'induced polarization' method of electrical integrity testing

8.2.6.4 *Induced polarization.* The circuit arrangement in this method is shown in Figure 8.12. The signal generator supplies a square wave pulse between the pile and the nearest electrode, and the decay of the electrical field that is produced is measured with respect to time using an oscilloscope to display the resulting waveform.

By using a series of different pulse amplitudes, a response curve for each pile may be produced and anomalous results used to detect defective piles.

8.3 Summary of methods

The wide diversity of testing methods may make selection difficult. The criteria which are applied may well include cost, whether or not pre-election can be tolerated, delay in testing before some methods may be used, and, in some cases, the possible type or location of a probable defect, which may render some tests more suitable than others. In common use are the double hole sonic test, the seismic (sonic echo) test and the dynamic reponse test. The first of these requires tubes to be installed in the pile, and is considerably more costly than the other two, which are carried out from the pile head and require little preparation. Although tests carried out from the pile head do not quantify features within a pile which repeat with depth, such features may only be of academic interest if, for example, discontinuity is present at a high level.

A summary of the methods and the considerations discussed is given in Table 8.1, which also gives an indication of costs.

9 Pile testing

9.1 Introduction

Load testing of piles is expensive, and the cost should be carefully weighed against the reduction in risk and assurance of satisfactory behaviour that the pile test provides. Piles cannot be readily inspected once they have been constructed, and variations in the bearing stratum are difficult to detect. Hence it is not easy for the engineer to be assured that the piled foundations comply with the specifications and drawings. It goes without saying that a comprehensive site investigation is essential.

The pile test programme should be considered as part of the design and construction process, and not carried out hurriedly in response to an immediate construction problem. Pile tests may be performed at various stages of construction; these are conveniently separated into tests carried out prior to the contract and tests carried out during construction. Pre-contract piles may be installed and tested to prove the suitability of the piling system and to confirm the design parameters inferred from the site investigation. Normally the installation of piles, particularly of the driven pre-formed types, will involve disturbance or re-moulding of soils, and high pore water pressures may be induced in the immediate vicinity of the pile. The bearing capacity of a pile may therefore increase or, occasionally, decrease with time after installation. It is common practice not to test piles immediately after placing but to specify a delay of perhaps 1 to 3 weeks or even more, depending on soil type and experience.

Testing of contract piles may involve integrity testing to check the construction technique and workmanship and for load testing to confirm the performance of the pile as a foundation element.

The extent of the test programme depends on the availability of experience of piling in a particular geologic environment and the capital cost of the works. It is thus difficult to be specific as to the number of piles to be tested and the extent to which the piles are instrumented. Pile testing may range from 'spot checks' on pile quality to comprehensively instrumented piles bordering on a research exercise. For small contracts it may not be possible to justify the cost of any pile tests and the engineer must rely heavily on his own or the piling contractor's experience.

There are two main aspects of the pile quality to be considered:

(i) the integrity of the pile and its ability to carry the applied loading as a structural unit
(ii) the load bearing and deformation characteristics of the soil/pile system.

It is noted that the errors in the position and alignment of the piles within a pile group may significantly influence the behaviour of the completed foundation.

The selection of, and the number of, piles to be tested obviously greatly influence the reliability of any conclusions drawn from the test results. Table 9.1 illustrates the effect of sample numbers in the likelihood of detecting a defective pile. From this table it is evident that to achieve a reliable assurance of the satisfactory quality of the group of piles, a large number of piles have to be tested. In general, load testing is unsatisfactory for quality control on the grounds of cost. It is usual to select at least one 'typical' pile for testing, together with any pile that, from its construction records, is thought to be unsatisfactory. Quality control of pile construction may be checked by adopting one of the non-destructive testing techniques which are now available. A great deal of effort has been put into the development of dynamic loading and integrity testing of piles; the former to check, as reliably and simply as possible, the bearing capacity and the latter to confirm soundness and structural continuity. These methods are now sufficiently well-developed to be considered useful tools in any quality control procedure.

The behaviour of a single test pile may well be very different from that of the pile working in a group. The results of observations of load distribution down the shaft of a pile at a bridge in Newhaven, England, are shown in Figure 9.1 (Reddaway and Elson, 1982). These results clearly demonstrate that under short-term test loading the soil resisted a large proportion of the load as friction in the alluvium, in spite of a bitumen slip coat which was applied to the upper 10 m of the pile. On completion of the bridge, the same pile (now working in a group of 32 piles), transmitted the entire loading to the underlying chalk. In such circumstances one would seriously question the usefulness of the pile test. Subsequent analysis of the situation, using an elastic continuum model, demonstrated that the observed behaviour was a function

Table 9.1 Probability of selecting at least 1 defective pile from a group containing 100 piles

No. of piles not meeting specification	No. of piles tested	Probability of selecting at least one low grade pile
2	2	0.04 (1 in 25)
2	5	0.1 (1 in 10)
2	10	0.18 (1 in 5.5)
2	20	0.33 (1 in 3.0)
10	2	0.18 (1 in 5.5)
10	5	0.41 (1 in 2.5)
10	10	0.65 (1 in 1.5)

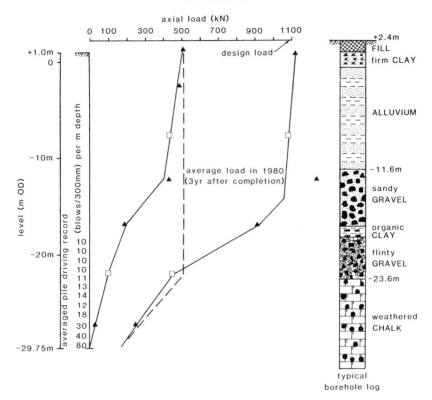

Figure 9.1 Newhaven Overbridge: axial load distribution in pile No. 122. Upper 10 m of pile slip coated. ☐, average of two active gauges; ▲, single gauge. Date of test 22 Nov 1976

of the very large difference in stiffness between the alluvium and the chalk. It should be noted that the designers had, not unreasonably, assumed that friction would be developed in the gravel which was moderately dense.

It is stressed that any pile test programme should be carefully designed, and that the purpose of each test should be clearly stated. A large amount of information can be obtained from carefully conducted and instrumental test piles. The information may well lead to refinements of the foundation design with a consequent possible cost saving, or certainly greater assurance of the satisfactory performance of the foundation. On the other hand poor-quality test pile data may lead to an uneconomic design and are obviously wasteful of resources and money.

9.2 Load testing of piles

9.2.1 *General comments*

Pile load testing, using the maintained load or the constant rate of penetration test procedures, is the most commonly adopted method of checking the performance of a pile.

For offshore installations it is usually not possible to carry out static load testing and reliance is placed on a dynamic 'wave equation' analysis to predict the relationship between static ultimate load capacity and the driving resistance. Such an analysis, at present, is of a semi-empirical nature and the input constants depend heavily on experience of piles driven in similar conditions.

Any test pile and loading system should be properly designed, bearing in mind the large forces and high strain energy in the system under full load. The test pile should be typical in all respects of the piles in the foundation. This is particularly important for pre-contract piles, as small variations in construction procedure and workmanship, can cause disproportionate effects on the behaviour of the pile. An obvious disadvantage of pre-contract piles is that the main piling contract should perforce be awarded to the pre-contract piling firm. Test piles which are not part of the permanent works, constructed and tested prior to the commencement of the main piling programme, avoid double mobilization charges. However the purpose of a preliminary test pile is often defeated if contract piling continues immediately after test pile installation and before the loading test can be carried out. Pile testing requires some considerable time and engineering input to carry out and to interpret the results; this process should not be rushed at the expense of quality of the data and reliability of the results.

It is important that the test piles are loaded in the way envisaged by the designer; in particular it may be necessary to sleeve the pile in any overburden. Such a free-standing length should then be designed as a long column. Any instrumentation built into the pile should be designed to have a minimal effect on the pile stiffness and ultimate load carrying capacity. This is not always easy to arrange. Inclinometer ducts or rod extensometers may affect the axial and lateral stiffness of the pile shaft, and built-in load cells may form physical discontinuities which appreciably alter the behaviour of the soil/pile system.

The objectives of a preliminary pile test programme are usually to determine

(i) the ultimate bearing capacity of the piles, relating this to the design parameters;
(ii) to separate the pile resistance contributed by the adhesion and end bearing capacity;
(iii) to determine the stiffness of the soil/pile system at design load. A back analysis of this data will enable the soil modulus to be evaluated, and hence the deformation of pile groups may be predicted with greatly increased confidence.

These various objectives will necessitate a carefully chosen test procedure and instrumentation programme.

The preparation of a test pile will usually involve cutting down piles to the required level, exposing the reinforcement, and casting a small pile cap or

Figure 9.2 A 2500 tonne pile load test using tension piles to provide reaction. Photograph courtesy of Cementation Piling and Foundations Ltd., Rickmansworth

bedding a 25 mm thick steel plate directly on to the pile. The surface of the pile cap or plate should be flat and should be perpendicular to the axis of the pile, to minimize eccentricity of loading. In particular circumstances it may be necessary to extend the pile by casting or welding on an additional length. Obviously the pile should be cut free of any blinding concrete. If the pile is to be loaded to its ultimate load, it may be necessary to provide additional reinforcement at the head of the pile to absorb eccentricities of loading and prevent bursting of the head of the pile. If additional reinforcement is required it is obviously impracticable to select test piles at random.

A pile test arrangement is shown in Figure 9.2, and the components of the equipment are discussed below.

9.2.2 *Test equipment*

9.2.2.1 *Provision of a reaction.*   The ultimate load of a pile may range from a few tens of tonnes to as much 2500 tonnes, and provision of a reaction to jack against requires careful consideration. The geometry of the pile-reaction arrangement should be such as to minimize interaction between the pile and the reaction, and to avoid movement of datum beams used as references to measure settlement.

Kentledge is commonly used to provide reaction, and more or less any material available in sufficient quantity can be used (Figure 9.3). Specially-cast concrete blocks or pigs of cast iron may be hired and transported to site. The

Figure 9.3 Test arrangement using kentledge. Illustration courtesy of CIRIA, London

cost of transport to and from the site is a significant factor. Regular-shaped blocks have the advantage that they may be stacked securely, and are unlikely to topple unexpectedly. Sheet steel piling, steel rail, bricks, or tanks full of sand or water have to be adopted as kentledge from time to time. The important criterion is that the mass of material is stable at all times during and after the test.

Clearance is required above the pile cap for the jack, load cell and reaction beam, and it is necessary to support the whole weight of the kentledge on a timber or other appropriate cribbage. The area of the cribbage should be sufficient to avoid bearing failure, a severe problem on soft ground. The kentledge is usually supported on a deck of steel beams and/or a timber mattress. The cribbage should also be spaced at least 3 or 4 diameters from the pile to avoid an unacceptably large interaction as load is transferred to the pile. This spacing determines the minimum length of the reaction beam required.

The reaction beam is subjected to high bending and buckling stresses and it should be designed to carry the maximum load safely. The maximum safe load should be clearly marked on the beam so that it is not inadvertently exceeded during a test. A thick spreader plate is normally welded to the centre of the beam to disperse the point load from the jack.

The whole system should be firmly wedged, cleated or bolted together to prevent slipping of any member. Frequently the piling contractor will have a ready-made set of equipment which is designed to suit his system of working. During the test it is important that none of the timber work or decking is unloaded completely, to avoid it becoming unstable. It is therefore usual to provide extra kentledge, and 20% above the ultimate load of the pile would be considered a satisfactory margin. Not so much overcapacity need be provided if the stack is built carefully from regular units. Common sense dictates that the kentledge is stacked uniformly, and interlocked if possible, and that the centre of gravity of the mass is as near as possible to the pile axis.

Raking piles are sometimes tested using kentledge as a reaction, provided the rake is not too great. A steel wedge of the correct angle is bolted to the reaction beam to provide a face parallel to the pile head. The horizontal reaction is provided by friction and the mass of the kentledge should be increased accordingly. This system of testing raking piles is not very satisfactory and it is difficult to ensure that eccentric loads are not applied to the pile head, leading to premature failure of the pile head.

Tension piles may be utilized to provide a satisfactory reaction, particularly for raking piles (Figure 9.4). It is obviously most convenient if adjacent permanent piles can be used. If only two adjacent piles are suitable they would often have to be lengthened to provide adequate pull-out resistance, and severe problems of lateral instability can occur under load. Some specifications do not permit this, while others insist that if it be done, uplift on the tension piles must be measured and must be of very limited magnitude.

It is important that the spacing of the piles is as large as practicable, as significant interaction between the test pile and the reaction pile can occur at spacings of less than 5 pile diameters. In practice, spacings of 3 to 4 diameters between the centres of the test pile and the anchor piles are commonly adopted. The effect of the pile interaction is to reduce the observed settlement of the pile, and it may be necessary to make corrections for this effect where close spacing is unavoidable. It is suggested that the pile layout is arranged to limit the effect of pile interaction to an absolute maximum of 20% of the settlement of the test pile.

Various systems are in common use to connect the tension pile to cross-

heads. The reinforcement of the anchor piles may be left projecting sufficiently
to allow a welded washer connection to be used (Figure 9.5). The fillet weld
should be designed to carry the anticipated load and each weld should be
carefully inspected before the system is loaded. High tensile steel reinforcement
should not be extended by welding, and bent or distorted bars should be

(a)

SECTION A–A (Reference beams and
temporary supports omitted for clarity)

Twin
RSJs
sleeved
and bolted
as shown

(b)

(c)

Figure 9.4   Test arrangement using tension piles. Illustration (a, b) courtesy of CIRIA, London;
(c), courtesy of Fondedile Foundations Ltd.

avoided. Accidental bending and subsequent straightening of such bars can
lead to brittle failures. Alternatively straight high tensile McCall or Dywidag
bars may be cast into the top of the piles, using patent anchor systems supplied
by these firms. Locking collars or nuts should be provided with such systems
to guard against slippage under high load. The top of the tension pile should be

Figure 9.5 Detail of washer connection to reinforcement (dimensions in mm)

adequately reinforced against the stresses imposed by the anchors. The tension piles should normally be reinforced for the full depth of the pile, though of course reinforcement in the pile may be curtailed as necessary with depth.

Connection to precast concrete anchor piles is not as easy as in the case of cast-in-place piles. It is necessary to break down the head of the pile to expose the reinforcement, and either cast on a special anchor block or extend the main steel using sleeved connectors or welding. Connection to steel H-section piles is straightforward, and suitable cleats may be designed and fixed to the piles by welding or bolting.

The use of *ground* or *rock anchors* to provide a reaction is sometimes convenient, depending on the particular geology of the site. The installation of the anchors usually requires the mobilization of a special drilling rig and its associated equipment. The main disadvantages of ground anchors as a reaction system are the axial flexibility of the anchor tendons and the lack of lateral stability of the system.

Multiple ground anchor systems are to be preferred (Figure 9.6) and each anchor should be proof tested to 130% of the maximum load before use. To reduce the extension of the cables during the test it is usual to pre-stress the anchors to as high a proportion of the maximum load as possible. Hence a suitable reaction frame and foundation must be provided to carry this load safely.

The tensile forces may be transmitted to the cross-head using a looped cable and steel saddle, or individual wires in the cable may be locked separately in an anchor block.

Figure 9.6   Test arrangement using ground anchors. Photograph courtesy of Sir Robert McAlpine & Sons Ltd., London

Owing to the high extensibility of the wires, very large amounts of energy are stored in the system, creating a potentially hazardous situation. Thus each component of the system should be designed and installed correctly.

9.2.2.2 Load application and measurement.   It is important that the application of the load to the test pile can be closely controlled, and a range of hydraulic jacks is readily available. Short, large-diameter low-pressure jacks are to be preferred to the high-pressure type as they are more stable. The capacity of the jack is not particularly important provided it exceeds the maximum applied load by a generous margin, and has sufficient travel of the ram (at least 100 mm). For very high loads it may be necessary to use a cluster of jacks. These should be set in a cradle to prevent relative movement of the jacks.

The jack may be actuated by a hand pump for loads up to approximately 300 tonnes. However, for very high loadings or constant rate of penetration tests, it is preferable to use a motorized pumping unit. Pressure gauges should be fitted to the system to allow the operator to check the load in the system at a glance, as he may be some way from the test pile.

The line of action of the load should be co-axial with the pile and time is well spent setting up the jack, load cell, etc., carefully (Figure 9.7). To take up slight misalignment or lack of parallelism a large-radius spherical seating should be inserted under the reaction beam. Under load, spherical seatings lock and will

Figure 9.7 Details of load train. Photograph courtesy of Sir Robert McAlpine & Sons Ltd., London

not correct any eccentricity of load during the test. Small-diameter ball seatings tend to lead to instability of the loading trains and increase the likelihood of violent collapse of the train under load. If an angular misalignment of more than a few degrees is noted, the reaction system should be repacked to eliminate the error.

In general the pressure gauges fitted to the hydraulic jack are not suitable for measuring the load applied to the test pile. Test-quality gauges may be fitted and calibrated with the jack in a compression loading frame. Such calibrated jacks may provide acceptable accuracy of load measurement for increasing load; however, the unloading cycle is usually non-linear owing to friction within the hydraulic jack. A typical calibration is shown in Figure 9.8; note the jack and gauge must be clearly marked and used together.

Load measurement is preferably carried out using a load cell, and several types are readily available. Commonly adopted load measuring devices, in order of accuracy, are

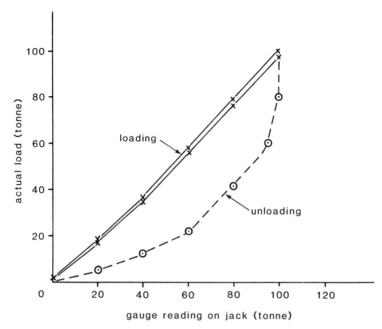

Figure 9.8 Typical calibration of hydraulic jack

(i) hydraulic load capsule, maximum capacity 450 tonne
(ii) load columns, maximum capacity 1000 tonne
(iii) proving rings, maximum capacity 200 tonne
(iv) strain gauged load cells of various types, maximum capacity 200 tonne.

For very large loads an array (preferably at least three) of load columns or strain-gauged load cells may be used. It is important that the load cell is calibrated regularly and that, for those devices sensitive to temperature, suitable corrections are determined.

Hydraulic load cell capsules are basically short, large-diameter hydraulic cylinders (Figure 9.9). The piston is bonded to the casing with a rubber element, sealing the oil-filled chamber. The oil pressure is measured with an accurate pressure gauge, or a pressure transducer may be substituted. The pressure gauge may be fitted with a capillary tube for reading clear of the test system. The accuracy of these devices is probably about ± 1%.

Load columns are short columns of steel or aluminium which have a hole machined in the central section to increase the deformation of the column (Figure 9.10). Deformation is measured using a sensitive dial indicator. The devices were developed by the National Physical Laboratory, but are now manufactured commercially. The devices are sensitive to changes in temperature; however, it should be possible to measure load to an accuracy of ± 1%.

Proving rings are machined from forgings of mild or high tensile steel

Figure 9.9   Hydraulic load capsule. Photograph courtesy of A. Macklow-Smith Ltd., Camberley

Figure 9.10   Load column. Photograph courtesy of A. Macklow-Smith Ltd., Camberley

Figure 9.11   Proving ring

Figure 9.12   Load cell

(Figure 9.11). Deformations are measured with a dial indicator or a displacement transducer. The deformation of the device under load is quite large, hence the sensitivity to temperature changes is reduced. Loads may be measured to an accuracy of 0.5%. However, the slender proportions of the ring can lead to instability of the load train.

Strain gauged load cells are basically machined steel columns, fitted with bonded electrical resistance strain gauges or vibrating wire strain gauges (Figure 9.12). The load cells, particularly the electrical resistance type, should be environmentally protected. The load cell developed by the Mining Research Development Establishment has been found to be particularly satisfactory. Special readout units are required for the load cells, although battery powered devices suitable for site observations are made by several firms. The overall accuracy of the system should be better than 0.5%. The load cells are usually temperature-compensated, although extreme changes of temperature should be avoided.

Figures 9.13 Measuring pile deformation in dial gauges. Illustration courtesy of CIRIA, London

9.2.2.3 *Measurement of settlement.* The measurement of deformation is perhaps the most difficult observation to make reliably. The use of dial gauges calibrated in 0.01 mm may give a false sense of accuracy to the observations. In practice, measurement to this precision is not necessary for pile testing and it is extremely difficult to maintain this degree of accuracy over a long period of time. The most commonly used systems of settlement measurement are dial test indicators or optical levelling.

Dial gauges are simple, robust, mechanical precision instruments, and as such they should be carefully stored and maintained. Dampness and dirt are the main causes of problems, and spare gauges should be available. For pile testing the most convenient gauge to use is one with a 50 mm travel and a least count of 0.01 mm. Gauges with a longer travel are made but they are not readily available.

Dial gauges should be clamped to a stable reference beam (Figure 9.13 shows a typical system). For short-term tests it is sometimes convenient to use magnetic clamps to mount the gauges. Normally four gauges are used for a pile test, so that any bending of the pile head can be detected.

The measuring points should be on the pile head or on brackets mounted on the side of the pile. Small machined steel or brass plates should be used as a datum for the dial gauge stem. On exposed sites, polythene bags provide a convenient method of protecting the gauges. Care must be taken to ensure that the reference plates are perpendicular to the pile axis, so that any sway of the pile has no vertical component.

Stable datum beams are quite difficult to set up, and the basic requirements are set out below. Firm foundations for the reference beams should be constructed, away from any loaded areas. If possible, adjacent piles should be used, or a separate concrete block supported by steel stakes constructed. If practicable, the datum beam foundations should be at least four pile diameters from the test pile. It is noted that a compromise must frequently be made between a long beam prone to vibration and temperature-induced deflections, and a shorter beam with foundations in the zone of influence of the test pile or kentledge.

The reference beam should be a heavy steel section or a rectangular hollow section tube clamped at one end. The other end is free to slide, but should be restrained against lateral and vertical movement (see Figure 9.14). Such a system will allow longitudinal movement of the beam caused by changes in temperature. The beam should be shaded from direct sunlight, and in hot climates it may be necessary to carry out the test at night when temperature fluctuations are less marked.

To check for gross errors in the observations or disturbance to the datum beams, it is good practice to level the test pile and datum beams at intervals throughout the test. If tension piles are being used to provide the reaction, these should also be levelled to check for undue movements.

The accuracy of any *optical levelling system* is poorer than that of dial gauges

Figure 9.14 Details of datum beam support. Building Research Establishment: reproduced by permission of the Controller, HMSO. Crown copyright

by a factor of at least 10. However the absolute accuracy of the system may be of a similar order, particularly in situations where it is difficult to establish a stable reference beam. Using optical methods it is difficult to arrange a constant rate of penetration test or to check the cessation of movement under a load increment owing to the coarseness of the least count. With an optical system it is easy to arrange for the instrument and reference point to be well outside the zone of disturbance.

A convenient system is to mount a good quality engineer's level on an adjacent pile for the duration of the test. A vernier sighting card (Figure 9.15) may be used to achieve an accuracy of 0.5 mm. Greater precision may be achieved by mounting invar scales on the sides of the piles and levelling these with a geodetic level fitted with an optical micrometer. An overall accuracy of ± 0.2 mm is attainable.

Electrical load and displacement transducers may readily be adapted for recording pile test data. However, similar care is required in setting up the equipment. Many transducers are based on the use of wire resistance strain gauges, and consequently are susceptible to loss of insulation resistance. If such devices are used, it is generally necessary to shield the whole of the test

TOP

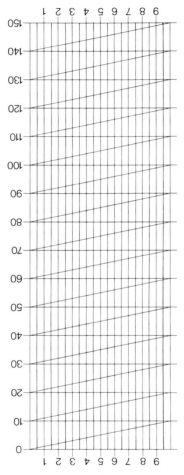

Figure 9.15 Sighting card for measuring pile deformation

arrangement from the weather. Automatic data acquisition may readily be achieved with a suitable logger. The output may be to paper tape or magnetic disk. Provided a compatible computer is chosen, it is possible to process the data on-site and plot up the relevant results. Indeed it is practicable to automate the whole of the test procedure, albeit at considerable cost, recognizing perhaps the need for human intervention in specific circumstances.

The degree of sophistication of the data-logging equipment provided depends to some extent on the volume of data to be recorded and on the number of piles to be tested. Some piling contractors have developed such equipment in what is effectively a mobile laboratory, and the quality of the output can be significantly better than that achieved with conventional test methods. For a few pile tests the equipment may be hired; however most available commercial data loggers are designed for laboratory use and may need to be adapted for outdoor use. Condensation or damp air frequently lead to instrumentation problems which may be difficult to solve for use under site conditions. A stable electrical supply is also required, particularly for computer equipment. To provide more information on the behaviour of the test pile, additional instrumentation is occasionally deemed necessary. This type of instrumentation is described in section 9.2.5.

9.2.3 *Pile load test procedures*

Two types of pile load test are commonly carried out; the maintained load test and the constant rate of penetration test. The maintained load test is convenient for testing end-bearing piles and for determining the load/settlement characteristics in clay soils. Usually, however, it is not as suitable for determining the true ultimate capacity of a pile.

The constant rate of penetration test procedure is best suited to determining the ultimate bearing capacity of a pile, as the method of testing is closely related to the test procedures used to obtain the shear strength of the soil. In clay soils, rapid pile testing may approximate to undrained loading conditions, whilst a constant load for several days may be necessary to allow full dissipation of pore pressures.

9.2.3.1 *Maintained load tests.*   The maintained load test requires a careful specification of loading increments and periods for which these increments are held constant. The ICE *Specification for Piling* sets out a suitable minimum scheme (Table 9.2), and the ASTM Test Designation D-1143 specifies a similar procedure. In addition, limits are placed on the rate of movement before the next load stage is added. The ICE *Specification for Piling* limits the rate of movement to 0.25 mm/h provided the rate is decreasing. Some engineers prefer a limit of 0.1 mm/h, and a frequent requirement is to hold the load constant for 24 h at design load. The particular details of the loading path followed are not of great significance, however it is important that the holding time at each load increment is the same so as to lead to the same degree of soil consolidation.

To determine these slow rates of movement a sensitive settlement measuring system is required. In granular soils or soft rocks the cessation of movement is easy to establish. However, in clay soils consolidation settlement will occur over an extended period. Plotting the results as the test proceeds is a useful way of resolving any ambiguities in the results (see Figure 9.17).

The maintained load test may be used to test contract piles to check acceptance criteria. Such piles are usually loaded to 100% of the design

Table 9.2 Suggested load increments and holding times.

Load*	Minimum time of holding load
25% DVL	30 min
50% DVL	30 min
75% DVL	30 min
100% DVL	1 h
75% DVL	10 min
50% DVL	10 min
25% DVL	10 min
0	1 h
100% DVL	6 h
100% DVL + 25% SWL	1 h
100% DVL + 50% SWL	6 h
100% DVL + 25% SWL	10 min
100% DVL	10 min
75% DVL	10 min
50% DVL	10 min
25% DVL	10 min
0	1 h

*SWL denotes specified working load; DVL denotes design verification load.

verification load (DVL) plus 50% of the specified working load to avoid overstressing the soil whilst still proving an adequate reserve of strength in the soil/pile system. The use of the DVL was introduced to take into account items such as downdrag friction load, which during a test may be in the reverse direction to that which may exist under long-term loading, and the fact that piles on a site may not be tested from the same level as may pertain for the piles in normal use. The ICE Specification also contains an extended proof load procedure which may be useful, particularly for piles bearing on cohesionless soil or on rock, and where further information may be desirable to verify design. On special test piles that will not form part of the permanent works, loading can be increased to failure, enabling a check to be made on the design parameters (usually by the CRP method).

For particular projects special loading paths may be called for, to simulate repeated loading for example. Such loading paths may easily be arranged but careful supervision is a necessity. Short-period, cyclic or sinusoidal loading requires the use of sophisticated servo-controlled equipment.

9.2.3.2 *Constant rate of penetration test.* The constant rate of penetration (CRP) form of testing is confined to special test piles, and is usually carried out at the pre-contract stage (Whittaker and Cooke, 1961). As the name implies, the loading is strain-controlled and a set rate of penetration of the pile load is specified. Rates of penetration between 0.5 and 2 mm/min are commonly used, the lower rate being used in clay soils.

Great care is required to maintain a steady rate of penetration, but this is difficult to achieve with manual pumping without some practice. If manual control is adopted time/penetration plots should be drawn as the test proceeds, so that deviations from the required path are easily identified. For this type of testing it is much more satisfactory to use a hydraulic power pack and control the rate of penetration with a dial gauge fitted with a pacing ring (Figure 9.16). Control of the jack movement is achieved by manual operation of the rate of oil flow. Fully automatic systems have been devised but are complex, and present difficulties in operation on site.

Loading of the CRP test should be continued to a deflection which goes to or beyond failure of the pile. This may be, typically, until a constant load has been recorded or until the penetration is at least 10% of the pile diameter (see Figure 9.18).

For this test, large travel of the jack and deflection gauges should be allowed for. It may be necessary to reset the gauges during the test, or alternatively, machined spacing blocks can be inserted under the gauge stems.

Once the equipment has been set up the test may be completed within an hour or so, and a rapid assessment of the ultimate capacity of the pile made. However, an experienced team of three or four people is required to carry out the test efficiently.

Figure 9.16   Dial gauge fitted with rotating bezel. Building Research Establishment: reproduced by permission of the controller, HMSO. Crown copyright

Tension piles may be tested using a constant rate of uplift procedure which is similar to the CRP test. The full tensile capacity of a pile is normally mobilized at a displacement of a few millimetres and the imposed rate of uplift should be correspondingly reduced to say 0.1 to 0.3 mm per minute.

9.2.3.3 *Recording the results.* It is essential that all the relevant data is recorded and to this end it is useful to use a standard pro-forma*. Similarly the load, settlement and time data should be clearly recorded throughout the test. It is important that any problems or unexpected occurrences are reported in the test records, as this information is most useful when interpretation of the results is made.

It is customary to present load/time, settlement/time and load/settlement graphs of the results. It is important that these last two graphs are plotted during a maintained loading procedure so that modifications to the test can be made if required.

Pile tests are expensive and the results of such tests can have a considerable influence on the construction of the permanent works. It is therefore imperative that adequate, qualified staff are available to set up and carry out the test. It is considered desirable that an engineer supervises the pile tests and is on site throughout the test period. In addition at least one technician should attend the tests, to control the hydraulic jack, etc. For constant rate of penetration tests a team of three or four people is necessary, albeit for a relatively short period. Similarly, interpretation of the results and preparation of a report on the test programme should not be stinted, and sufficient staff time should be allocated to complete the report speedily.

Special pile tests, involving a higher level of instrumentation will involve a commensurately larger team to carry out successfully. Setting up the instrumentation should be carried out by experienced technicians or engineers. Staff time involved in carrying such special pile tests can be considerable, and represents a significant cost in the overall test programme.

9.2.4 *Interpretation of the results*

A considerable amount of data may be computed from a pile test, and with more sophisticated instrumentation a greater understanding of the soil/pile interaction may be achieved. As discussed in section 9.1, it is important that the designer is satisfied that the behaviour of an isolated test pile under relatively short-term loading stresses the bearing stratum in a similar manner to the piles in the foundation. Interpretation of the pile test data may be carried out on several levels:

(i) qualitative inspection of the load settlement curves

*See Table 9.4

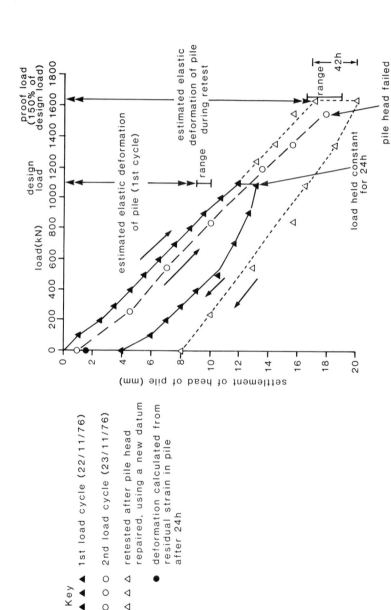

Figure 9.17 Typical pile proof load test result

(ii) a check for compliance with load and deformation specifications
(iii) a back analysis of the data to provide information on the soil stiffness and
strength criteria.

9.2.4.1 *Load settlement curves.*   Typical shapes of load settlement curves are
given by Tomlinson (1977). In particular unusual load settlement curves
should be thoroughly investigated as they may indicate unexpected geological
problems, such as soft sensitive clay, defects in the pile shaft, or poor
construction techniques.

From straightforward pile tests in which the load settlement characteristics
of the head of a pile are measured, there is as yet no satisfactory published
method which allows separation of the components of shaft friction or
adhesion and end-bearing with a satisfactory degree of confidence. Van
Wheele (1957) described a technique for this but it is thought not to be very
satisfactory. Recent research by Fleming suggests that, subject to a pile
having been made to settle by several per cent of its diameter, it is possible
to effect reasonably such a separation. The results of this work are currently
in preparation for publication.

In the past it has generally been found that, in order to study the division
of load between a pile shaft and base, it is necessary to install a system of
strain gauges. This can have the advantage of also providing guidance on
the distribution of load along a pile shaft and contributing to design
knowledge, subject to adequate provision of gauges. However, this requires
considerable skill and care (see section 9.2.5).

9.2.4.2 *Estimation of ultimate load.*   The ultimate load of a pile is usually not
well defined; typical load/deformation curves are shown in Figures 9.17 and
9.18. In practice an exact definition of the ultimate load is not all that
important, provided an adequate factor of safety is clearly demonstrated. Two
simple criteria which cover most situations are

(i) the load at which settlement continues to increase without further increase
in load
(ii) the load causing a settlement of 10% of the pile diameter (base diameter).
The latter limit is likely to give a low estimate of the ultimate load as it is
unlikely that general yielding of the soil around the pile will have been
initiated.

Other definitions of ultimate load are summarized by Tomlinson (1977),
and Fellenius (1980) compares a number of methods of estimating the ultimate
load.

Unfortunately many of the proposed techniques are either empirical or
are based on set deformation criteria. Several of the methods are also sensitive
to the shape of the load/deformation curve, and it is preferable to use a
considerable number of load increments to define the shape clearly.

Figure 9.18   Results of a CRP test. After Fellenius (1980)

Two methods of estimating the ultimate pile load are shown in Figures 9.19 and 9.20, after Chin (1970) and Brinch Hansen (1963). Chin's method assumes the form of the load/deformation curve is hyperbolic and is an empirical method. However, it is useful in that if the results are plotted at the time of the test, deviations from the expected curve are easily seen and the test procedure can be checked. It may in some cases be used to obtain an estimate of shaft or base load separately. Brinch Hansen's method is very simple to use, and in practice has been found to give consistent results.

9.2.4.3 *Settlement criteria.*   Settlement limits at design load or 1.5 times the design load are frequently specified, on the basis of previous experience (e.g. a maximum of 10 mm settlement at design load). Such specified settlement limits should be realistic and should take into account the likely elastic compression of the pile shaft. It is sometimes difficult to discern how the specified criteria for a single pile relate to the behaviour of the pile group. A much more satisfactory approach is to use the pile test data to assess the stiffness of the soil mass using the solutions outlined in Chapter 4. This data may then be applied to the pile group to estimate the overall performance of the foundation under the design loads. This approach is particularly useful in the case of piled structures carrying high lateral loads, as the stiffness of the soil is difficult to establish and may well be affected by the piling operation.

Contract test piles are frequently loaded to $1\frac{1}{2}$ times the design load as a performance test. Such a load test should avoid serious permanent deformations in the surrounding soil and the significant alteration of the load/deformation characteristics of the pile so tested. As these tests are not

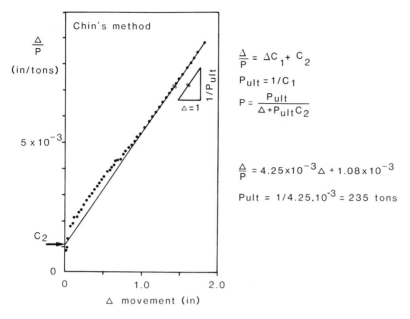

Figure 9.19 Chin's method for estimation of ultimate load. After Chin (1970)

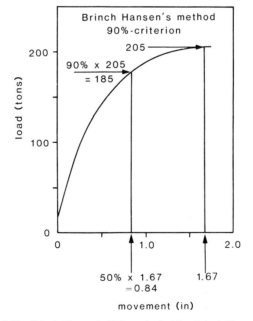

Figure 9.20 Brinch Hansen's 90% method. After Brinch Hansen (1963)

taken to failure it is sometimes difficult to assess the acceptability of the pile. In such situations it is useful to estimate the likely elastic settlement of the pile and then to draw up a realistic and unambiguous specification for test pile performance. Such an assessment can quickly be made by summing the compression of the shaft and settlement of the toe. For such calculations it is necessary to estimate the shaft adhesion from the site investigation data. In practice, by making appropriate assumptions, upper and lower limits to the settlement may be established.

9.2.5 *Special pile instrumentation*

To obtain a greater understanding of the soil/pile behaviour, it is sometimes desirable to install further instrumentation in the test piles. It should be stressed that such instrumentation is expensive in cost and manpower, if reliable results are to be obtained. In general, for site use simple robust equipment should be used to obtain the necessary information, albeit with some loss of accuracy. There are several case histories reported in the literature (Reese *et al.*, 1975) for fully instrumented piles; these are really research projects and are not considered in detail.

Much of the equipment for full pile instrumentation is not available 'off the shelf' and delivery periods for specialized apparatus is usually several months. Some of the readout equipment may not be specifically designed for site use and an instrument hut may have to be provided.

9.2.5.1 *Distribution of load.* Special high-capacity load cells have been developed by the BRE (Whitaker, 1963). The cells are steel multi-column devices, the strain being measured by electric resistance strain gauges (Figure 9.21).

A simpler modular load cell system has recently been developed by the BRE (Price and Wardle, 1983). The load transducer comprises a steel tube fitted with an internal vibrating wire gauge. Load is transferred to the transducer, across a discontinuity in the pile formed by a soft rubber sheet, by steel bars bonded into the concrete. Load cells of the desired capacity may be made using an array of transducers (Figure 9.22). These cells are robust and have the advantage that the structural collapse of the pile cannot occur, following corrosion of the transducer elements.

The measurement of base load in a driven pile is not as easy as for cast-in-place piles. Normally special load cells occupying the full cross-section of the pile cannot be used, although with some segmental piles the load cell can be inserted at the joints. A more satisfactory technique is to cast vibrating wire strain gauges into the pile (Figure 9.23). The gauge installation may be calibrated by load-testing a short length of pile. Stability of the system is reasonable for short-term loading, and the gauges should not be damaged by driving. Long-term load measurement requires correction for creep strains in

Figure 9.21 BRE pile load cells. Building Research Establishment: reproduced by permission of the Controller, HMSO. Crown copyright

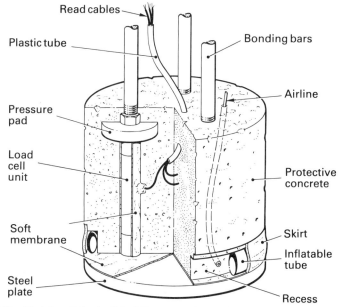

Figure 9.22 Modular pile load cell. Building Research establishment: reproduced by permission of the Controller, HMSO. Crown copyright

Figure 9.23   Vibrating wire strain gauge installation. Photograph courtesy of CIRIA, London

the concrete, and a consequent uncertainty in the estimation of the pile load. Several gauges should be installed at any level to allow for inevitable losses. The cables should be continuous and may be led out through a duct below the head of the pile. If a segmental pile system is being used, forming waterproof cable joints at each segment junction causes severe problems.

A similar system may be used for bored piles (O'Riorden, 1982), but unfortunately the sensitivity of the system is reduced owing to lower stresses in the concrete. Variation of the modulus of the concrete will also add to the uncertainty in relating strain to load. Protecting gauges during concreting is difficult, and a considerable number of gauges may be damaged. Successful installations of vibrating wire strain gauges are much easier to achieve at the head of bored piles, where concreting can be carefully monitored.

An alternative, cheap technique to separate shaft and base resistance of a cast-in-place pile is to form a compressible base to the shaft. High-density polystyrene has been successfully used, and the pile is deformed sufficiently to compress the packing in the later stages of the test to mobilize the base resistance. Water-filled rubber bags have been used to similar effect. A similar technique is to test two piles, one constructed with a void at the base.

Steel piles may be instrumented using vibrating wire or wire resistance strain gauges. With the latter system, the sealed gauges which can be welded to the steel are to be preferred, and waterproof cabling must be used. The strain gauges effectively turn the steel pile into a load measuring column and computation of the pile load is straightforward. With tubular piles the gauges may be fixed to the inside and sealed in. Further mechanical protection is not necessary except to restrain the cables during driving. With H-section piles the strain gauges and cables need to be protected with steel ducts (a difficult operation which can seriously affect the stiffness of the pile and cause stability problems during driving).

In general, wire resistance strain gauges should be used only in situations where they can be sealed into a dry compartment. With this system small electrical voltages are measured and cable earth leakage must be small. Similarly, the readout unit must be set up in a dry atmosphere, not an easy thing to arrange on a civil engineering site. If dynamic response of piles to driving is to be investigated, this type of gauge is the only suitable system, but special ultraviolet recorders and continuous excitation of the gauges should be adopted.

9.2.5.2 *Deformation measurement.* Axial deformation of large-diameter piles may be measured using a simple rod extensometer (Figure 9.24). Such extensometers should be free to move in the enclosing tubes, and care should be taken to exclude debris. Using the information provided by these devices a crude estimate of shaft adhesion and base resistance may be made. Similarly, magnetic ring extensometers may be cast into the pile.

Figure 9.24 Typical pile test using rod extensometers

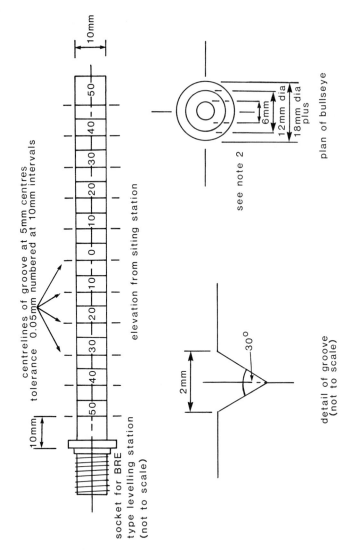

Figure 9.25 Survey targets. *Notes*: 1, material stainless steel; 2, bullseye to be target of concentric grooves 1 mm wide around 1 mm drilled centre point. The target to be the end of standard socket for **BRE** type station

Gross lateral deformation of piles may be measured by incorporating inclinometer tubes into the pile and surveying these with a suitable instrument. The computed profile will be accurate to a few tens of millimetres, unfortunately not sufficiently accurate for bending stresses to be estimated. Strain gauges may be used in steel piling to determine the variation of bending strain with depth and hence to compute stresses and deformations. Many sets of gauges are required to perform this reliably, and such installations are used only for research purposes (Reese et al., 1975).

In lateral pile load tests it is desirable to measure rotation of the pile head. This may be accomplished by using a sensitive spirit level mounted on the pile, or by using an electro-level. Such observations are also valuable if groups of piles are tested to determine the global stiffness of the group.

Lateral deformations may be measured using a datum beam and dial gauges. These measurements suffer the same problems as for vertical measurements (section 9.2.2.3). For long-term observations on pile groups, optical methods are to be preferred. Suitable targets are shown in Figure 9.25. A single second theodolite fitted with a parallel plate micrometer may be used for the observations. An accuracy of about ± 0.2 mm may be achieved using this system over a long period of time. However, reliable, stable datum marks are a necessity and these may be difficult to establish, especially in areas underlain by thick alluvial deposits. Geodetic levelling techniques may be used to observe vertical deformation to a similar order of accuracy. The absolute accuracy of such level surveys is again dependent on the establishment of a reliable, stable datum (Cheney, 1973).

9.2.6 Other pile tests

9.2.6.1 Tension pile tests. This test method is similar to compression testing of piles, and maintained load or constant rate of uplift procedures may be used. The reaction system may be provided by a strong-back resting on adjacent piles or spread foundations. Such a test arrangement is shown diagrammatically in Figure 9.26. Two jacking points on either side of the pile are used to avoid bending the test pile, and care is required to load the jacks evenly. Some cracking of reinforced concrete tension piles is to be expected if loads are high unless the reinforcement is debonded except in the bottom section of pile.

9.2.6.2 Lateral load tests. For structures carrying major lateral loads it is desirable to carry out lateral load tests to confirm the design assumptions. The applied loads are relatively small and it is convenient to use an adjoining pile as a reaction (Figure 9.27). Provision must be made for a stable datum for deformation measurement as for vertical tests. If two identical piles are being tested it may be more convenient to measure the relative movement between

Figure 9.26 Tension pile test

Figure 9.27   Lateral pile load test. Photograph courtesy of Westpile Ltd., Uxbridge

them using an extensometer. To increase the reliability of the back-analysis
it is worthwhile measuring the rotation of the pile head. Foundations are
frequently subjected to lateral, vertical and moment loading. It is usually
not possible to arrange for the pile test to mimic such combined loads.
Instead, application of a lateral load near to the ground-level existing at the
time is often chosen for such tests. Similarly it is difficult to simulate fixed
head conditions. Data from such a simplified load test may be analysed
using the principles outlined in Chapter 4, and the computed soil parameters
may then be used to predict the response of the pile or pile group to a more
complicated loading regime. Bearing in mind the process of analysis which
has to be used in such circumstances, where possible the testing should be
planned and loading levels arranged to give the minimum degree of
extrapolation in the analysis.

9.3 Dynamic testing of piles

The use of dynamic testing methods for estimating the ultimate capacity and
integrity of piles is now widespread. Assessment of pile capacity during
driving has been undertaken routinely by the use of empirical driving
formulae such as those discussed in section 4.1.4. There are many limitations
to such formulae, particularly in respect of uncertainties in the proportion
of the available driving energy that is transmitted to the pile. The accuracy
of dynamic methods of estimating pile capacity has been significantly
improved using modern instrumentation and fast data acquisition systems

that allow measurement of the dynamic force and velocity waves in the pile during driving. The data may be analysed at different levels of sophistication, as described in the following sections, to give estimates of the static pile capacity at time of driving. Similar techniques, with lower energy input, may be used to assess the integrity of piles (particularly cast-in-situ piles), as discussed in Chapter 8.

9.3.1 *Wave equation analysis*

The propagation of driving energy along a pile, allowing for interaction with the surrounding soil, may be analysed with sufficient accuracy using a 'one-dimensional' idealization. In this idealization, only vertical (strictly speaking 'axial'), displacement of the pile is considered, and the governing differential equation is

$$(AE)_p \frac{\partial^2 w}{\partial z^2} = (A\rho)_p \frac{\partial^2 w}{\partial t^2} - f \tag{9.1}$$

or

$$\frac{\partial^2 w}{\partial z^2} = \frac{1}{c^2} \frac{\partial^2 w}{\partial t^2} - \frac{f}{(AE)_p} \tag{9.2}$$

where $(AE)_p$ is the cross-sectional stiffness of the pile,
        $(A\rho)_p$ is the mass per unit length of the pile,
        c is the wave propagation speed in the pile ($= \sqrt{E/\rho}$),
        w is the vertical displacement of the pile,
        z is the distance down the pile,
        t is the time variable, and
        f is the mobilized soil resistance per unit length of pile.

Historically, this equation has been implemented using finite difference or finite element techniques, with the pile being modelled as a discrete assembly of mass points interconnected by springs. This model, originating in the work of Smith (1960), forms the basis of a range of computer programs for studying pile drivability.

More recently, equation (9.2) has been solved numerically by using the characteristic solutions, which are of the form

$$w = g(z - ct) + h(x + ct) \tag{9.3}$$

where g and h are unspecified functions which represent downward (increasing z) and upward travelling waves respectively. Taking downward displacement and compressive strain and stress as positive, the force, F, and particle velocity, v, in the pile are given by

$$F = -(AE)_p \frac{\partial w}{\partial z} = -(AE)_p(g' + h') \tag{9.4}$$

and

$$v = \frac{\partial w}{\partial t} = -c(g' - h') \tag{9.5}$$

where the prime denotes the derivative of the function with respect to its argument.

The velocity and force can each be considered as made up of two components, one due to the downward travelling wave (represented by the function g) and one due to the upward travelling wave (represented by the function h). Using subscripts d and u for these two components, the velocity is

$$v = v_d + v_u = -cg' + ch' \tag{9.6}$$

The force F is similarly expressed as:

$$F = F_d + F_u = -(AE)_p g' - (AE)_p h' = Z(v_d - v_u) \tag{9.7}$$

where $Z = (AE)_p/c$ is referred to as the pile impedance. [Note, some authors have referred to the pile impedance as $Z = E/c$, relating axial stress and velocity rather than force and velocity. The more common definition of pile impedance as $Z = (AE)_p/c$ will be adopted here.]

The relationships given above may be used to model the passage of waves down and up piles of varying cross-section, allowing for interaction with the surrounding soil. It is helpful to consider the pile as made up of a number of elements, each of length Δz, with the soil resistance acting at nodes at the mid-point of each element (see Figure 9.28). Numerical implementation of the characteristic solutions involves tracing the passage of the downward and upward travelling waves from one element interface to the next. The time increment, Δt, is chosen such that each wave travels across one element in the time increment (giving $\Delta t = \Delta z/c$).

Between nodes i and $i + 1$, the soil resistance may be taken as T_i, the value of which will depend on the local soil displacement and velocity (see later). Taking T_i as positive when acting upwards on the pile (that is, with the soil resisting downward motion of the pile), the soil resistance will lead to upward and downward waves of magnitude

$$\Delta F_u = -\Delta F_d = T_i/2 \tag{9.8}$$

These waves will lead to modification of the waves propagating up and down the pile.

The procedure for calculating new values of wave velocities at each node is shown schematically in Figure 9.29. Thus, consider the downward and upward waves at nodes $i - 1$ and $i + 1$, at time t. The new downward travelling wave at node i at time $t + \Delta t$ is given by

$$(v_d)_i[t + \Delta t] = (v_d)_{i-1}[t] - T_{i-1}[t + \Delta t]/(2Z) \tag{9.9}$$

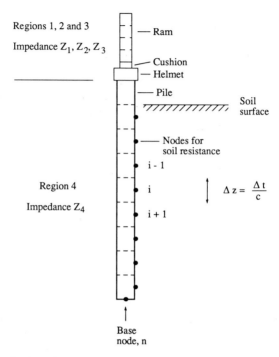

Figure 9.28 Idealization of pile as elastic rod with soil interaction at discrete nodes

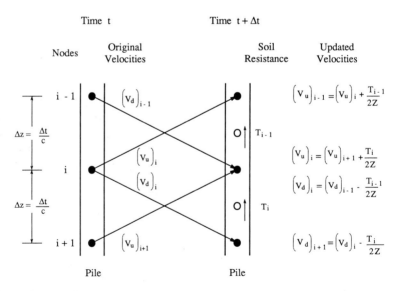

Figure 9.29 Modification of downward and upward waves due to soil interaction. After Middendorp and van Weele (1986)

while the new upward travelling wave fractionally above node i is

$$(v_u)_i[t + \Delta t] = (v_u)_{i+1}[t] + T_i[t + \Delta t]/(2Z) \qquad (9.10)$$

At the base of the pile, the downward travelling wave will be reflected, with the magnitude of the reflected wave dependent on the base resistance, Q_b, offered by the soil. The axial force in the pile must balance the base resistance, which leads to an expression for the reflected (upward travelling) wave velocity of

$$(v_u)_n[t + \Delta t] = (v_d)_n[t + \Delta t] - Q_b[t + \Delta t]/Z \qquad (9.11)$$

The base velocity (the nth node) is

$$v_n = 2v_u + Q_b/Z = 2v_d - Q_b/Z \qquad (9.12)$$

where all quantities refer to time $t + \Delta t$.

For a force F_d arriving at the pile tip, equation (9.11) implies a reflected force of

$$F_u = - Zv_u = Q_b - F_d \qquad (9.13)$$

The magnitude of the reflected wave thus varies from $- F_d$, where the tip resistance is zero, to F_d, where the base velocity is zero and the base resistance is twice the magnitude of the incident force (see equation (9.12)).

It should be noted that the characteristic solutions of the wave equation may be used to model the pile hammer assembly as well as the pile, in order to perform drivability studies. The geometry and mass density of each component (ram, cushion, helmet) is matched, and the ram is given an initial velocity to model the impact. As discussed by Middendorp and van Weele (1986), a relatively crude model of the hammer will generally suffice to give adequate results.

Dynamic soil model along pile shaft. Accurate prediction of the performance of piles during driving requires modelling of the dynamic response of the soil around (and, for open-ended pipe piles, inside) the pile, both along the shaft and at the base. Following traditional approaches for the analysis of machine foundations, the soil response can generally be represented by a combination of a spring and dashpot. However, it is also necessary to consider limiting values of soil resistance where, along the shaft, the pile will slip past the soil and, at the tip, the pile will penetrate the soil plastically.

In the original work of Smith (1960), which still forms the basis of many commercially available pile driving programs, the soil response was modelled conceptually as a spring and plastic slider, in parallel with a dashpot (Figure 9.30(a)). For such a model, the soil resistance (per unit length of pile) may be written as

$$T = T_s + T_d = Kw + Cv \qquad (9.14)$$

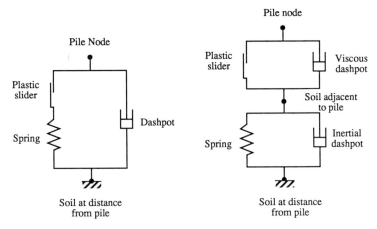

Figure 9.30   Soil models for dynamic response: (a) original soil model of Smith (1960); (b) soil model of Randolph and Simons (1986)

where the subscripts s and d refer to static and dynamic resistances respectively, and v is the local velocity. Smith (1960) simplified this response to

$$T = \frac{w}{Q} T_{max}(1 + Jv) \qquad (9.15)$$

subject to the term outside the bracket being limited to T_{max}, the limiting static resistance. The parameters Q and J are referred to as the 'quake' (the pile displacement for full mobilization of the static resistance T_{max}) and 'damping constant' (with units of s/m or equivalent). Typical values for Q and J recommended by Smith (1960) are 2 to 3 mm, and 0.1 to 0.2 s/m.

A more sophisticated model, based on elastodynamic theory, has been introduced by Randolph and Simons (1986) (see also Randolph (1990)). As shown in Figure 9.30, the model separates out inertial or radiation damping due to the surrounding soil (which operates prior to slip between pile and soil), from viscous damping that may lead to an increase in the limiting resistance, T_{max}. The spring constant, K, and inertial dashpot constant, C, are related to the shear modulus, G, and mass density, ρ_s, of the soil, by

$$K = 2.75G \qquad (9.16)$$

$$C = \frac{2\pi r_0 G}{V_s} = 2\pi r_0 \sqrt{G\rho_s} \qquad (9.17)$$

where V_s is the shear wave velocity in the soil and r_0 is the pile radius.

Any viscous enhancement of the static resistance is modelled by taking the limiting dynamic resistance as

$$T_d = T_{max}[1 + \alpha(\Delta v/v_0)^\beta] \qquad (9.18)$$

where $v_0 = 1$ m/s and Δv is the *relative* velocity between the pile and the soil. It is more correct to use the relative velocity in this equation, rather than the absolute pile velocity, since the main viscous effects will be confined to the zone of high shear strain rate immediately adjacent to the pile. Typical values for the viscous parameters may be taken as $\beta = 0.2$ (following Gibson and Coyle (1968) and Litkouhi and Poskitt (1980)) and α in the range 0 for dry sand, up to 1 or possibly higher for clay coils.

Dynamic soil model at pile base. Similar models may be used to describe the response of the soil at the pile base. Unlike along the pile shaft, the soil directly beneath the pile tip does not 'slip' past the pile in the way that occurs along the shaft, and it is less easy to separate the viscous and inertial damping. In practice, the original model of Smith is generally used (Figure 9.30(a)), but with the spring and dashpot constants chosen in the light of elastodynamic theory, using the analogue of Lysmer and Richart (1966).

Thus, the dynamic resistance at the pile base may be written

$$Q_b = Kw + Cv \qquad (9.19)$$

where

$$K = \frac{4Gr_0}{(1 - v)} \qquad (9.20)$$

and

$$C = \frac{3.4r_0^2}{(1 - v)}\frac{G}{V_s} = \frac{3.4r_0^2\sqrt{G\rho_s}}{(1 - v)} \qquad (9.21)$$

In order to allow for plastic penetration of the pile tip, it is customary to limit the *static* component of Q_b to the bearing capacity of the pile tip. Thus the quantity Kw should be limited to

$$Kw \leqslant Q_{max} = A_b q_b \qquad (9.22)$$

where A_b is the area of the pile base (the area of steel for an H-section or open-ended pipe pile) and q_b is the limiting end-bearing pressure. It should be noted that only the static component is limited to Q_{max}, since energy will continue to be radiated into the far field during plastic penetration. Thus there will be soil resistance from the dashpot representing the inertial resistance of the soil, in addition to the limiting static end-bearing capacity.

9.3.2 *Dynamic stress wave data*

Wave equation analysis, as described in the previous section, is used in determining the drivability of piles, using commercially available software such as the WEAP family (Goble and Rausche, 1976, 1986; Rausche *et al.*, 1988). However, a potentially more significant application lies in the interpretation of dynamic stress wave data measured during pile driving. The data form the basis of a dynamic pile test, which may obviate the need for more costly static loading tests.

Measurement of stress wave data is achieved by the use of light-weight strain cells and accelerometers that are attached to the pile a few diameters below the hammer. A description of the instrumentation, and one of the most widely-used monitoring systems (the Pile Driving Analyser, or PDA) has been given by Likins (1984).

The first stage of interpretation of stress wave data is performed in real-time in the field by equipment such as the Pile Driving Analyser. Strain and acceleration data are processed, generally through electronic hardware, to obtain force and velocity data. From these data, various parameters may be derived. Thus, integration with time of the product of force and velocity up to the time at which the product becomes negative leads to a figure for the maximum energy transmitted to the pile. This allows the overall operating efficiency of the hammer to be assessed in terms of its rated energy. If additional information is available on the ram velocity at impact, then the energy losses may be subdivided into mechanical losses in the hammer, and losses in the impact process due to inelasticity and bounce of the components.

In traditional pile driving formulae, one of the largest sources of error in estimating the overall pile capacity is uncertainty in the energy transmitted to the pile. Measurement of the actually transmitted energy allows use of simple pile driving formulae with increased confidence. Such formulae provide a means whereby information obtained on instrumented piles may be extrapolated in order to assess the quality of uninstrumented piles driven on the same site. Of course, for piles where stress wave data are obtained, more sophisticated techniques may be used to assess the pile capacity.

The relationships presented in section 9.3.1 allow the dynamic and static soil resistance to be estimated from the stress wave data. Equation (9.8) implies that, as the stress wave travels down the pile, the magnitude of the force will decrease by half the total (dynamic plus static) shaft resistance, T. Thus, at the bottom of the pile, the downward travelling force is

$$F_d = F_{do} - T/2 \qquad (9.23)$$

where F_{do} is the original value at the pile head. Similarly, equation (9.13) may be used to obtain the upward travelling force after reflection at the pile tip as

$$F_u = Q_b - F_d = Q_b + T/2 - F_{do} \qquad (9.24)$$

On the way back up the pile, *provided the particle velocity at each position is still downwards, implying upward forces from the soil on the pile*, the upward travelling wave will be augmented by half the shaft resistance, to give a final return wave of

$$F_{ur} = Q_b + T - F_{do} \qquad (9.25)$$

where the subscript r refers to the return (upward travelling) wave at a time $2l/c$ later than the time at which the value of F_{do} was obtained (l being the length of pile below the instrumentation point). The total dynamic pile capacity is then

$$R = Q_b + T = F_{do} + F_{ur} \qquad (9.26)$$

Equations (9.6) and (9.7) may be used to derive the upward and downward components of force from the net force and particle velocity at the instrumentation level, so that equation (9.26) may be re-written:

$$R = 0.5(F_0 + Zv_0) + 0.5(F_r - Zv_r) \qquad (9.27)$$

where subscripts 0 and r refer to times t_0 (generally close to the peak transmitted force) and $t_r = t_0 + 2l/c$. This equation is the basis for estimating the total dynamic pile capacity directly from the stress wave measurements. A search may be made for the value of t_0 which gives the largest value of capacity.

Case method of analysis. Since the dynamic capacity will be greater than the current static capacity, a simple method is needed to estimate the static capacity in the field, without the need for a full numerical analysis of the pile. In the Case approach, which has gained widespread acceptance, this estimate is made on the basis that all the dynamic enhancement of capacity occurs at the pile tip, with a dynamic component of resistance that is proportional to the pile tip velocity, v_b. Thus the dynamic tip resistance is taken as

$$Q_d = j_c Z v_b \qquad (9.28)$$

where j_c is the Case damping coefficient. These simplifying assumptions lead to an expression for the static pile capacity, R_s

$$R_s = 0.5(1 - j_c)(F_0 + Zv_0) + 0.5(1 + j_c)(F_r - Zv_r) \qquad (9.29)$$

The assumptions regarding the dynamic soil resistance are clearly an oversimplification, and the deduced static pile capacity can be very sensitive to the value adopted for the damping parameter, j_c. However, the above expression can provide useful guidance on the static pile capacity where it is possible to calibrate the parameter j_c for a particular site. Where no static load tests are carried out, guidelines for j_c as given below may be adopted (Rausche et al., 1985).

Table 9.3   Suggested values for case damping coefficient

Soil type in bearing strata	Suggested range of j_c	Correlation value of j_c
Sand	0.05–0.20	0.05
Silty sand/sandy silt	0.15–0.30	0.15
Silt	0.20–0.45	0.30
Silty clay/clayey silt	0.40–0.70	0.55
Clay	0.60–1.10	1.10

Figure 9.31 shows example stress wave data, for a 300 mm square concrete pile driven with a 6 tonne hammer. In the early part of the blow, before any reflections from the soil resistance arrive at the instrumentation point, the force and velocity are directly proportional ($F = Zv$) since only a downward travelling wave component is present. As reflections from the soil arrive, the force curve separates from the velocity curve, with the separation showing the magnitude of the shaft resistance. At a time $2l/c$ after impact (where l is the length of pile below the instrumentation point), reflection from the pile tip occurs, and the forms of the force and velocity curves change. In this particular example (taken from a re-drive), little of the blow reaches the pile tip, and the reflection from the base is minimal.

Implementation of equation (9.29) to give the Case estimate of pile capacity is shown in the latter part of the blow (after the return wave arrives back at the instrumentation point). While the total dynamic resistance (using $j_c = 0$) reaches a peak of 2.67 MN, the use of $j_c = 0.75$ gives an estimate of the maximum 'static' resistance of about 1.73 MN.

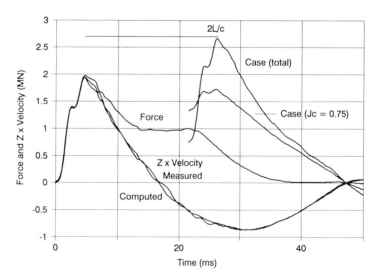

Figure 9.31   Examples of stress wave data

Detailed matching of stress wave data. The Case formula (equation (9.29)) offers only an approximate estimate of pile capacity, and a more reliable estimate requires detailed matching of the stress wave data. This process is an iterative one, where the soil parameters for each element down the pile are varied until an acceptable fit is obtained between measurements and computed results. In order to avoid uncertainties in modelling the hammer, either the measured force signal or the measured velocity signal is used as an upper boundary condition in the computer model. The fit is then obtained in terms of the other variable. The iterative process, with examples of the effect of varying different parameters, has been described by Goble *et al.* (1980).

It is possible to automate the matching process, with the computer optimizing the soil parameters in order to minimize some measure of the difference between measured and computed response (Dolwin and Poskitt, 1982). However, it has been found that computation time can become excessive, particularly for long piles, unless the search zone for each parameter is restricted by operator intervention. It is rather more straightforward to carry out the matching process manually. Experience soon enables assessment of where values of soil resistance, damping or stiffness need to be adjusted in order to achieve an improved fit. A satisfactory fit may generally be achieved after 5–10 iterations of adjusting the parameters and re-computing the response.

Limitations in the soil models used for pile driving analysis entail that the computer simulation will not exactly match the real situation. A consequence is that the final distribution of soil parameters should not be considered as unique, but rather as a best fit obtained by one particular operator. Generally, the total static resistance computed will show little variation provided a reasonable fit is obtained. However, the distribution of resistance down the pile, and the proportion of the resistance at the pile base, may show considerable variation (Middendorp and van Weele, 1986).

An interesting investigation of operator dependence in the analysis of stress wave measurements has been reported by Fellenius (1988). Eighteen operators were given four sets of stress wave data to analyse, covering a range of pile types and soil conditions. One of the sets of data was from a re-drive of a pile that was subjected to a static load test the following day. All the operators were using the same computer program, CAPWAP, which is one of the most widely utilized programs for such analyses, originating from the work of Goble and his co-workers (Goble and Rausche, 1979).

The study by Fellenius shows a good measure of agreement among the participants in predicting the static pile resistance, with the coefficient of variation being under 10% for three out of the four cases considered. For the one pile that was subjected to a static load test, the predictions spanned the measured static capacity.

An example of matching stress wave data is shown in Figure 9.31. The

346   PILING ENGINEERING

measured force has been used as input data, and measured and computed velocities are compared. The agreement is reasonably good. The fit has been obtained for a static shaft capacity of 1.78 MN, and assumed base capacity of 2.24 MN (giving 4.02 MN total capacity). However, the blow was not sufficient to mobilize the full base resistance. Analysis of the data showed a residual force at the pile base of 1.17 MN, and a maximum base resistance during the blow of 1.50 kN. The minimum static pile capacity then lies between $1.78 + 1.17 = 2.95$ MN, and $1.78 + 1.50 = 3.28$ MN. In this case, without the use of a heavier hammer, it is not possible to be more specific concerning the ultimate capacity of the pile. Note, however, that the minimum pile capacity exceeds the simple Case estimate, due to partial rebound of the pile during the return passage of the stress wave.

9.3.3 General comments

There are a number of different systems available commercially to undertake dynamic pile monitoring and interpretation. Probably the two most widely used are (1) the Pile Driving Analyser and CAPWAP program from Pile Dynamics Inc., USA, and (2) the TNO equipment (Figure 9.32) and TNO-WAVE program, from the Institute TNO for Building Materials and Building Structures, Holland.

Regardless of which system is used, there are a number of intrinsic limitations in assessing the static pile capacity from dynamic data:

(1) The capacity of a driven pile generally increases with time following installation. This phenomenon, referred to as 'set-up', is generally attributed to dissipation of excess pore pressures generated during installation. Case studies have shown that, particularly in soft cohesive

Figure 9.32   The TNO dynamic pile test equipment. Photograph courtesy of Institute TNO, Delft

soil, the pile capacity may increase by a factor of 4 to 5 following installation, over a time period of several weeks or months, depending on the consolidation characteristics of the soil and the pile diameter (Randolph *et al.*, 1979). Where the pile capacity is to be estimated from dynamic measurements, it is necessary to allow for such set-up. This may be achieved by 're-striking' the pile after an appropriate time delay. It is important to ensure that the full pile capacity is mobilized in such a re-strike within the first few blows, so as not to reduce the long-term performance of the pile.

(2) One of the major pitfalls to be avoided is trying to estimate the capacity from hammer blows of insufficient energy to fail the pile. This is quite common during re-drive tests and, in such cases, analysis of the data will only yield an estimate of the maximum static resistance mobilized during that blow.

(3) Use of the Case formula to estimate the static capacity of the pile is not recommended without a full numerical matching of the stress wave data as corroboration. Even where this has been done, and a reasonable estimate of damping parameter, j_c, is available, care should be taken to ensure that the conditions assumed in deriving the relationship are met in practice. Two particular conditions are (a) that there are no major changes in cross-section (or impedance) of the pile along its length, and (b) that the pile velocity remains positive (downwards) over the major part of the return time of the stress wave. In many cases, the latter condition is not fulfilled, and the Case formula will underpredict the pile capacity. A correction may be made, taking account of the proportion of pile that was rebounding during the return passage of the stress wave, and using the earlier part of the blow to estimate the soil resistance in that region.

(4) The two main causes of differences between static and dynamic performance of the pile are (a) viscous damping and (b) inertial damping. The dynamic shaft capacity of a pile may exceed the static capacity by a factor of 2 or more, due to viscous effects. Similarly, the dynamic base capacity may be 2 to 4 times the static capacity due to inertial effects alone. In both cases, differences between dynamic and static capacity are greater in soft cohesive soil, than in stiffer or coarser material. Allowance for damping is made by appropriate choice of damping parameters in the dynamic analysis. However, the deduced static capacity is relatively sensitive to the choice of damping constant, and it is clear that further research is needed in order to provide better guidance on damping parameters for different soil types. At this stage, it is strongly recommended that at least one static load test be performed on a given site in order to calibrate the dynamic analyses.

(5) One area which has received insufficient attention is the different response of H-section piles and open-ended pipe piles under dynamic

and static conditions. It may be shown that both types of pile tend to drive in an 'unplugged' condition, with soil moving up the inside of the pipe, or filling the space between the flanges of the H-pile. However, during a static load test the reverse is true. The frictional resistance of the soil plug is such that both types of pile will tend to fail as a solid body. Thus, during a dynamic test, these piles will show relatively high shaft friction but low end-bearing. By contrast, during a static test, the shaft capacity will be just that on the outside of the pile, while the end-bearing resistance will act over the gross area of the pile. It is essential that estimates of the static capacity of such piles take account of differences in the failure modes during dynamic and static penetration.

(6) Residual stresses acting down the length of the pile can have a significant effect on the calculated pile response under dynamic conditions. These stresses will generally build up along the pile, with some locked in end-bearing stress, balanced by negative shear stresses acting along the pile shaft (particularly the lower part). Accurate assessment of pile capacity and, in particular, the relative shaft and base capacity, necessitates modelling of the residual forces locked into the pile between each blow.

Dynamic methods for estimating the static resistance of piles are evolving rapidly at present. It is suggested that such methods should not be relied upon as the sole means of assessing the acceptability of piles, except in areas where the geology is similar and a large body of experience has been accumulated. For general applications, it is suggested that these dynamic methods should be calibrated against static load tests, before extensive testing of contract piles is undertaken. Used in this way, dynamic monitoring of piles during driving will enable economies to be made in the static pile testing programme, and allow a larger proportion of contract piles to be tested, giving increased confidence in the foundation performance.

9.4 Safety aspects of pile testing

Substantial loads are mobilized during a pile test and a large amount of strain energy is stored in the equipment. It is therefore imperative that all parts of the pile testing apparatus are carefully designed and that short-cuts or poor workmanship are not entertained for what is after all a temporary structure. Failures usually result in the explosive release of energy, for example the head of a concrete pile may fly apart under the action of eccentric loads.

Major parts of the equipment should be clearly marked with the design load and care should be taken not to overload the equipment. Reaction beams, in particular, should be used with the spreader plates, cross-heads, or grillage to support kentledge for which they were designed, and grillage beams should be at the correct spacing. It is good practice for working drawings of the test layout to be sent on site with the equipment.

Eccentricities of load can cause unexpected high bending moments in the pile and loading frame. They may, in extreme cases, lead to twisting and collapse of the load column, a very dangerous situation. Thus care should be taken to align the elements of the load column and ensure all surfaces mate properly. The reaction beam should be carefully levelled and set with the web vertical. To increase the stability of the system it is better to use double beams.

Kentledge constitutes a heavy structure and stability of the stack is important. Cribbages should be soundly constructed and the timbers clipped together with dogs if necessary. Cribs should be of sufficient plan area to prevent bearing failure of the soil. On soft ground timber mattresses may be used to spread the load. The kentledge should be carefully stacked and interlocked if possible. Stability of the outer blocks can be increased by suitably placed lacing. The stack should be symmetrical about the pile and the mass of kentledge should exceed the test load by a reasonable margin.

Tension piles are reasonably straightforward to design, and should be reinforced for the full length of the shaft. High stresses in the concrete will occur at the heads of the pile around tie bar anchors; reinforcement for these areas should be carefully designed. Any splices in the reinforcement should be designed with a generous lap. If extended main steel is to be used as anchor straps, these should be in mild steel and the welds for the washed stops should be designed and not left to the welder on site to do as he thinks best. During the test the heads of the tension piles should be levelled from time to time to check that soil failure is not occurring.

Ground anchors have little inherent stiffness against lateral movement, and tests should be carried out with multiple anchors. Good alignment of the equipment is particularly important for this type of reaction. When stressing the tendons care should be taken to ensure that each strand carries an equal load.

High pressure jacks present a special problem, and checks should be made to ensure that pressure lines are not frayed and that couplings do not leak. The hydraulic fluid should be clean, and the various filters in the pumping system should be in place.

During the test all unnecessary personnel should keep clear of the test area, especially when load is being increased. The hydraulic pump should be sited clear of the loaded area, and as little time as necessary should be taken to read the dial gauges. It is important that regular visual inspections are made of the equipment to check on its satisfactory performance. Obvious signs of distress are:

(i)   excessive deflection of beams
(ii) yielding of tension bars of piles
(iii) horizontal deflection of frames
(iv) movement of kentledge blocks
(v)   difficulty in maintaining load on the pile.

The test should not be left unattended whilst under load. In some circumstances it may be necessary to fence the area. Also, lighting should be adequate at night so that men and equipment can be moved safely, and the area should be kept tidy and free from rubbish.

Table 9.4   Essential information to accompany pile test sheets

Item	Information needed
Dates	Date pile was installed. Date of test
Location	Sufficient detail to permit the site to be located and also the position of the pile relative to the works. The pile number should always be stated
Pile type	Various categories of pile type are set out in Weltman and Little (1977) (undergoing revision)
Pile installation details	If a bored pile (i.e. non-displacement), the depth of temporary casing, how much concrete, mix details, how placed (e.g. trémie or chute into dry bore), and any special circumstances, particularly regarding groundwater. If driven (i.e. displacement), type of hammer, weight, drop, final set. The driving record should accompany pile test report
Pile dimensions	Nominal diameter (or section) weight per m run as applicable. Size of underream, bulb, 'wings' etc. Length of pile—include the entire pile length
Installed level (driven) Concreted level (bored) Trimmed level and toe level	Give full data so that there is no doubt regarding the reduced level of the pile toe, the ground level *at the time of the test* and the level of the top of the pile either at the end of installation, after trimming or both (above or below ground level)
Orientation	State whether vertical or raking (with degree of rake if applicable)
Design Load of pile	State Design Load or indicate on plot
Type of test set up and settlement measuring system	Whether kentledge or tension pile test. Leading plan dimensions of the set-up including reference system. Method of measuring settlement, subsidiary levelling of reference beams, temperature corrections
Weather	Brief comment on weather conditions and extremes of temperature during test
Soil information	It is not generally feasible to provide full soil information. If a report exists, the name of the Company who produced the report should be given, together with a reference number. The position of relevant site investigation boreholes should be given on the pile location plan. Summary logs of nearby boreholes with SPT N values and cohesion values alongside are useful. In some cases, a Bored Piling Contractor will log the boring—such logs should be given with an indication of their source. The *reduced levels* of boreholes must be given

10 Choice of piling method and economics of design

10.1 General

It is pertinent, before beginning to design a piled foundation, to contemplate the underlying reasons why piles may be chosen as the supporting elements in that particular case. Some of the reasons advanced are related to engineering requirements, but others are concerned more with convenience and speed of action. These issues should not be confused and sheer expediency should not be allowed to override valid engineering requirements.

Engineering reasons are often expressed as a need to place foundations into firm ground, below soils of great variability or doubtful long-term competence, or to place foundations at such a level that settlements, and more particularly differential settlements, will be kept to acceptable values. In addition there is value in being able to provide a foundation system which can be fairly readily tailored so as to match each load concentration with a local support concentration, perhaps avoiding the use of large ground-bearing slabs which may be difficult to accommodate. It is also economically advantageous, and from an engineering point of view desirable, to avoid where possible having to carry out deep excavations simply for the purpose of installing footings of a conventional kind, perhaps below water or in circumstances where adjacent buildings might be put at increased risk.

From the expediency point of view there are of course advantages in using a competent specialist service for work below ground, since the trade skills necessary at the foundation stage do not match very well those necessary for structural work. Piling often can be used therefore to save time and to simplify the work of following trades and there are in certain cases advantages to be gained where one specialist foundation contractor can provide a combined service of piling, retaining wall construction and perhaps ground anchoring or other related works within a single contractual package.

All engineering design involves the choice of a solution to a particular problem which is effective, practical and economic, and these objectives apply to piling as to any other item.

To be effective the piles must carry the loads which the supported structure imparts to them, together with any additional forces which may result from deformations of the soil mass in which they are embedded. They must also be sound, durable and free from significant defects, and their design must recognize fully the properties of the ground and the implications of ground-water movements so that deformations or settlements will be such as not to cause unacceptable strains in the supported or adjacent structures.

To be practical, the piles must be of such a type as will permit access for piling equipment to the locations where they are required, the design must recognize the limits of what is possible in current practice in terms of the equipment available, and the method of construction must recognize and seek to minimize difficulties related to ground conditions which could impede proper construction.

To be economic, the design must be such as to maximize the bearing capacity of each pile while at the same time providing for an adequate margin of safety against failure or excessive deformation of either individual piles or pile groups. The materials of the pile need also to be reasonably stressed and not used wastefully.

It is sometimes thought that, for any particular ground conditions and structural application, there is but one best choice of pile type which will result in sound and durable results. However it is more often than not the case that any one of a number of methods could be employed to yield a satisfactory end product and the real choice may rest on such items as access for equipment, environmental disturbance, or on the important consideration of overall cost. So many variables exist in practice, not least being the detail of ground conditions, that it is not possible to produce a set of rules which will invariably lead to easy identification of a best solution.

It is clear that the factors involved in making a best choice require knowledge of ground conditions, equipment installation procedures, and materials, and that several skills are needed on the part of both the designer and the contractor carrying out the work. In so far as the usual practice of carrying out pile design, selection and the implementation of the work is concerned, several procedures are adopted. Some larger firms of consulting engineers with geotechnical skills carry out pile design within their own companies and they frequently discuss their requirements with specialist contractors in the process of formulating a solution. Other structural engineers set out guidelines for pile design, perhaps stipulating settlement or defor-mation requirements and occasionally specifying the type of pile which they consider appropriate, and then leave detailed design in the hands of specialist contractors with a request for competitive quotations. Yet others provide a pile design for the works, but leave specialist contractors the option of submitting an alternative scheme based on any other proposal which will achieve the same ends. Each of these approaches can work satisfactorily in practice, but they all must recognize that certain skills and experience are necessary on the part of those involved, and full information to allow both proper design and construction must be made available to all those involved.

10.2 Limitations on method choice imposed by ground conditions

It is clear that a proper choice of piling method can only be based on adequate site investigation and on a knowledge of the geology and history of use of the site in question. Without such an appreciation serious difficulties may be encountered at the construction stage or subsequently, perhaps entailing a change of method during the work, a requirement for a different type of plant, and in certain conditions even remedial work. Some examples of items which can give rise to difficulties are outlined in the following paragraphs.

The presence of groundwater and its possible fluctuations can be significant in certain construction procedures in relation to bored piles and care should be taken to obtain as much relevant information as possible. It is particularly relevant to pile head casting levels, lengths of temporary casing necessary to maintain bore stability, the use of stabilizing fluids such as bentonite suspensions, concrete quality and the method of placing concrete.

Hard strata, such as dense granular soils and rocks, must be identified and their properties quantified. If for example a driven pile is chosen in circumstances where considerable penetration of dense upper soils is required, then time and energy will be consumed in achieving this and piles may be damaged during the driving process. It is not in general sensible to seek to achieve significant penetrations into hard rock-bearing strata by any straightforward driving process and the specification of very small sets and prolonged hard driving is often counterproductive. In so far as bored piles are concerned, the costs of hard rock penetration are high, and lack of adequate information may add considerably to the expense of the work because larger penetrations than necessary are called for.

Boulder clays pose a particularly intractable problem because conventional site investigation cannot easily lead to identification of either true boulder dimensions or frequency, and it has sometimes been recommended that, where practicable in such ground, large-diameter boreholes and direct inspection should be used rather than the small-diameter standard investigation holes. Boulders can severely hinder many pile installation procedures and can exert a very important influence on the cost of piling. Piles, particularly of preformed types, can be damaged, and for all types of pile delays in construction can occur. In the presence of large quantities of boulders it is worth considering how the boring length in the difficult strata may be minimized.

It is not without reason that it has sometimes been said that most claims for extra payment in piling result either from battles with adverse groundwater conditions or in seeking to penetrate rock or other hard strata.

Other ground-related difficulties which can present a problem include downdrag on piles due to consolidation of upper soil layers after pile installation, or heave which may result from swelling of clays around piles when trees have been removed from a site prior to construction. A similar condition may arise because of a deep excavation following pile installation and both these the effects are often underestimated in terms of the forces which are caused in the piles.

The forces resulting on piles in either conditions of downdrag or heave may be reduced by providing a slip layer around the pile over the affected length, or the pile may be reinforced and lengthened if necesssary to cater for the forces as they occur. Where the heave is related to deep basement excavation only, these measures may not be necessary, depending on analysis of the amount of movement expected since some small degree of cracking in a finished pile may be acceptable.

The general stability of excavations or sloping ground surrounding a piling site can lead to slip failures and movements which produce large forces on the piles. These forces are sometimes difficult to quantify, but must be taken into consideration when preparing a pile design. Piles have little ability to resist the forces which result from major slip failures in the soil around them, since these forces are often very large.

Large diameter piles may have their bases enlarged in stiff stable water-free ground by underreaming, but if there are severe water inflows or bands of cohesionless material at the level chosen for the pile bases, then it may be impossible to ensure completely stable base excavations or a clean bearing surface. The conditions where instability is likely are often indicated by silt or sand seams in a clay or by significant reductions in moisture content as compared with those of the general clay mass.

In the re-use of sites which have previously been occupied by chemical works or gas works, it is frequently found that the ground contains residual chemicals which could adversely affect the durability of one or several types of pile. Protective coatings may be required in the case of preformed piles, or in the case of cast-in-place piles a protective lining or sleeve may be needed.

It is clear from these few but important examples of difficulties which result from particular soil conditions that the means of avoiding or minimizing undesirable effects can differ between one pile type and another and that in certain cases particular types of pile may be ruled out as unsuitable. In practice there are many ground-condition influences which bear on the economy and proper choice of piles for any job and one ignores or underestimates the importance of carrying out a proper and appropriate site investigation at one's peril. In general terms, the larger the job in question, the more dramatic and expensive the end consequences of insufficient information become.

10.3 Structural consideration of pile use

Piling is used for the support of wide range of structures and the choice of pile must be appropriate to the requirements of each type of structure. A major and prestigious structure, for example, will warrant perhaps the use of large and deep piles merely to keep the possibilities of structural movements in the completed work to an absolute minimum, whereas for some purely functional industrial structure, settlement may be of little consequence and simple bearing capacity may become the overriding feature. It is often not apparent at first sight how important the control of settlement and differential settlement may be. For example, fine tolerances are sometimes specified for items such as machinery bases and automatic stacking systems in warehouses. In structures such as bridges, the sensitivity of the system may be very dependent on the assumptions of deck design, whether it is a case of a simply supported span or spans in a continuous beam arrangement.

One should recognize that structural design and foundation design should not be carried out in isolation. The loads from the structure will lead to deformations of the ground which depend on soil types, and these deformations will in turn lead to internal stress changes within the structure, thus changing the loads on the piles. Many cases exist where such considerations may be unimportant, perhaps where the structure rests on piles bearing onto hard rock, and one can therefore afford not to become too involved in detailed analysis. However, cases exist in plenty where soil/structure interaction effects are important in terms of both the finished structure and adjacent existing structures. These considerations may have a determining influence on the selection of pile types.

10.3.1 The use of piles for vertical load bearing

The majority of piles are designed to carry vertical loads and the secondary lateral loads to which they may be subjected are small. These loads derive in the main from structures with a certain degree of redundancy and cannot by their nature be exact. Derived loads usually represent combinations of dead loads, live loads and wind loads, and the final structure is usually designed on the worst loading condition, say for each column. It is important in considering real cases, and in particular probable structural settlements, to take into account realistic conditions, although each pile must of course still be capable of carrying safely the required maximum structural load.

The starting point of most structural engineering designs is an assumption of a non-deforming structure, although it is possible for simpler types of structure to incorporate the concept of a settling foundation in order to assess how loads may change in such a circumstance. This step in the procedure is then often followed by the acceptance of some differential settlement of the foundations on a purely empirical basis. Such empirical deformation criteria in relation to structural damage are discussed by Skempton and MacDonald (1956) and by Bjerrum (1963).

In order to satisfy the designer of a piled foundation who seeks the nearest possible approximation to a non-deforming foundation, a great deal of attention is devoted to ensuring that the deformation of each individual pile, and of all the supporting piles acting as a group, is minimized. The choice of depth required for piles, particularly where load is carried mainly by friction, is often based on this requirement, and in general the deeper the piles are the more likely is the settlement to be restricted. This means that in many cases Factors of Safety are raised in order to achieve this result.

In the past it was common practice, in the case of piles which carried most of their loads by friction or adhesion, to make all the piles on a site of the same length, but this practice has been largely abandoned, provided piles in the same group are kept to similar lengths and that very large variations are avoided between adjacent piles at close centre-to-centre spacings. It is necessary in such cases still to found all piles in the same basic stratum, or at least in strata which have broadly similar deformation characteristics. Most driven piles carry a high proportion of load by end bearing on dense strata or rock, and the possibility of substantial length variations does not therefore arise except where, perhaps because of underlying geological faults, the bearing stratum changes radically in level across a site.

The interaction between piles and rafts has been discussed in Chapter 5 (sections 5.2 and 5.3) and it is clear that where ground conditions permit, advantage can be taken of load sharing effects to improve the economy of design. The piles may also be used as settlement reducers and substantial control is theoretically possible over the deformation of the raft and the supported structure. The practice of using a raft and piles as a combined system has been used on a somewhat empirical basis on a small scale in the past, but the possibilities of using this type of approach in a much more satisfactory way are now worth consideration when the soils underlying the raft are competent.

An alternative approach to pile/structure interaction has been suggested by Fleming (1984) in which the load/settlement characteristics of piles are

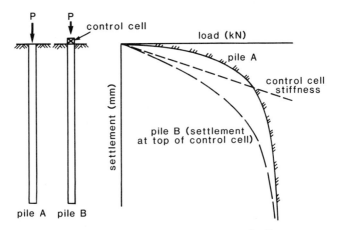

Figure 10.1   Structural settlement control cell

regarded only as boundary conditions within which alterations to performance may be made by placing a pile settlement control element on each pile head.

The principle involved is indicated in Figures 10.1 and 10.2. Figure 10.1 indicates the result of a load test on a single pile. If say a suitable rubber block is placed on the pile head and load and settlement are then measured at the upper surface of the block, the load/settlement relationship can be modified. By selection of the thickness of the block, or its deformation behaviour, the characteristic of the pile may be tuned to any suitable but similar form within the boundary.

In Figure 10.2 three representations of a structure are shown. The first diagram describes a 'flexible' building resting on a clay soil with equal load acting on each pile from a column. As the soil deforms or 'dishes', the structure will deform without alteration of pile loads. The second diagram shows a 'rigid' building in the same situation, but in this case as the soil layers 'dish', the edges of the structure and supporting piles are pushed more deeply into the soil mass. Hence the load on external piles will go up and on internal piles will go down. The structure is now called upon to accept internal shear stresses, and it is these additional stresses which place the design of the structure at greater risk. The third diagram shows how the desirable features of constancy of support which occur in the flexible case can be combined with the presence of a rigid structure, minimizing the internal stresses to which the structure may be subject as a result of forced distortion. The piles deform more or less as in the flexible case but the rigid structure experiences the correct support loads corresponding to the design assumptions.

Since for many structures on clay soils settlement is time-dependent, it may not be possible to choose the stiffness of the blocks or control cells to place on

Figure 10.2   Deformation of piled structure on clay soil

the pile heads to cope exactly with the longer-term consolidation movements. A major part, perhaps of the order of 60% of the deformation of the soil mass around piles in stiff clay, is immediate settlement which occurs during construction and the deformation of each block will be more or less concurrent with the development of settlement, so that there should be little problem in choosing stiffnesses up to this stage. Even in the later consolidation stage when settlement increases but load stays constant, much smaller load shifts between piles would be necessary to ensure equal settlements at pile heads than would occur in any conventional structural support arrangement. Some compromise choice of stiffness might prove best—say basing deformation calculations on immediate settlement plus half of the remaining consolidation settlement. It seems generally to be true, making reference to several papers in the 'Settlement of Structures' Conference (British Geotechnical Society, 1974) that settlement of structures on piles in clay soils occurs more rapidly than normal one-dimensional consolidation theory would imply and it has been suggested that this may at least in part be due to the piles also acting as a vertical drainage system. This deformation control system has not been exploited up to the present time but it could have clear advantages since it does not seek the absolute minimization of pile settlements and could be adapted to situations such as where tall buildings adjoin low level surrounding structures. Until now two factors have inhibited use of the system, namely the precision with which calculation of settlement at each pile position could be carried out and the development of simple, durable and low-cost materials for the control medium. However, computer calculation methods are now available which will enable the former problem to be solved to an adequate degree of accuracy, and basic materials are becoming available which may give the life and durability required for the pile head capping materials. Such a system would if properly used, justify the structural engineer's assumption of a non-differentially-deforming structure, would place the problem of ensuring that this condition is met in the hands of the geotechnical engineer, and would remove to a large extent the need for present empirical differential-settlement-limiting criteria.

It is clear that there is scope for the improvement of pile support systems in some of the ways discussed above as well as though more advanced methods of individual pile capacity calculation, and that the engineer who would seek the most economical design of any major foundation needs to have an intimate knowledge of the many different options which may be open to him.

10.3.2 The use of piles for lateral load bearing

The design of piles subject to lateral loading has been discussed in Chapters 4 and 5. In practice one has to make a choice between the use of vertical piles used singly or in groups to carry such loads or of groups incorporating at least some piles installed to a angle of rake. The capacity of a pile as a structural unit to carry shear loads at its head depends on the strength of the section, and when the forces become high one is impelled to find some structurally acceptable solution which keeps stresses within reasonable limits.

However, in choosing raking piles one should be aware of the problems and limitations which may be involved. Some of the factors involved are as follows:

(i)   Raking piles are usually more expensive than vertical piles. This is partly involved with extra time taken to set up and maintain the equipment in position, the less efficient use of hammers in the case of driven piles, and the difficulties of concrete placing in bored piles.

(ii) The standards of tolerance which can be maintained in the installation of raking piles are not as good as for vertical piles. Most analyses of pile groups of this kind ignore the effect of tolerances, but if tolerances are properly taken into account in such groups they can have a significant if limited effect on calculated pile loads.

(iii) Where the upper part of a raking pile is embedded in a soil which is likely to suffer time-dependent settlement, the pile will in due course be subject to bending stresses unrelated to the structural design load conditions. This may require increase of strength of the section which is in turn reflected in costs.

(iv) Many machines used for pile installation carry the pile driving or forming equipment on a long mast, so that they become intrinsically less stable, particularly as the line of the pile gets further from the vertical position. In certain cases, when working close to river banks or railway lines for example, there is a major limitation on how machinery can be positioned to produce the desired end result.

(v)   Design of groups involving raking and vertical piles and with loads which are both vertical and horizontal should have regard to the constancy of the relationship between these. If for example the vertical load is near constant, but the horizontal force varies greatly, then it is better to employ groupings with rakers balanced in two opposed directions rather than to have an arrangement of vertical piles plus piles raking in one direction only. This is simply to minimize the shears in the heads of the piles when horizontal load falls to a minimum value (see Chapter 5, section 5.4.2).

(vi) The use of raking piles to 'spread' load under vertically loaded foundations, where the piles are fully embedded in the soil mass and where the whole foundation is expected to undergo significant consolidation settlement, must lead to large bending stresses being developed in the piles. In certain cases this can lead to such stress levels in the piles that the section will suffer damage, which may in turn lead to problems in the supported structure.

It should, however, be said that where groups of raking piles derive their axial capacity from strata which are hard and relatively non-deformable, they provide a stiffness in terms of laterally applied forces which can be very desirable. The main issue in design is to avoid large and unquantifiable secondary stresses, and provided this can be achieved all will be well.

Where there are very heavy lateral loads to be carried and neither raking piles nor single piles other than perhaps those of very large diameter are suitable, then diaphragm piers or 'barrettes' have a useful application. They can be given high stiffness in the direction of applied horizontal loading without fear of the problem of major secondary stresses.

10.3.3 Tolerances in the installation of piles and their structural effects

It is not possible to install piles consistently in the right position nor to precisely the alignment which is required by a design. The achievable tolerances are strongly related to such factors as the accurate setting up of the equipment, the fixity in machine parts, the presence of obstructions in the ground and wide variations in the properties of the soil especially near the point of entry of the pile into the surface, the inclination of strata, and operator error. The tolerances which can normally and reasonably achieved are quoted for example in the publication Specification for Piling (Institution of Civil Engineers, 1988) and in the specifications issued by the Federation of Piling Specialists.*

The proper positioning of the machine used to drive or bore the piles is a primary requirement, but once a pile or boring tool has penetrated some short way into the soil, correction of an initial error becomes more and more difficult and eventually is virtually impossible in the case of a driven pile and very costly and difficult in the case of a bored pile.

If a single pile is installed to carry a single load, positional errors mean that the load is applied eccentrically, and if an additive alignment error also occurs and upper soils are very weak, then the resulting bending moment applied at the pile head will increase for some short distance below the pile head until soil constraints come into effect and prevent further increases. In the case of a rigidly capped pair of piles the same will be true in the direction at right angles to the line joining pile centres, and for positional and alignment errors in the same direction. However, for rigidly capped groups of 3 or more piles or where the error is in the direction of the long axis of the cap in the case of a pair of piles, the main effect is not to produce significant moments in the piles but rather to adjust the vertical loads in the individual members of the group.

Where alignment errors are known to occur consistently with driven piles it has been suggested that preboring, at least through the upper soil layers, can reduce the problem, but this adds to cost and of course some error may still be inherent in the boring operation.

Whereas in the case of driven precast concrete piles, reinforcement is often fairly substantial to cater for handling stresses, for cast-in-place piles,

*15 Tooks Court, London EC4.

reinforcement is often nominal only and tolerance errors may require additional reinforcement, as will laterally applied loads. One of the advantages of cast-in-place piles is that the reinforcement may be tailored to fit, though not on a pile-by-pile basis except where some deviation or oddity is discovered before concrete placing. The pile also starts its life in an unstressed state. Reinforcement specified is therefore often based on the achievement of standard tolerances.

It is evident, since tolerances are generally the same for piles of large or small diameter, that by far the most severe conditions are associated with piles of small cross-section. If steel reinforcement becomes large in quantity, the cost of piling will be increased and the spacing of reinforcement bars within the section may be reduced to unacceptable values. It is generally held that the clear distance should be maintained at 100 mm between bars in cast-in-place piles if possible and it should never be less than about 80 mm where normal high slump concrete is used with 20 mm aggregate.

Where lateral loads are applied to piles, the bending effects may add to the tolerance effects, and reinforcement has to be provided for the combined moments.

From a theoretical point of view the case of an axially loaded bent pile has been considered in order to determine the maximum load which can be taken by the pile section, the maximum bending moment in the pile, and the maximum lateral soils reaction induced along the pile in relation to permissible values (Broms, 1963). It is apparent from this type of analysis that the possibility of overstressing really only appears with very slender piles in soft ground and with fairly severe deviations of alignment. Work by Francis *et al.* (1965) bears out that a pile with initial alignment imperfections has the same buckling load as a straight pile, though it may deflect more.

The methods described in Chapter 4 (section 4.3) for the calculation of bending moments in laterally loaded piles may equally be used in the consideration of tolerance effects, taking the eccentricity of load as applying a moment and a direct stress at the pile head, and resolving the forces in the axial and lateral directions for the alignment error.

The analysis of positional and alignment errors by a simple method has also been described by Fleming and Lane (1970) and this method has been used extensively in routine pile design conditions. The method is based on the work of Broms (1964). Reinforcement for the section is easily calculated using standard concrete design procedures.

The use of the methods demonstrates several important principles, as follows:

(a) Maximum moments occur just below the top of the restraining stratum if the effects of alignment deviation and eccentricity are additive. If the effects are in opposed direction, the maximum moment occurs nearer to the pile head.

(b) Maximum effort should be devoted to ensuring that a pile which is not fixed in position at the head is located accurately.

(c) A higher standard of alignment accuracy is desirable in that part of the piles which passes through soft upper soils than at greater depths. Deviations near the pile base are unimportant where the piles are reasonably long.

(d) Checking of pile positions prior to placing concrete in cast-in-place piles is highly desirable. If an error is then detected, steps may then be taken to cater for extra bending stresses through the provision of extra reinforcing steel.

In general tie beams between pile heads to minimize lateral deflections are only necessary where the upper ground is soft and the piles are slender, and if tolerance effects have not been taken into account adequately at the pile design stage.

For large-diameter piles or for most conventional piles where the upper ground is firm or dense, calculation indicates and experience shows that tie beams are generally unnecessary. The fixing of pile heads into a stiff cap reduces moments to about half of what they would be in a free headed case.

Reinforcement may also be required in driven piles to cater for uplift forces where piles are driven into soils such as stiff clays. This heave or uplift is largely consequent upon volume displacement.

The above considerations demonstrate how tolerances can have significance in relation to the selection of pile type on any particular contract, and how they influence subsequent structural work at ground level and the amount of the steel reinforcement which is necessary.

10.4 Constructional consideration of pile use

10.4.1 *Site access*

The problems encountered in gaining access to sites are numerous, and many invitations to tender for piling work are accompanied by the sensible requirement that before tendering the specialist contractor must visit the site and satisfy himself as to the means of getting in and out and that there is adequate working space.

Naturally if a small and inaccessible site is considered, perhaps in the basement of an existing building, one must look for a machine which will work in those conditions, and this often means the selection of bored piling equipment rather than driven, because small boring machines are more readily available. Some driven steel or iron piles are constructed in this way, for example in Sweden, but the system does not appear to be used much elsewhere.

When working in enclosed areas it has to be remembered that if standard diesel engines are used the exhaust fumes must be ducted away to open air.

Otherwise the equipment for such work can be driven either by remotely provided compressed air or electricity.

Whatever the size of a piling site, the physical dimensions of entrances and the slope and stability of access ramps require consideration. Ramps should not in general slope more steeply than about 1 :10. It is rarely sensible to have to undertake a partial dismantling job on a piling machine just to get it into a site, but there are times when such action becomes necessary. Occasionally piling equipment is picked up bodily and lowered into deep basements by heavy cranes.

Several questions must be asked in relation to the disposition of piles within a site:

(i) Is there enough space for all the equipment needed? This may include extra craneage, various materials which must be stored in readiness for use, and items of ancillary plant. It is not just a matter of fitting the equipment into the site, since manoeuvring space is needed, and double-handling materials as work proceeds in order to make space is not economical. Inadequate working space can also constitute a safety hazard.

(ii) Is the headroom adequate to allow the intended equipment to function? Roofs, overhead electrical cables and pipes, and overhangs on buildings all impose constraints. Occasionally cut-down equipment is made for special low headroom circumstances, but this nearly always results in some other functional penalty which is reflected in a lower rate of output than normal.

(iii) Are the clearances from existing structures sufficient to allow pile construction? There are many types of equipment, each with its own dimensional characteristics, but as a generalization the centres of piles must lie no closer to adjacent walls than about 0.75 m, although a few machines can work closer than this. It is particularly difficult to work into the corners of a site, where buildings surround the corner on each side, and minimum distance from each wall in such a case may have to be 1.0 m to the pile centre or perhaps a little more (see Figure 10.3).

(iv) Is the site surface sufficient to bear the weight of the piling equipment? Many piling machines have self-weights in the range 35 to 100 tonnes including the self-weight of the base unit upon which equipment is mounted. With constant and heavy usage, undrained and badly prepared surfaces can quickly become quagmires. This in turn delays pile installation and can endanger the stability of the machines. Particular care should be exercised to ensure that old excavations are properly backfilled, for the same reason. To attempt to work on poorly prepared or waterlogged ground is false economy. In general a carpet of hardcore is laid to enable piling to be carried out without hindrance.

Sloping site surfaces are not generally satisfactory to work on except for the smallest and lightest types of equipment. Traditional tripod percussive boring equipment is light and adaptable and may be used in cases where no alternative may be possible. Small diameter rotary drills of the kind used for rock drilling may also be appropriate.

10.4.2 Contract size

In the case of cast-in-place piles, when a site is large enough and where the number of piles to be installed justifies it, several machines may be used to speed the work. The order of working has always to be thought about because it may be unwise to construct immediately adjacent piles in sequence. The risk is that the installation of a pile nearby may damage an already-cast pile, either by displacement in the case of a driven pile or by allowing concrete to flow between piles in the case of a bored pile, up until the time when the concrete in the first pile has achieved an initial set. Attempts to use too many machines on a site which is really too small for them can lead to logistical difficulties and problems with spoil removal and concrete supply, and the complications may grow, bearing in mind that on large jobs a main contractor may be seeking to follow up piling as closely as possible with cap and beam construction.

The selection of a piling method for a particular site is to some extent governed by the size of the job to be carried out. Transportation cost is a function of piling machine size and the distance to be travelled to reach the site. Small machines are far less costly to transport than large machines, and if there are not many piles to be constructed, they begin with a distinct advantage in terms of overall cost. Economy implies that small equipment and small numbers of piles go together, though such a convenient arrangement is not always possible. In the United Kingdom regulations govern the movement of heavy loads by road and in certain cases up to seven days' notice must be given of impending travel. Rapid changes of intention for whatever reason generally involve delays and, as would be expected, cost is increased.

Large contracts do however offer a piling specialist contractor continuity of use of expensive equipment and some freedom from plant time loss in moves. They are therefore in general competitively priced and every effort is made to keep output high. Hence, just as small machines are best for small jobs, so large machines are most appropriate for large jobs.

10.4.3 Limitations on various types of piling equipment

Every piling machine is designed to operate within certain limits of pile length and cross-sectional area, and the power of machines varies to cater for the heaviest work which may correspondingly be expected. Table 10.1 indicates approximately the normal range of lengths available for a selection of types of pile often used for work on land-based sites.

Similarly, maximum loadbearing capacity varies with the pile cross-

Table 10.1

Pile type	Normal maximum length (m)
Steel 'H'	30 to 50*
Steel pipe	30 to 40*
Precast concrete	30 m without join*
Precast concrete—joined sections	30 to 40 m**
Driven cast-in-place—permanent concrete shell	25 to 30 m*
Driven cast-in-place—withdrawn tube	20 to 30 m
Bored—tripod small-diameter	20 to 25 m*
Bored—rotary small-diameter	25 to 30 m*
Bored—rotary large-diameter	50 to 60 m
Bored—continuous-flight auger	25 to 30 m

Note: *exceptionally longer.
       **very exceptionally much longer.

sectional area, subject to the ground being of sufficient competence. Some examples of approximate maximum capacities are summarized in Table 10.2, and fuller details of pile dimensions and appropriate loadings will be found in Chapter 3. Precast concrete piles which are formed of lengths jointed together by proprietary jointing systems usually employ high strength concrete. It should also be noted that it is more difficult to accommodate bending stresses in piles when the concrete is stressed to a level near the maximum permissible values than in the case of piles with lower axial stresses.

In normal usage, the term 'small-diameter' as applied to bored piles includes piles up to 600 mm diameter, and the term 'large-diameter' refers to piles of larger diameter. In the large-diameter range, the sizes generally on offer range from 750 mm diameter rising in stages of 150 mm to 1500 mm diameter and thereafter rising in stages of 300 mm.

In stiff clay soils, or in soft rock such as chalk and marl in dry conditions, bases may be enlarged by underreaming to approximately three times the

Table 10.2

Pile type	Size (mm)	Typical design load (kN)
Steel 'H'	up to 350 × 400	600–4400
Steel pipe	up to 900	400–7300
Precast concrete	200–600 sq	600–2000
Precast concrete—	up to 300 sq	550–1500
jointed sections	up to 390 hex.	700–2000
Driven cast-in-place—permanent concrete shell or light steel casing	400–750	500–2000 shell 600–2500 light steel casing
Driven cast-in-place, withdrawn tube	275–550	350–2000
Bored small-diameter (tripod or rotary)	150–600	100–1500
Bored—large-diameter	750–2100	2000–30000+
Bored—continuous-flight auger	300–750	400–2600

Figure 10.3   Rotary piling machine working in restricted corner position

shaft diameter, but in the United Kingdom bases of more than 5400 mm diameter have rarely been used.

The bases of certain driven cast-in-place piles may also be enlarged within limits in certain systems, notably in the case of the Franki pile, and this can be used to provide piles of required capacity at a higher level on a cohesionless bearing stratum for example than would otherwise be possible.

It should be pointed out that although the maximum lengths given in Table 10.1 are possible, ground conditions often mean that such lengths are neither possible nor necessary.

For large-diameter piles, the larger sizes in the available range require powerful and substantial machines, fitted with appropriate robust augers or drilling buckets. The changing of these tools when required is relatively simple and quick, so that piles of several diameters may be selected to match the loading variations of any one site. However, if every available size should be chosen, so many boring tools and temporary casings might be required that not only would on-site establishment costs rise, but a great deal of storage space would be needed.

Advancement of bores into hard rock, for example to provide frictional sockets, may call for special equipment. Aggressive augers with special teeth are most commonly used for this purpose, but core barrels, down-the-hole hammers, heavy chisels and unusually, rock roller bits may be employed. There is only a limited range of sizes available for down-the-hole hammers and rock-roller bits, and for both rock roller bits and core barrel use, a direct or reverse circulation technique is usual.

Penetration of rotary bores into boulder-laden clays may be obstructed by large boulders which will not pass into the drilling tool, and these are usually dealt with by chiselling; in general terms the heavier the chisel the more effective it will be.

10.4.4 *Production*

The time taken to complete a contract is an important consideration in most instances, and this in turn is dependent on the type of pile selected and the equipment used in its installation. On large contracts driven piling can be very rapid where upper soils are soft down to a good bearing stratum, with output from efficient equipment being several hundreds of metres per day.

Likewise, in ideal conditions and where very little temporary casing is needed to stabilize soft soils near the ground surface, rotary boring of small diameter piles can be very rapid.

However, driving piles through relatively stiff cohesive strata, or boring piles to substantial depths through soft and unstable strata where long temporary casings are needed, leads to reduced output and therefore increased costs. Fortunately the conditions which slow down bored piling are often those in which driven piling thrives and vice versa, so that economic choice becomes fairly obvious.

In the case of large bored piles, there is a slowing down of the processes as the diameter increases, but the use of long temporary casings is probably the factor of greatest influence in production. These casings may be installed relatively quickly if they are short, and a 'mudding in' process which involves creating a slurry column of soil in bentonite suspension may be used to aid insertion down to a depth of about 10 m. To place much longer casings through cohesionless soils, a large vibrator may be attached to the casing and by this method casings of the order of 25 m have been inserted. However, the effectiveness of such vibrators diminishes rapidly in the presence of clay soils and their use for situations where interbedded layers of cohesive and cohesionless soils occur is limited. Therefore, if large bored piles are required to go to great depths in unstable soils, bentonite suspensions remain the most practical way to sustain an open borehole throughout the process of pile formation. It is clear that the more elaborate the method becomes, the more plant will be required and in general the slower the process will become.

The continuous-flight auger method, with grout or concrete injection through the stem of the auger as it is withdrawn, represents a neat way of using bored piles to avoid casing problems and in circumstances where, but for environmental considerations, a driven pile would be used. It provides a rapid installation rate, perhaps just a little slower than for a comparable solution using driven piles.

The use of on-board computers to monitor the construction of continuous-flight auger piles is now common among major specialist contractors in the United Kingdom following their introduction by Cementation. Full records of construction can be printed on-site and analysed off-site by a central computer using a program designed to scan for such items as under-supply of concrete in relation to auger lift and other possible untoward events. It is believed that this procedure can aid considerably in the production of reliable piles.

Advice on production rates is always best sought from an experienced specialist contractor who is in possession of the full facts of the job in question.

10.4.5 Environmental factors

Increasing attention has been directed to this subject in recent years and some discussion has been included in Chapter 7 (section 7.1.4) with regard to driven piles. Although the duration of a piling contract may be short in comparison with the whole contract period, noise and vibration perception may be more acute during the piling phase.

In the United Kingdom, the Control of Pollution Act (1974) provides a legislative framework for, amongst other things, the control of construction site noise. The Act defines noise as including vibration and it provides for the publication and approval of Codes of Practice, the approved code being British Standard 5228. A section of the Code (Part 4) deals specifically with piling noise. This Code is being revised to include guidance on vibration and is expected to be published in 1992.

Because some of the traditionally noisy methods of piling have become unacceptable in many urban and suburban locations, plant manufactures have responded by seeking quieter alternative techniques.

The noise from pile driving displays high peaks associated with each hammer blow, and some manufactures concerned with sheet pile driving have for example enclosed the hammer in an acoustic box. This leads to some appreciable noise reduction but can affect the efficiency of the process of installation.

In the case of bored piling equipment, much of the noise emanates from the engine providing the power, and it is possible to reduce the steady noise level by improving the soundproofing qualities of the engine enclosure, but this does increase the difficulties of routine maintenance. Some of the noise is associated with the driving and extraction of temporary casings and this is not easy to reduce. However, continuous flight auger equipment does not involve a casing process and therefore this equipment is currently the quietest method available for pile installation.

Driven cast-in-place piling can be made less noisy by the use of an internal drop hammer acting on a plug at the bottom of the driving tube and by attention to the soundproofing of the machine's engine enclosure, but the process is generally slower than using a top hammer configuration.

Excessive noise presents a health hazard by inducing premature deafness, although this is more a problem for the site operatives than for neighbouring residents (Noise at Work Regulation, 1989). Operatives now wear ear defenders on many sites.

It is widely recognized that noise and vibration, although related, are not amenable to similar treatment. In the main, noise from site is airborne and consequently the prediction of noise levels is relatively straightforward, given the noise characteristics and pattern of use of the equipment. On the other hand, the transmission of vibration is largely determined by site soil conditions and the particular nature of the structures involved. There are cases cited in which vibrations have not materially affected immediately adjacent structures, but have affected structures at a greater distance from the source. Some general guidance may be derived from the study of case histories of similar situations (Construction Research and Information Association, in preparation; Attewell and Farmer, 1973). Useful references on the whole subject of ground vibrations are provided also by Skipp (1984) and by Building Research Establishment Digest No. 353 (1990).

Piling processes produce vibrations of two basic types, namely free and forced vibrations. Free vibrations are normally associated with impact loading, such as when a hammer strikes a pile or a steel casing, and the frequency is then determined by the pile, soil and structure characteristics. Forced vibrations occur when a vibratory tool, such as a heavy casing vibrator, is used, producing continuous vibrations with well defined frequency characteristics. Continuous forced vibrations are more likely to set up resonant conditions in structural members of certain dimensions, which in consequence may vibrate at much greater amplitudes than other non-resonant members.

In general human perception of vibrations occurs at levels which are low in comparison with thresholds of risk for structural damage. The British Standards Institution have recently issued 'Guide to the evaluation of human exposure to vibration in buildings (1 Hz to 80 Hz)' (BS 6472: 1984) which sets out acceptable criteria for vibration levels in various different types of accommodation. The vast majority of piling processes currently in use would give rise to vibrational energy within the stated 1 Hz to 80 Hz frequency range.

The International Standards Organisation is working on criteria for structural damage caused by vibration, measured at foundation level in terms of peak particle velocities, for various categories of structure, above which there is initially a risk of architectural damage in the form of minor cracks in plaster, etc. Compliance with the proposed criteria can normally be achieved, and where structural damage occurs below the relevant thresholds it is usually because some other cause is also present. It is frequently feared by adjacent building occupiers that vibrations caused by piling will adversely affect computer operation on their premises. This matter is dealt with in an article by Boyle (1990).

Apart from the direct effect of transmitted vibrations, it is also possible to initiate settlement of the ground by reason of the compaction effect on loose soils such as fine sands or silty sands. However other materials also deform as a result of repeated loading.

Various methods of reducing vibration transmission from piling have been advocated with conflicting claims as to their success. It may for example help to use low-displacement piles rather than piles of large displacement, but in the last resort is may be necessary to change the piling equipment and adopt a different piling technique. This is a costly exercise when the problems of the particular site are discovered too late.

The incidence of damage due to piling vibrations in the United Kingdom is very low, but it is clear that because of human sensitivity, many people confronted with the problem for the first time become understandably fearful for the safety of their property. Confidence will often be restored by undertaking a programme of monitoring and programming work so as to minimize the periods of annoyance.

10.5 Cost considerations

It must be obvious that there are so many variations of pile type, ground conditions, structural requirements and other factors involved, that there is no simple way of determining cost effectiveness in a particular case, apart from competitive price comparisons and perhaps several trial designs.

In the United Kingdom the piling market is fairly evenly divided between driven or displacement piles and bored or replacement types, although the ground conditions of some areas favour one type more than another.

Where soft strata overlie a definite granular or rock stratum and there is a well defined depth at which substantial end bearing capacities can be achieved, a driven pile will usually prove most attractive for conventional uses, provided environmental or other unusual circumstances do not intervene. If the bearing stratum is within say 20 m of ground surface, then a simple unlined concrete cast-in-place type of pile is often the least expensive choice, and although it increases cost, some systems of this kind make provision for possible base enlargement. Expansion of the pile foot is often not necessary, but where it is effective, it can sometimes lead to significant saving in pile length and hence overall economy may be improved.

If the bearing stratum is at a greater depth, say normally down to 40 m, then a segmental jointed precast pile has much to commend it. A balance has to be struck between the segment length, joint cost, and driving time in order to make best use of the system.

Where the depth of pile required becomes much greater, or where harder driving can be expected on the way to a deeper founding stratum, driven steel H sections may prove economic. They can easily be raked and have high tensile as well as compressive bearing capacity. In work in ports and harbours steel box piles are frequently used, since they are compatible with sheet pile driving and have similar durability. The H pile, because of its low displacement volume, can be used to minimize heave problems which might otherwise result from the use of larger-section piles, but to be used efficiently they need to be stressed to high levels.

Intrinsically steel is an expensive material and therefore steel piles are mainly used for the special purposes as described.

Permanently lined cast-in-place piles formed in steel tubes are invariably more expensive than piles of similar size without such linings, but piles formed within concrete shells can be competitive with the simpler forms of driven cast-in-place pile. The shells can be treated appropriately in aggressive ground or where downdrag is anticipated, and because they are manufactured in short lengths, they can when necessary be driven using low headroom equipment.

In so far as bored piles are concerned, in general they find their best use in clay soils where open boreholes can be formed without the use of long temporary casings. In terms of cost per metre, they can seldom achieve the low prices of standard driven cast-in-place piles of similar diameter, when the driven pile is used in its ideal conditions. They are less efficient than driven piles as load transfer units but they can be installed rapidly, they are efficient in terms of installation energy use, they are relatively vibration-free particularly when a rotary system is used, and they have the advantage that the soil is at least seen and its properties can be appreciated during the process of installation.

As the length of casing needed to stabilize upper soft soils increases, so also does the cost of any given diameter of pile, and where there is a need for very deep casings or in those circumstances where casing has to be abandoned in favour of bentonite, the cost per metre run of pile will rise rapidly to perhaps twice the cost of the simplest unlined similar pile.

Piles bored by percussion methods cannot generally be constructed as fast as when rotary boring equipment is used, and are therefore more expensive per metre of comparable size, but nevertheless these piles still find substantial use on small sites. The large-diameter version of the percussion pile has merit in the construction of secant walls when required.

There is therefore a wide range of pile prices which cannot be generalized because of the many different circumstances of use. Any variation from standard procedures in the construction of piles in general, increases cost. Raking piles are notably more expensive than vertical piles, permanently lined piles than unlined piles, and heavily reinforced than nominally reinforced piles.

As stated previously, the costs of establishing large equipment on site are much greater than for small equipment, and the choice of plant depends very much on the volume of work to be carried out. However, having chosen the best equipment and method there are still several factors which are of strong influence on the price of the contract.

Load tests on piles are expensive; they are often a requirement of a controlling local authority and sometimes they are necessary and sometimes not. As an alternative to such testing, engineers from time to time negotiate with local authorities on the basis of avoiding a test by providing a higher factor of safety in design calculations. This is particularly sensible on small contracts and where there is an extensive history of similar work. In such cases not only does the cost of a load test become an inordinately large fraction of total cost, but the physical act of carrying out a pile test may disrupt and delay work by obstructing the site for perhaps two or more weeks. On sites where large-diameter piles are used it is often simply impractical and unreasonably costly to apply a load test to the larger piles, and subject to previous history of use, a similar procedure to avoid testing is often adopted.

Integrity testing is inexpensive, and is now sufficiently well established in the hands of experienced companies to provide evidence of major defects in piles without the great cost of load testing. To take most advantage of its use on large and potentially troublesome sites, it is best to start testing early in a contract rather than to leave all testing to the end. The statistical chance of finding a faulty pile by a load test on any site is usually very small indeed, because of the small percentage of piles which can be tested at any reasonable and acceptable cost.

Another cost to be considered directly with piling is the cost of cutting back piles to a final trimmed level, and it is here that a great deal of damage can be done by the use of crude methods. Various improvements in this process are currently appearing and these include the use of hydraulic fracturing equipment placed in holes drilled in each pile head at appropriate positions or the use of non-explosive demolition agents such as 'Bristar'. In the case where the latter process is used, holes are formed in each pile head at suitably chosen positions and subsequently filled or part-filled with a mixture of an inorganic silicate compound and an organic compound. This sets in the hole after a short time and then expands to shatter surrounding concrete. These methods may appear to be more costly than other simpler procedures, but when the damage which can be caused to pile/structure connections from using unrefined methods is fully appreciated, they may in the end provide a good and economic solution to this problem.

The cost of pile caps and beams is closely related to the early design decisions in a contract. A large-diameter pile may be used instead of numbers of smaller piles, and it is then necessary to consider the combined cost of the piles and any capping arrangements in assessing the cost effectiveness of one solution as opposed to the other. Large-diameter piles rarely need substantial caps nor ground restraint arrangements at the pile head. Some useful notes on the design of pile caps are provided by Whittle and Beattie (1972).

Outside the normal limits of site costs, there are certain risk factors which must be evaluated and against which insurances will normally afford some protection. These include the problems of damage to services and to adjacent structures. Although a contractor may be insured against such unfortunate occurrences, it should not be forgotton that when the unforeseen event occurs, it absorbs a great deal of time and effort which is costly to the contractor and the other parties to the contract, and it is unlikely that these costs can ever be fully recovered.

Unforeseen ground conditions represent perhaps the most common source of claim from specialist contractors on piling contracts and these are sometimes inevitable and invariably unwelcome. It is very difficult to choose a piling method in the first place which would minimize such claims, but when something does go wrong the most important factor operating in the case is often delay. Promptness in isolating the cause of a problem and providing a remedy should be a primary objective.

References

Abdrabbo, C.M. (1976). 'A model scale study of axially loaded pile foundations'. Ph.D. Thesis, University of Southampton.

American Society for Testing and Materials (1974). Standard method of testing piles under axial compressive load. *ASTM Standards Yearbook*, Pt. 19, D 1143–74, pp. 178–188.

Attewell, P.B. and Farmer, I.W. (1973). Attenuation of ground vibrations from pile driving. *Ground Engineering* **6**(4), (July).

Baguelin, F. and Frank, R. (1979). 'Theoretical studies of piles using the finite element method', in *Int. Conf. on Numerical Methods in Offshore Piling*, ICE, London. pp. 33–91.

Banerjee, P.K. and Butterfield, R. (1981). *Boundary Element Method in Engineering Science.* McGraw-Hill, Maidenhead.

Banerjee, P.K. and Davies, T.G. (1978). The behaviour of axially and laterally loaded single piles embedded in nonhomogeneous soils. *Géotechnique* **28**(3), 309–326.

Banerjee, P.K. and Driscoll, R.M.C. (1978). Program for the analysis of pile groups of any geometry subjected to horizontal and vertical loads and moments, PGROUP, (2.1). HECB/B/7, Department of Transport, HECB, London.

Bartholomew, R.F. (1979). 'The protection of concrete piles in aggressive ground conditions: an international appreciation'. in *Proc. Conf. Recent Piling Developments in the Design and Construction of Piles*. Institution of Civil Engineers, London.

Barton, Y.O. (1982). 'Laterally loaded model piles in sand: Centrifuge tests and finite element analyses. Ph.D. Thesis, University of Cambridge.

Begemann, H. (1974). 'General Report for Central and Western Europe', in *Proc. Eur. Symp. on Penetration Testing*, Stockholm, Vol. 2.1, pp. 29–39.

Berezantzev, V.C., Khristoforov, V. and Golubkov, V. (1961). 'Load bearing capacity and deformation of piled foundations', in *Proc. 5th Int. Conf. on Soil Mech. and Found. Eng.*, Vol, 2, pp. 11–15.

Bishop, A.W. (1972). 'Shear strength parameters for undisturbed and remoulded soil specimens', in *Stress Strain Behaviour of Soils*, (Proc. Roscoe Mem. Symp.) ed. Parry, R.H.G., G.T. Foulis (Blackie Publishing Group), pp. 3–58.

Bishop, A.W., Green, G.E., Garga, V.K., Andresen, A. and Brown, J.D. (1971). A new ring shear apparatus and its application to the measurement of residual strength. *Géotechnique* **21** (4), 273–328.

Bishop, A.W., Webb, D.L. and Lewin, P.I. (1965). Undisturbed samples of London Clay from the Ashford Common Shaft: Strength—effective stress relationships. *Géotechnique* **15**(1), 1–31.

Bjerrum, L. (1963). Discussion: 'Compressibility of soils, stress strain properties, pore pressure prediction', in *Proc. European Conf. Soil Mechs. Found. Eng.*, Wiesbaden, 1963, **2**. 16–17.

Bobroski, J., Bardhan-Roy, B.K., Magiera, R.H. and Lowe, R.H. (1975). The structural integrity of large diameter bored piles', in *Behaviour of Piles: Proc. Conf. ICE.*, London.

Bolognesi, A.T.L. and Moretto, O. (1973). 'Stage grouting preloading of large piles in sand', in *Proc. 8th Int. Cont. on Soil Mech. and Found. Eng.*, Moscow, Vol. 2.1.

Bolton, M.D. (1984). The strength and dilatancy of sands. Cambridge University Engineering Department Research Report, CUED/D-Soils TR 152.

Bourges, F. and Mieussens, C. (1979). Déplacements latéraux à proximité des remblais sur sols compressibles—méthode de prévision. Bull. Liaison Lab. Ponts et Chaussées, No 101.

Bourges, F., Frank, R. and Mieussens, C. (1980). Calcul des efforts et des déplacements engendrés par des poussées latérales de sol sur les pieux. Note Technique du Dept. Sols et Fondations, Lab. Central des Ponts et Chaussées, Paris.

Bowles, J.E. (1974). *Analytical and Computer Methods in Foundation Engineering.* McGraw-Hill, New York.

Bransby, P.L. and Milligan, G.W.E. (1975). Soil deformations near cantilever sheet pile walls. *Géotechnique* 25(2), 175–195.

Brinch Hansen, J. (1963). Discussion: hyperbolic stress-strain response cohesive soils. *J. Soil Mech. and Found. Div., ASCE,* **89** (SM4), 241–242.

British Standard 6472: 1984. Guide to evaluation of human exposure to vibration and shock in buildings. British Standards Institution, London.

British Standard 5930: 1981. 'Site Investigations'. British Standards Institution, London.

British Standards Institution (1972). BS 8110: Structural use of Concrete.

British Standards Institution (1982). Recommendations for ground anchorages. A draft for development. (Ref. DD81.) London.

Broms, B.B. (1963). Allowable bearing capacity of initially bent piles. *Proc. ASCE,* **89** (SM5), 73.

Broms, B.B. (1964). Lateral resistance of piles in cohesive soils. *J. Soil Mech. and Found. Div., ASCE.* **90** (SM2), 27–63.

Broms, B.B. (1964). Lateral resistance of piles in cohesionless soils. *J. Soil Mech and Found. Div., ASCE,* **90** (SM3), 123–156.

Broms, B.B. (1965). Discussion to paper by Y Yoshimi. *J. Soil Mech. and Found. Div., ASCE,* **91** (SM4), 199–205.

Broms, B.B. (1978). *Precast Piling Practice.* Balken Piling Ltd.

Brooker, E.W. and Ireland, H.O. (1965). Earth pressures at rest related to stress history. *Can. Geotech. J.* **11**(1).

Brown, P.T. and Gibson, R.E. (1972). Surface settlement of a deep elastic stratum whose modulus increases linearly with depth. *Can. Geotech. J.* **9**(4), 467–470.

Building Research Establishment Digest 250 (1981) Building Research Establishment, Watford.

Burgess, I.W. (1976). The stability of slender piles during driving. *Géotechnique* 26(2), 281–292.

Burgess, I.W. (1980). Reply to discussion by Ly B.L., *Géotechnique* 30(3), 322–323.

Burland, J.B., Broms, B.B. and de Mello, V.F.B. (1977).'Behaviour of foundations and structures', in *Proc. 9th Int. Conf. on Soil Mech. and Found. Eng.,* Tokyo, Vol. 2, pp. 495–546.

Burland, J.B. (1973). Shaft friction of piles in clay—a simple fundamental approach. *Ground Engineering* **6**(3), 30–42.

Burland, J.E. and Cooke, R.W. (1974). The design of bored piles in stiff clays. *Ground Engineering* **7**(4), 28–30, 33–35.

Burland, J.B., Simpson, B. and St. John, H.D. (1979). 'Movements around excavations in London Clay', in *Proc. 7th European Conf. on Soil Mech. and Found. Eng.,* Brighton 1, 13–29.

Butler, F.G. and Morton. K. (1970). 'Specification and performance of test piles in clay', in *Proc. Symp. on Behaviour of Piles,* ICE, London, p. 67.

Butterfield, R. and Banerjee, P.K. (1971). The elastic analysis of compressible piles and pile groups. *Géotechnique* **21**(1), 43–60.

Butterfield, R. and Banerjee, P.K. (1971). The problem of pile group—pile cap interaction. *Géotechnique* **21**(2), 135–142.

Butterfield, R. and Douglas, R.A. (1981). Flexibility coefficients for the design of piles and pile groups. CIRIA Technical Note 108.

Carter, J.P. (1982). 'A numerical method for pile deformations due to nearby surface loadings', in *Proc. 4th Int. Conf. on Numerical Methods in Geomechanics,* Edmonton, Vol 2, pp. 811–817.

Chandler, K. Corrosion control. *Engineering,* March 1976, **216**(No. 3). Technical File 27.

Chandler, R.J. (1968). The shaft friction of piles in cohesive soils in terms of effective stresses. *Civ. Eng. Public Wks. Rev.* **63**, 48–51.

Chandler, R.J. (1969). The effect of weathering on the shear strength properties of Keuper Marl. *Géotechnique* 19(3), 321–334.

Chandler, R.J. (1972). Lias Clay: weathering processes and their effect on shear strength. *Géotechnique* 22(3), 403–431.

Chandler, R.J. and Martins, J.P. (1982). An experimental study of skin friction around piles in clay. *Géotechnique* 32(2), 119–132.

Cheney, J.E. (1973). 'Techniques and equipment using the surveyor's level for accurate measurement of building movement', in *Proc. Symp. Field Instrumentation in Geotech. Eng.,* Butterworth, London.

Chin, F.K. (1970). 'Estimation of the ultimate load of piles not carried to failure', in *Proc. 2nd S.E. Asia Conf. on Soil Eng.* pp. 81–90.

Claessen, A.I.M. and Horvat, E. (1974). Reducing negative friction with bitumen slip layers. *J. Geotech. Eng. Div., ASCE,* **100**(GT8), 925–944.

Cole, K.W. Uplift of piles due to driving displacement. *Civ. Eng. Public Wks.* March 1972, 263–269.

Cole, K.W. (1980). 'The South abutment of Kessock Bridge. Scotland', in *Proc. IABSE Conf.,* Vienna.

Combarieu, O. (1979). 'Comportement d'une fondation mixte semelle-pieux', in *Proc. 7th Eur. Conf. on Soil Mech. and Found. Eng.,* Brighton. Vol. 1, pp. 105–110.

Combarieu, O. and Evrard, H. (1979). Les fondations mixtes, semelle-pieux. Bull Liaison Laboratoires Ponts et Chaussées, Vol. 102, July–August.

Combarieu, O. and Morbois, A. (1982). Fondations mixtes semelle-pieux. *Annales de l'Inst. Tech. du Bat. et des Trav. Pub.,* No. 410, December.

Cooke, R.W. (1974). 'Settlement of friction pile foundations', in *Proc. Conf. on Tall Buildings,* Kuala Lumpur, pp. 7–19.

Cooke, R.W. and Price, G. (1978). Strains and displacements around friction piles. Building Research Station CP 28/78, October.

Cooke, R.W., Price, G. and Tarr, K.W. (1979). Friction piles under vertical working load conditions—load transfer and settlement. *Géotechnique* **29**(2), 113–147.

Cooke, R.W., Price, G. and Tarr, K.W. (1980). Jacked piles in London clay: interaction and group behaviour under working conditions. *Géotechnique* **30**(2), 449–471.

Cooke, R.W., Bryden Smith, D.W., Gooch, M.N. and Sillet, D.F. (1981). Some observations of the foundation loading and settlement of a multi-storey building on a piled raft foundation in London clay. *Proc. ICE,* Vol. 107 (Part 1), 433–460.

Coyle, H.M. and Castello, R.R. (1981). New design correlations for piles in sand. *J. Geotech. Eng. Div., ASCE,* **107**(GT7), 965–986.

Davisson, M.T. (1963). 'Estimating buckling loads for piles', in *Proc. 2nd Pan-Amer. Conf. on Soil Mech. and Found. Eng.,* Brazil, Vol. 1, pp. 351–371.

Davisson, M.T. and Gill, H.L. (1963). Laterally loaded piles in a layered soil system. *J. Soil Mech. and Found. Div., ASCE,* **89**(SM3) 63–94.

Davisson, M.T. and Robinson, K.E. (1965). 'Bending and buckling of partially embedded piles', in *Proc. 6th Int. Conf. on Soil Mech. and Found. Eng.,* Vol. 2, pp. 243–246.

De Beer, E.E. and Wallays, M. (1972). 'Forces induced in piles by unsymmetrical surcharges on the soil around the piles', in *Proc. 5th Eur. Conf. on Soil Mech. and Found. Eng.,* Madrid, Vol. 1, pp. 325–332.

De Mello, V.F.B. (1969). 'Foundations of buildings on clay', State of the Art Report, in *Proc. 7th Int. Conf. on Soil Mech. and Found. Eng.,* Vol. 1, pp. 49–136.

Dumbleton, M.J. and West, G. (1976). Preliminary sources of information for site investigation in Britain. Department of the Environment, Transport and Road Research Laboratory Report LR403, revised edition.

Elson, W.K. (1984). Design of laterally loaded piles. CIRIA Report 103, London.

Evangelista, A. and Viggiani, C. (1976). 'Accuracy of numerical solutions for laterally loaded piles in elastic half-space', in *Proc. 2nd Int. Conf. on Numerical Methods in Geomechanics,* Blacksburg, Vol. 3, pp. 1367–1370.

Fahey, M. and Randolph, M.F. (1984). Effects of disturbance on parameters derived from self-boring pressuremeter tests in sand. *Géotechnique* **34**(1), 81–97.

Fellenius, B.H. (1972). Down drag on piles in clay due to negative skin friction. *Can. Geotech. J.* **9**(4), 323–337.

Fellenius, B.H. (1980). The analysis of results from routine pile load tests. *Ground Engineering* **6**, Sept. 19–31.

Fischer, H.C. (1984). 'Stress wave theory for pile driving applications', in *Proc. 2nd Int. Conf. on Application of Stress Wave Theory on piles,* Stockholm.

Fleming, W.G.K. (1984). Discussion on State of the Art paper, Advances in Piling and Ground Treatment for Foundations' conference, ICE, London.

Fleming, W.G.K. and Thorburn, S. (1983). 'Recent piling advances', State of the Art Report', in *Proc. Conf. on Advances in Piling and Ground Treatment for Foundations,* ICE, London.

Fleming, W.K. and Sliwinski, Z.J. (1977). The use and influence of bentonite in bored pile construction. D.O.E/CIRIA Piling Development Group Report PG. 3.

Fleming, W.G.K. and Lane, P.F. (1970). 'Tolerance requirements and construction problems in piling', in 'Behaviour of Piles' conference, ICE, London (1971) Paper 17.

Fookes, P.G. and Collis, L. Cracking and the Middle East. Concrete, 1176, **10**, February, 14–19.

Francescon, M. (1983). 'Model pile tests in clay: stresses and displacements due to installation and axial loading'. Ph.D. Thesis, University of Cambridge.

Francis, A.J., Stevens, L.K. and Hoadley, P.J. Paper to *Symposium of Soft Ground Engineering*, Brisbane. Institution of Civil Engineers Australia, 1965.

Frank, R.A. (1974). 'Etude théorique du comportement des pieux sous charge verticale: Introduction de la dilatance', Dr-Ing-Thesis, Universite Pierre et Marie Curie (Paris VI), also Rapport de Recherche No 46, Laboratoire Central des Ponts et Chaussës, Paris.

Geddes, W.G.N., Martin, G.P. and Land, D.D. (1966). Clyde Dry Dock Project, Greenock. *Proc. ICE.* **33**, 615.

Geiger, J. (1959). Estimating the risk of damage to buildings from vibrations. *Bauingenieur* **34**, 1959.

Gens, A. and Hight, D.W. (1979). 'The laboratory measurement of design parameters for a glacial till', in *Proc. 7th European Conf. on Soil Mech. and Found. Eng.*, Vol. 2, 57–65.

George, A.B., Sherrell, F.W. and Tomlinson, M.J. (1977). 'The behaviour of steel H-piles in slaty mudstone', in *Proc. Conf. on Piles in weak rock*, ICE, London, 95–104 (also CIRIA Technical Note 66, May 1976).

German Code of Practice for Piling: DIN 4026.

German Code of Practice. Protection against vibration in building construction DIN 4150: 1970.

Ghosh, N. (1975). 'A model scale investigation of the working load stiffness of single piles and groups of piles in clay under centric and eccentric vertical loads', Ph.D. Thesis, University of Southampton.

Goble, G.G. and Rausche, F. (1979). 'Pile drivability predictions by CAPWAP', in *Proc. Int. Conf. on Numerical Methods in Offshore Piling*, ICE, London, pp. 29–36.

Gravare, C.J., Gable, G.G. Rausche, F. and Likins, G. (1980). Pile driving construction and control by the CASE method. *Ground Engineering* **13**, March, 20–25.

Hagerty, D. and Peck, R.B. (1971). Heave and lateral movements due to pile driving. *J. Soil Mech. and Found. Div., ASCE*, (SM11) Nov. 1971, 1513–1531.

Hain, S.J. and Lee, I.K. (1978). The analysis of flexible pile raft systems. *Géotechnique* **28**(1), 65–83.

Hambly, E.C. (1976). A review of current practice in the design of bridge foundations. Report for the Department of the Environment, Building Research Establishment, UK.

Hanna, T.H. (1968). The bending of long H-section piles. *Canadian Geotech. J.* **5**(3), 150–172.

Hansbo, S. and Jenderby, L. (1983). A case study of two alternative foundation principles: conventional friction piling and creep piling. Vag-och Vattenbyggaren **7**(8), 29–31.

Healy, P.R. and Weltman, A.J. (1980). Survey of problems associated with the installation of displacement piles. D.O.E/CIRIA Piling Development Group Report PG. 8.

Heerema, E.P. and de Jong, A. (1979). 'An advanced wave equation computer program which simulates dynamic pile plugging through a coupled mass-spring system', in *Proc. Int. Conf. on Numerical Methods in Offshore Piling*, ICE, London, pp. 37–42.

Hiley, A. (1925). A rational pile-driving formula and its application in piling practice explained. *Engineering* (London) **119**, 657, 721.

Hobbs, N.B. and Healy, P.R. (1979). Piling in Chalk DOE/CIRIA Piling Development Group Report PG. 6. CIRIA, London.

Hobbs, N.B. and Robins, P. (1976). Compression and tension tests on driven piles in chalk. *Géotechnique* **26**(1), 33–46.

Hodges, W.G.H. and Pink, S. (1971). 'The use of penetrometer soundings in the estimation of pile bearing capacity and settlement for driven piles in highly weathered chalk', in *Stress Strain Behaviour of Soils*, (Proc. Roscoe Mem. Symp.) ed. Parry, R.H.G. G.T. Foulis (Blackie Publishing Group) pp. 693–723.

Hughes, J.M.O., Wroth, C.P. and Windle, D. (1977). Pressuremeter tests in sand. *Géotechnique* **27**(4), 455–477.

Institution of Civil Engineers (1988). Specification for Piling. ICE, London.

Institution of Civil Engineers (1979). Recent developments in the design and construction of piles. ICE, London.

Ionising Radiations (Sealed Sources) Regulations, 1969. HMSO, London.

Ito, T. and Matsui, T. (1975). Methods to estimate lateral force acting on stablizing piles. *Soils and Foundations* **15**(4), 43–59.

Johanessen, I.J. and Bjerrum, L. (1965). 'Measurement of the compression of a steel pile to rock due to settlement of the surrounding clay', in *Proc. 6th Int. Conf. on Soil Mech. and Found. Eng.*, Vol. 2, pp. 261–264.

Kennet, R. (1971). Geophysical borehole logs as an aid to ground engineering. *Ground Engineering* **4**(5).

Kim, J.B. and Brungraber, R.J. (1976). Full scale lateral load tests on pile groups. *Proc. Geotech. Div. ASCE*, (GTI) 87–101.

Kovacs, W.D. (1975). Discussion of paper by Y. Lhsaki and R. Iwasaki, *Soils and Foundations* **13**(4), 61–73.

Kraft, L.M. (1982). Effective stress capacity model for piles in clay. *J. Geotech. Eng. Div., ASCE*, **108**(GT11), 1387–1404.

Kuhlemeyer, R.L. (1979). Static and dynamic laterally loaded floating piles *J. Geotech. Eng. Div., ASCE*, **105**(GT2), 289–304.

Lo, M.B. (1967). Discussion to paper by Y.O. Beredugo, *Can. Geotech. J.* **4**, 353–354.

Lord, J.A. (1976). A comparison of three types of driven cast-in-situ pile in chalk. *Géotechnique* **26**(1), 73–93.

Lupini, J.F., Skinner, A.E. and Vaughan, P.R. (1981). The drained residual strength of cohesive soil. *Géotechnique* **31**(2), 181–213.

Ly, B.L. (1980). Discussion of Burgess, I.W., *Géotechnique* **30**(3), 321–322.

McClelland, B. (1974). Design of deep penetration piles for ocean structures. *J. Geotech. Eng. Div., ASCE*, **100**(GT7), 705–747.

McClelland, B., Focht, J.A. and Emrich, W.J. (1969). Problems in design and installation of offshore piles. *J. Soil Mech. and Found. Div., ASCE*, **6**, 1491–1513.

Marsland, A. (1976). In-situ and laboratory tests on glacial clays at Redcar Building Research Establishment Current Paper CP 65/76 HMSO (London).

Marsland, A. and Butler, M.E. (1967). 'Strength measurements on stiff fissured Barton Clay from Fawley, Hampshire', in *Proc. Geotech. Conf.*, Oslo, **1**, pp. 139–145.

Marsland, A. and Randolph, M.F. (1977). Comparison of the results from pressuremeter tests and large in situ plate tests in London clay. *Géotechnique* **27**(2), 217–243.

Matlock, H.S. and Reese, L.C. (1960). Generalised solutions for laterally loaded piles. *J. Soil Mech. and Found. Div., ASCE*, **86**(SM5), 63–91.

Matlock, H.S. (1970). 'Correlations for design of laterally loaded piles in soft clay', in *2nd Annual Offshore Technology Conf.*, Houston.

Mattes, N.S. and Poulos, H.G. (1969). Settlement of single compressible pile. *J. Soil Mech. and Found. Div., ASCE*, **95**(SM1), 198–207.

Meyerhof, G.G. (1956). Penetration tests and bearing capacity of cohesionless soils. *J. Soil Mech. Found. Div., ASCE*, **82**(SM1), Paper 866.

Meyerhof, G.G. and Ranjan, G. (1972). The bearing capacity of rigid piles under inclined loads in sand: I—Vertical piles. *Can. Geotech. J.* **9**(4), 430–446 (see also *Can. Geotech. J.* **10**(1), 71–85 for Part II).

Meyerhof, G.G. (1976). Bearing capacity and settlement of pile foundations. *J. Geot. Engng. Div., ASCE*, **102**(GT3), 197–228.

Milligan, G.W.E. (1983). Soil deformations near anchored sheet pile walls. *Géotechnique* **33**(1).

More, J.N., Kightley, M.L., Gravare, C.J. and Hermansson, I. (1983). 'CAPWAP—an economic and comprehensive alternative to traditional methods of load testing piles', in *Proc. Sym. Advances in Piling and Ground Treatment for Foundations*, ICE, March, pp. 167–174.

Murff, J.D. (1980). 'Pile capacity in a softening soil'. *Int. J. Num. and Anal. Methods in Geomechanics*, **4**, 185–189.

Navy, Dept. of (1971). *Design Manual—Navfac DM*-7. U.S. Government Printing Office, Washington D.C.

Naylor, D.J. (1983). Finite element study of embankment loading on piles. Report to the Department of Transport (H.E.C.B.).

Nordlund, R.L. (1963). Bearing capacity of piles in cohesionless soils. *J. Soil Mech and Found Eng. Div., ASCE*, **89**(SM3), 1–35.

Olson, R.E. and Dennis, N.D. (1982). Review and compilation of pile test results: Axial pile capacity. PRAC Project 81–29. American Petroleum Institute, Dallas.

Omar, R.M. (1978). Discussion of Burgess I.W., *Géotechnique* **28**(2), 234–239.

O'Neill, M.W., Hawkins, R.A. and Mahar, L.J. (1982). Load transfer mechanisms in piles and pile groups. *J. Geotech. Eng. Div., ASCE*, **108**(GT12), 1605–1623.

O'Neill, M.W., Ghazzaly, O.I. and Ha, H.B. (1977). 'Analysis of three dimensional pile groups with non-linear soil response and pile-soil-pile interaction', in *Proc. 9th Offshore Tech. Conf.*, Vol. 2, pp. 245–256.

O'Riordan, N.J. (1982). The mobilisation of shaft adhesion down a bored, cast-in-situ pile in the Woolwich and Reading beds. *Ground Engineering* **15**(3), 17–26.

Packshaw, S. (1951). Pile driving in difficult conditions. ICE Introductory Note.

Padfield, C.J. and Sharrock, M.J. (1983). Settlement of structures on clay soils. CIRIA Special Publication 27, London.

Palmer, D.J. (1985). Ground borne vibrations from piling: a review. CIRIA report (in preparation).

Parry, R.H.G. (1980). A study of pile capacity for the Heather platform. *Ground Engineering* **13**(2), 26–28, 31, 37.

Parry, R.H.G. (1972). Some Properties of heavily overconsolidated Lower Oxford Clay at a site near Bedford. *Géotechnique* **22**(3), 485–507.

Parry, R.H.G. and Swain, C.W. (1977). Effective stress methods of calculating skin friction of driven piles in soft clay. *Ground Engineering* **10**(3), 24–26.

Parry, R.H.G. and Swain, C.W. (1977). A study of skin friction on piles in stiff clay. *Ground Engineering* **10**(8), 33–37.

Peck, R.B. (1969). State of the Art Report. *7th Int. Conf. on Soil Mech. and Found. Eng.*, Mexico.

Peck, R.B., Hanson, W.E. and Thornburn, T.H. (1974). *Foundation Engineering*. John Wiley & Sons, New York, etc.

Piling, Specification for Piling (1988). Institution of Civil Engineers, London.

Pise, P.J. (1982). Laterally loaded piles in a two-layer soil system. *J. Geotech. Eng. Div., ASCE*, **108**(GT9), 1177–1181.

Poulos, H.G. and Davis, E.H. (1968). The settlement behaviour of single axially loaded incompressible piles and piers. *Géotechnique* **18**, 351–371.

Poulos, H.G. (1971). Behaviour of laterally loaded piles: I—single piles, and II—pile groups. *J. Soil Mech. and Found. Div., ASCE*, **97**(SM5), 711–731, 733–751.

Poulos, H.G. and Davis, E.H. (1974). *Elastic Solutions for Soil and Rock Mechanics*. John Wiley & Sons, New York, etc.

Poulos, H.G. and Davis, E.H. (1980). *Pile Foundation Analysis and Design*. John Wiley & Sons, New York, etc.

Poulos, H.G. (1980). User's guide to program DEFPIG—deformation analysis of pile groups. School of Civil Engineering, University of Sydney.

Poulos, H.G. (1982). The influence of shaft length on pile load capacity in clays. *Géotechnique* **32**(2), 145–148.

Poulos, H.G. and Randolph, M.F. (1983). Pile group analysis: a study of two methods. *J. Geotech. Eng., ASCE*, **109**(3), 355–372.

Preiss, K. and Caisserman, F. (1975). Non-destructive integrity testing of bored piles by gamma ray scattering. *Ground Engineering* **8**(3).

Price, G. and Wardle, I.F. (1983). 'Recent developments in pile/soil instrumentation systems', in *Proc. Conf. Field Measurements in Geomechanics*, Zurich.

Randolph, M.F. (1977). 'A theoretical study of the performance of piles', Ph.D. Thesis, University of Cambridge.

Randolph, M.F. (1980). PIGLET: A computer program for the analysis and design of pile groups under general loading conditions. Cambridge University Engineering Department Research Report, Soils TR 91.

Randolph, M.F. (1981). The response of flexible piles to lateral loading Geotechnique **31**(2), 247–259.

Randolph, M.F. (1981). Piles subjected to torsion. *J. Geotech. Eng. Div., ASCE*, **107**(GT8), 1095–1111.

Randolph, M.F. (1983). 'Design considerations for offshore piles', in *Proc. Conf. on Geotech. Practice in Offshore Engng., ASCE*, Austin, pp. 422–439.

Randolph, M.F. (1983). 'Settlement considerations in the design of axially loaded piles'. *Ground Engineering* **16**(4), 28–32.

Randolph, M.F., Carter, J.P. and Wroth, C.P. (1979). Driven piles in clay—the effects of installation and subsequent consolidation. *Géotechnique* **29**(4), 361–393.

Randolph, M.F. and Houlsby, G.T. (1984). The limiting pressure on a circular pile loaded laterally in cohesive soil. *Géotechnique* **34**, 4.

Randolph, M.F. and Poulos, H.G. (1982). 'Estimating the flexibility of offshore pile groups', in *Proc 2nd Int. Conf. on Numerical Methods in Offshore Piling*, Austin, pp. 313–328.

Randolph, M.F. and Wroth, C.P. (1978). Analysis of deformation of vertically loaded piles. *J. Geotech. Engng. Div., ASCE*, **104**(GT12), 1465–1488.

Randolph, M.F. and Wroth, C.P. (1978). 'A simple approach to pile design and the evaluation of pile tests', in *Behaviour of Deep Foundations*, ed. R. Lundgren, ASTM STP 670, pp. 484–499.

Randolph, M.F. and Wroth, C.P. (1979). An analytical solution for the consolidation around a driven pile. *Int. J. Num. and Anal. Methods in Geomechanics* **3**, 217–229.

Randolph, M.F. and Wroth, C.P. (1979). An analysis of the vertical deformation of pile groups. *Géotechnique* **29**(4), 423–439.

Randolph, M.F. and Wroth, C.P. (1981). Application of the failure state in undrained simple shear to the shaft capacity of driven piles. *Géotechnique* **31**(1), 143–157.

Randolph, M.F. and Wroth, C.P. (1982). Recent developments in understanding the axial capacity of piles in clay. *Ground Engineering* **15**(7), 17–25, 32.

Reddaway, A.L. and Elson, W.K. (1982). The performance of a piled bridge abutment at Newhaven. CIRIA Technical Note 109, London.

Reddy, A.S. and Valsangkar, A.J. (1970). Buckling of fully and partially embedded piles. *J. Soil Mech. and Found. Div., ASCE*, **96**(SM6), 1951–1965.

Reese, L.C. (1958). Discussion of paper by B. McClelland and J.A. Focht. *Trans. ASCE*, **123**, 1071–1074.

Reese, L.C. (1977). Laterally loaded piles: program documentation. *J. Geotech. Eng. Div. ASCE*, (103), GT4.

Reese, L.C. and Matlock, H.S. (1956). 'Non-dimensional solutions for laterally loaded piles with soil modulus proportional to depth', in *Proc. 8th Texas Conf. on Soil Mech and Found. Eng.*, pp. 1–41.

Reese, L.C., Cox, W.R. and Koop, F.D. (1974). 'Analysis of laterally loaded piles in sand', in *6th Annual Offshore Technology Conf.*, Houston, pp.

Reese, L.C., Cox, W.R. and Koop, F.D. (1975). 'Field testing and analysis of laterally loaded piles in stiff clay', in *Proc. 7th Offshore Tech. Conf.*, Vol. 2, pp. 671–690.

Reese, L.C., Touma, F.T. and O'Neill, M.W. (1976). Behaviour of drilled piers under axial loading. *Proc. J. Geotech. Eng. Div., ASCE*, **102**(GT5).

Reid, W.M. and Buchanan, N.W. (1983). 'Bridge approach support piling', in *Proc. Conf. on Advances in Piling and Ground Treatment for Foundations*, ICE, London.

Samuels, S.G. (1975). Some properties of the Gault Clay from the Ely-Ouse Essex Water Tunnel. Build. Res. Est. Current Paper 63/75, 19 pp.

Scott, R.F. (1965). *Principles of Soil Mechanics*. Addison Wesley, Reading, Mass.

Settlement of Structures. Conf. of British Geotechnical Society, Cambridge. Pentech Press, London 1974.

Schofield, A.N. and Wroth, C.P. (1968). *Critical State Soil Mechanics*, McGraw-Hill, Maidenhead.

Seed, H.B. and Reese, L.C. (1955). 'The action of soft clay along friction piles'. *Proc. ASCE*, **81**, Paper 842.

Semple, R.M. and Rigden, W.J. (1984). 'Shaft capacity of driven piles in clay', in *Proc. ASCE National Convention*, San Francisco.

Simpson, B.; Calabresi, G., Sommer, H. and Wallays, M. (1979). 'Design parameters for stiff clays', in *Proc. 7th Eur. Conf. on Soil Mech. and Found. Eng.*, Brighton, Vol. 5, pp. 91–125.

Skempton, A.W. (1951). 'The bearing capacity of clays', in *Building Research Congress*, ICE, London, pp. 180–189.

Skempton, A.W. (1959). Cast in situ bored piles in London clay. *Géotechnique* **9**, 153–173.

Skempton, A.W. (1970). First time slides in over consolidated clays. *Géotechnique* **20**(3), 320–324.

Skempton, A.W. (1974). 'Slope stability of cuttings in brown London Clay'. Special Lecture, *9th Int. Conf. on Soil Mechanics and Found. Eng.*, Tokyo, Vol. 3.

Skempton, A.W. and Brown, J.D. (1961). A landslide in boulder clay at Selset, Yorkshire. *Géotechnique* **11**(4), 280–293.

Skempton, A.W. and MacDonald, D.H. (1956). The allowable settlements of buildings. *Proc. ICE*, 5 (Dec.), 727–767.

Skempton, A.W. and Northey, R.D. (1952). The sensitivity of clays. *Géotechnique* **3**, 30–53.

Skipp, B.O. (1984). 'Dynamic ground movements—man-made vibrations', in *Ground Movements and their Effects on Structures*, eds. P.B. Attewell and R.K. Taylor, Blackie, Glasgow and London, pp. 381–434.

Sutherland, H.B. (1974). Review Paper, Session 1: granular material. *Conf. on Settlement of Structures*, Cambridge, 1974.

Smith, E.A.L. (1960). Pile driving analysis by the wave equation. *J. Soil Mech. and Found. Div., ASCE*, **86** (SM4), 35–61.

Smith, I.M. and Chow, Y.K. (1982). 'Three-dimensional analysis of pile drivability', in *Proc. 2nd Int. Conf. on Numerical Methods in Offshore Pilling*, Austin, pp. 1–10.

Sokolovskii, V.V. (1965). *The Statics of Granular Media*, Translated by J.K. Lusher; Pergamon, Oxford.

Stroud, M.A. (1974). 'The standard penetration test in insensitive clays and soft rocks', in *Proc. Eur. Symp. on Penetration Testing*, Stockholm, Vol. 2.2, pp. 367–375.

Stroud, M.A. and Butler, F.G. (1975). 'The standard penetration test and the engineering properties of glacial materials', in *Proc. Symp. on the engineering behaviour of glacial materials*, held at the University of Birmingham, 1975.

Sullivan, W.R., Reese, L.C. and Fenske, C.W. (1979). 'Unified method for analysis of laterally loaded piles in clay', in *Proc. Int. Conf. on Numerical Methods in Offshore Piling*, ICE, London, pp. 135–146.

Swain, A. (1976). 'Model ground anchors in clay', Ph.D. Thesis, University of Cambridge.

Teng, W.C. (1962). *Foundation Design*. Prentice-Hall Inc., New Jersey.

Terzaghi, K. (1954). Anchored bulkheads. *Trans. ASCE*, 119.

Terzaghi, K. and Peck, R.B. (1967). *Soil Mechanics in Engineering Practice*. John Wiley & Sons, New York, etc.

Thorburn, S. (1963). 'Tentative correction chart for the standard penetration test in non-cohesive soils'. *Civ. Eng. Public Wks. Rev.* **58**, 752–753.

Thorburn, S., Laird, C. and Randolph, M.F. (1983). 'Storage tanks founded on soft soils reinforced with driven piles', in *Proc. Conf. on Recent Advances in Piling and Ground Treatment for Foundations*, ICE, London.

Thorburn, S. and MacVicar, R.S.L. (1971). 'Pile load tests to failure in the Clyde alluvium' (see also Discussion to Session A)', in *Proc. Conf. on Behaviour of Piles*, ICE, London, pp. 1–7, 53–54.

Thorburn, S. and Rigden, W.J. (1980). 'A practical study of pile behaviour', in *Proc. 12th Annual Offshore Technology conf.*, Houston.

Thorburn, S. and Thorburn, J.Q. (1977). Review of problems associated with the construction of cast-in-place piles. D.O.E/CIRIA Piling Development Group Report, PG. 2.

Timoshenko, S.P. and Goodier, J.N. (1970). *Theory of Elasticity* (3rd edn.), McGraw-Hill, New York.

Tomlinson, M.J. (1957). 'The adhesion of piles driven in clay soils', in *Proc. 4th Int. Conf. on Soil Mech. and Found. Eng.*, Vol. 2, pp. 66–71.

Tomlinson, M.J. (1970). Adhesion of piles in stiff clay. CIRIA Report 26, London.

Tomlinson, M.J. (1977). *Pile Design and Construction Practice*. Viewpoint Publications, London.

Toolan, F.E. and Fox, D.A. (1977). Geotechnical planning of piled foundations for offshore platforms. *Proc. ICE*, **62** (Part 1), 221–244.

US Navy Bureau of Yards and Docks (1962). 'Navdocks', *Design Manual DM-7*, US. Govt. Printing Office, Washington D.C.

USS Sheet Steel Piling Design Manual (1975). United States Steel International Inc., Pittsburgh.

Van Wheele, A.F. (1957). 'A method of separating the bearing capacity of a test pile into skin friction and point resistance', in *Proc. 4th Int. Conf. Soil Mech. and Found. Eng.*, ICE, London, Vol. 2.

Vaughan, P.R., Lovenbury, H.T. and Horswill, P. (1975). The design, construction and performance of Cow Green Embankment Dam. *Géotechnique* **25** (3), 555–580.

Vesic, A.S. (1969). Experiments with instrumented pile groups in sand. ASTM STP 444, 177–222.

Vesic, A.S. (1977). Design of Pile Foundations, National Co-operative Highway Research Program, Synthesis of Highway Practice No 42, Transportation Research Board, National Research Council, Washington D.C.

Vijayvergiya, V.N. and Focht, J.A. (1972). 'A new way to predict the capacity of piles in clay', in *Proc. 4th Annual Offshore Tech. Conf.*, Houston, Vol. 2, pp. 865–874.

Walker, L.K. and Darvall, P. Le P. (1973). 'Dragdown on coated and uncoated piles', in *Proc. 8th Int. Conf. on Soil Mech. and Found. Eng.*, Vol. 2.1, pp. 257–262.

Weltman, A.J. (1980). Pile Load Testing Procedures. DoE/CIRIA Piling Development Group Report PG. 7.

Weltman, A.J. and Head, J.M. (1983). *Site Investigation Manual*. DoE/CIRIA Special Publication 25, CIRIA, London.

Weltman, A.J. and Healy, P.R. (1978). Piling in boulder clay and other glacial tills, DoE/CIRIA Report PG. 5, London.

Weltman, A.J. and Little, J.A. (1977). A review of bearing pile types DoE/CIRIA Piling Development Group Report PG. 1.

West, A.S. (1972). *Piling Practice*. Butterworth, London.

Whitaker, T. (1963). Load cells for measuring the base loads in bored piles and cylinder foundations. BRS Current Paper, Engineering Series No. 11.

Whitaker, T. and Cooke, R.W. (1966). 'An investigation of the shaft and base resistance of large bored piles in London clay', in *Proc. Symp. on Large Bored Piles*, ICE, London, pp. 7–49.

Whitaker, T. (1975). *The Design of Piled Foundations* (2nd edn.) Pergamon, Oxford.

Whittle, R.T. and Beattie, D. (1972). 'Standard pile caps' I and II: *Concrete* 6 (1 Jan.), 34–36, and 2 February, 39–41.

Williams, A.F. and Pells, P.J.N. (1981). Side resistance of rock sockets in sandstone, mudstone and shale. *Can. Geotech. J.*, **18**.

Wroth, C.P. (1972). Discussion in *Specialty Conf. on Performance of Earth and Earth-Supporting Structures*, ASCE, New York, Vol. 3, pp. 231–234.

Wroth, C.P., Randolph, M.F., Houlsby, G.T. and Fahey, M. (1979). A review of the engineering properties of soils, with particular reference to the shear modulus. Cambridge University Research Report, CUED/D—Soils TR 75.

ADDITIONAL REFERENCES

Bauer, G.E. (1975). Instrumentation of a 50 ft deep braced excavation. A.S.T.M. Special Technical Publication 584.

Bica, A.V.D. and Clayton, C.R.I. (1989). Limit equilibrium design methods for free embedded cantilever walls in granular materials. *Proc. Instn. Civ. Engrs*, Part 1, October, 879–898.

Bolton, M.D. (1986). The strength and dilatancy of sands. *Géotechnique* **36** (1), 65–78.

Bolton, M.D. and Powrie, W. (1988). Behaviour of diaphragm walls in clay prior to collapse. *Géotechnique* **38** (2), 167–189.

Bolton, M.D., Powrie, W. and Symons, I.F. (1989). The design of stiff in-situ walls retaining overconsolidated clay. *Ground Engineering*, Nov 1989 (Part I); Jan–Feb 1990 (Part I—with enlarged diagrams); March 1990 (Part II).

Boyle, S. (1990). The effect of piling operations in the vicinity of computing systems. *Ground Engineering*, June, 23–27.

Burland, J.B. and Twine, D. (1988). 'The shaft friction of bored piles in terms of effective strength', *Proc. 1st Int. Geot. Sem. on Deep Foundations on Bored and Auger Piles*, Ghent, pp 411–420.

Busbridge, J.R. (1975). Discussion, *Diaphragm Walls and Anchorages*, ICE, London.

Carter, J.P. and Kulhawy, F.H. (1988). Analysis and design of drilled shaft foundations socketed into rock. Report to Electric Power Research Institute, California, Project RP1493-4, Report EL-5918.

Cole, K.W. and Burland, J.B. (1972). 'Observations of a retaining wall movements associated with a large excavation', *5th ICSMFE Conf.*, Madrid.

Corbett, B.O., Davies, R.V. and Langford, A.D. (1975). 'A load bearing diaphragm wall at Kensington and Chelsea Town Hall, London,' *Diaphragm Walls and Anchorages*, ICE, London.

Cundall, P.A. (1976). 'Explicit finite difference methods in geomechanics,' in *Numerical Methods in Engineering* (*Proc. EF Conference on Numerical Methods in Geomechanics, Blacksburg, V A*), Vol. 1, pp. 132–150.

Dolwin, J.D. and Poskitt, T. (1982). 'An optimisation method for pile driving analysis', *Proc. 2nd Int. Conf. on Num. Methods in Offshore Piling*, Austin, 91–106.

Fellenius, B.H. (1988). 'Variation of CAPWAP results as a function of the operator', *Proc. 3rd Int. Conf. on Application of Stress–wave Theory to Piles*, Ottawa, 814–825.

Fellenius, B.H. (1989). Unified design of piles and pile group. Transportation Research board, Washington, TRB Record 1169, pp 75–82.

Gibbs, H.J. and Holtz, W.G. (1957). Research on determining the density of sands by spoon penetration testing. *Proc. 4th Int. Conf. on Soil Mech. and Found. Eng.*, No 1, London, 35–39.

Gibson, G.C. and Coyle, H.M. (1968). Soil damping constants related to common soil properties in sands and clays, Report No. 125–1, Texas Transport Institute, Texas A and M University, Houston.

Gnaedinger, J.P., Gill, S.A. and Keim, P.A. (1975). Performance of tied back excavation at Water Tower Place, A.S.T.M. Special Technical Publication 584.

Goble, G.G. and Rausche, F. (1976). Wave equation analysis of pile driving—WEAP program, US Department of Transportation, Federal Highway Administration, Implementation Division, Office of Research and Development, Washington D.C. 20590.

Goble, G.G. and Rausche, F. (1979). 'Pile drivability predictions by CAPWAP', in *Proc. Int. Conf. on Num. Methods in Offshore Piling*, ICE, London, pp 29–36.

Goble, G.G., Rausche, F. and Liking, G.E. (1980). 'The analysis of pile driving—A state-of-the-art', *Proc. Int. Conf. on Stress–Wave Theory on Piles*, Stockholm, 131–161.

Goble, G.G. and Rausche, F. (1986). *WEAP86 Program Documentation in 4 Vols.* Federal Highway Administration, Office of Implementation, Washington DC.

Griffiths, D.V., Clancy, P. and Randolph, M.F. (1991), 'Piled raft foundation analysis by finite elements', in *Proc. 7th Int. Conf. on Num. Methods in Geom.*, Cairns.

Gysi, H.J., Linder, A. and Leoni, R. (1977). 'Behaviour of a prestressed diaphragm wall', 9th ICSMFE, Tokyo (2), 83–90.

Hewlett, W.J. and Randolph, M.F. (1988). Analysis of piled embankments. *Ground Engineering* **22** (3), 12–18.

Hodgson, F.T. (1975). 'Design and construction of a diaphragm wall at Victoria Street, London,' *Diaphragm Walls and Anchorages*, ICE, London.

Horvath, R.G. (1978). Field load tests on concrete to rock bond strength for drilled pier foundations. University of Toronto Publication 78–07.

Horvath, R.G., Trow, W.A. and Kenney, T.C. (1980). 'Results of tests to determine shaft resistance of rock-socketed drilled piers'. *Proc. Int. Conf. on Structural Foundations on Rock*, Sydney, **1**, 349–361.

Houlsby, G.T., Evans, K.M. and Sweeney, M.A. (1988). 'End bearing capacity of model piles in layered carbonate soils', in *Proc. Int. Conf. on Calcareous Sediments*, Perth, Vol 1, pp 209–214.

Jaeger, J.C., and Starfield, A.M. (1974). *An Introduction to Applied Mathematics* (2nd Ed.) Clarendon Press, Oxford.

Johnston, I.W. and Lam, T.S.K. (1989). Shear behaviour of regular traingular concrete/rock joints—Analysis. *J. Geot. Eng. Div., ASCE*, **115** (GT5), 711–727.

Kishida, H. and Uesugi, M. (1987). Tests of interface between sand and steel in the simple shear apparatus. *Géotechnique* **37** (1), 46–52.

Kulhawy, F.H. (1984). 'Limiting tip and side resistance: Fact or fallacy', in *Proc. of Symp. on Analysis and Design of Pile Foundations, ASCE Spec. Conf.*, San Francisco, pp 80–98.

Kuwabara, F. and Poulos, H.G. (1989). Downdrag forces in groups of piles. *J. Geot. Eng. Div., ASCE*, **115** (GT6), 806–818.

Likins, G.E. (1984) 'Field measurements and the pile driving analyser', *Proc. 2nd Int. Conf. on Application of Stress Wave Theory*, Stockholm, May 1984.

Litkouhi, S. and Poskitt, T.J. (1980). Damping constant for pile driveability calculations. *Géotechnique*, **30** (1), 77–86.

Littlejohn, G.S. and Macfarlane, I.M. (1975). 'A case history study of multi-tied diaphragm walls', *Diaphragm Walls and Anchorages*, ICE, London.

Lo, M.B. (1967), Discussion to paper by Y.O. Beredugo, *Can. Geotech. J.* **4**, 353–354.

Lysmer, T. and Richart F.E. (1966). Dynamic response of footing to vertical loading, *J. Soil Mech. and Found. Eng. Div., ASCE*, **98**, 85–105.

Mair, R.J. and Woods, D.M. (1987). *Pressuremeter Testing, Methods and Interpretation*, CIRIA and Butterworth, London.

Marcuson, W.F. and Bieganousky, W.A. (1977). SPT and relative density in coarse sands. *J. Geotech. Eng. Div.*, ASCE **103** (GTII), 1295–1309.

Marti, J. and Cundall, P.A. (1982). Mixed discretization procedure for accurate solution of plasticity problems. *Int. J. Num. Methods in Eng.* **6**, 129–139.

Meigh, A.C. (1987). *Cone Penetration Testing, Methods and Interpretation.* CIRIA and Butterworth, London.

Middendorp, P. and van Weele, A.F. (1986) 'Application of characteristic stress wave method, in offshore practice' *Proc. 3rd Int. Conf. on Num. Methods in Offshore Piling*, Nantes, Supplement, 6–18.

Murff, J.D. (1987). Pile capacity in calcareous sands: State of the Art. *J. Geot. Eng. Div., ASCE*, **113** (GT5), 490–507.

Neely, W.J. (1988). Bearing capacity of expanded-base piles in sand. *J. Geotech. Eng. Div., ASCE*, **116** (GT1), 73–87.

Pappin J.W., Simpson, B., Felton P.J. and Raison, C. (1986). Numerical analysis of flexible retaining walls. *Symposium on Computer Applications in Geotechnical Engineering.* The Midland Geotechnical Society, UK.

Randolph, M.F. (1988), 'The axial capacity of deep foundations in calcareous soil', *Proc. Int. Conf. on Calcareous Sediments*, Perth, Vol 2, pp 837–857.

Randolph, M.F. (1990). 'Analysis of the dynamics of pile driving'. In *Developments in Soil Mechanics IV: Advanced Geotechnical Analyses*, Eds P.K. Banerjee and R. Butterfield, Elsevier Applied Science Publishers Ltd.

Randolph, M.F. and Simons, D.A. (1980). An improved soil model for uni-dimensional pile driving analysis', *Proc. 3rd Int. Conf. on Num. Methiods in Offshore Piling*, Nantes, 1–17.

Randolph, M.F. and Murphy, B.S. (1985). 'Shaft capacity of driven piles in clay', in *Proc. 17th Offshore Technology Conf.*, Houston, Paper OTC 4883, pp 371–378.

Randolph, M.F. and Springman, S.M. (1991), 'Analysis of pile response due to external loads and soil movement', *Proc. 10th Eur. Conf. on Soil Mech. and Found. Eng.*, Florence.

Randolph, M.F., Carter, J.P. and Wroth, C.P. (1979). Driven piles in clay—the effects of installation and subsequent consolidation, *Géotechnique*, **29** (4), 361–393.

Rausche, F., Goble, G.G. and Likins, G.E. (1985). Dynamic determination of pile capacity. *J. Geot. Eng. Div., ASCE*, **111**, 367–383.

Rausche, F., Goble, G.G. and Likins, G.E. (1988), 'Recent WEAP developments', in *Proc. 3rd Int. Conf. on Application of Stress Wave Theory to Piles*, Ottawa, pp 164–173.

Rowe, R.K. and Armitage, H.H. (1987). A design method for drilled piers in soft rock. Canadian *Geotechnical Journal* **24** (1), 126–142.

Saxena, S.K. (1975). 'Measured performance of a rigid concrete wall at the World Trade Centre', *Diaphragm Walls and Anchorages*, ICE, London.

Sills, G.C., Burland, J.B. and Czechowski, M.K. (1977). 'Behaviour of an anchored diaphragm wall in stiff clay', *9th ICSMFE Conf.*, Tokyo (2), 147–153.

Skempton, A.W. (1986). Standard penetration test procedures and the effects in sands of overburden pressure, relative density, particle size, ageing and overconsolidation. *Géotechnique* **36** (2), 425–447.

Sliwinski, Z.J. and Fleming, W.G.K. (1975). 'Practical considerations affecting the construction of diaphragm walls', *Diaphragm Walls and Anchorages*, ICE, London.

Springman, S.M., Randolph, M.F. and Bolton, M.D. (1991), 'Modelling the behaviour of piles subjected to surcharge loading', in *Proc. Int. Conf. on Centrifuge Modelling*, Boulder, Colorado.

St. John, H.D. (1975). Discussion, *Diaphragm Walls and Anchorages*, ICE, London.

Stroud, M.A. (1989) 'The standard penetration test—its application and interpretation,' in *ICE Conf. Penetration Testing*, Birmingham. Thomas Telford, London.

Terzaghi, K. and Peck, R.B. (1948). *Soil Mechanics in Engineering Practice.* John Wiley and Sons, Inc., New York.

Tomlinson, M.J. (1986) *Foundation Design and Concrete.* 5th edn. Longman Scientific and Technical.

Wilkins, M. (1963) Calculation of Elastic-Plastic Flow. Lawrence Radiation Laboratory Report UCRL 7322.

Wood, L.A. and Perrin, A.J. (1984). 'Observation of a strutted diaphragm wall in London Clay', *Geotechnique Symposium*, in press.

Youd, T.D. (1972). 'Factors controlling maximum and minimum densities of sand', in *Proc. Symp. Eual. Dens. ASTM* STP 523, 98–112.

Index